Selected Titles in This Series

An Introduction to
Infinite Ergodic
Theory

Mathematical
Surveys
and
Monographs

Volume 50

An Introduction to Infinite Ergodic Theory

Jon Aaronson

American Mathematical Society

Editorial Board

ABSTRACT. The book is about measure preserving transformations of infinite measure spaces. It
could be of interest to mathematicians working in ergodic theory, probability and/or dynamical
systems and should be accessible to graduate students.

QA
313
A23
1997

Library of Congress Cataloging-in-Publication Data

Aaronson, Jon, 1949–
    An introduction to infinite ergodic theory / Jon Aaronson.
        p. cm. — (Mathematical surveys and monographs, ISSN 0076-5376 ; v. 50)
    Includes bibliographical references (p.   –   ) and index.
    ISBN 0-8218-0494-4 (alk. paper)
    1. Ergodic theory.   I. Title.   II. Series: Mathematical surveys and monographs ; no. 50.
QA313.A23   1997
515′.42—dc21   96-54848
                                                                                CIP

For Nilli

Contents

Preface

Infinite ergodic theory is the study of measure preserving transformations of infinite measure spaces (early references being [**Hop1**] and [**St**]). It is part of "non-singular ergodic theory", the more general study of non-singular transformations (since a measure preserving transformation is also a non-singular transformation).

Non-singular ergodic theory arose as an attempt to generalise the classical ergodic theory of probability preserving transformations. Its major success was the ratio ergodic theorem. Another side to the theory also developed concentrating on facts which are valid "in the absence of invariant probabilities".

This book is more concerned with properties specific to infinite measure preserving transformations.

It should be readable by anyone initiated to metric space topology and measure theoretic probability.

Some readers may like to begin by following an example and perhaps one of the simplest in the book is Boole's transformation $T : \mathbb{R} \to \mathbb{R}$ defined by $Tx = x - \frac{1}{x}$.

This is a conservative, exact measure preserving transformation of $\mathbb{R}$ equipped with Lebesgue measure; and for each absolutely continuous probability P on $\mathbb{R}$ and non-negative, integrable function $f : \mathbb{R} \to \mathbb{R}$ with unit integral,

$$P\left(\left[\sum_{k=0}^{n-1} f \circ T^k \leq \frac{\sqrt{2n}}{\pi} t\right]\right) \to \frac{2}{\pi} \int_0^t e^{-\frac{s^2}{\pi}} ds$$

as $n \to \infty$.

The book begins with an introduction to basic non-singular ergodic theory (chapters 1 and 2), including recurrence behaviour, existence of invariant measures, ergodic theorems and spectral theory. One of the results in §2.4 is the collapse of absolutely normalised pointwise ergodic convergence for ergodic measure preserving transformations of infinite measure spaces.

This leaves a wide range of possible "ergodic behaviour" which is catalogued in chapter 3 mainly according to the yardsticks of intrinsic normalising constants, laws of large numbers and return sequences (the return sequence of Boole's transformation is $\frac{\sqrt{2n}}{\pi}$).

The rest of the book (excepting chapter 5) consists of illustrations of these phenomena by examples.

Markov maps which arise both in probability theory and in smooth dynamics are treated in chapter 4. They illustrate distributional convergence phenomena (mentioned above) as do the inner functions of chapter 6. Geodesic flows on hyperbolic surfaces were one of the first examples considered ([**Hop1**]), and these are treated in chapter 7. Some of the extremely pathological examples in the subject can be found in the chapter on cocycles and skew products (chapter 8). In chapter 5, there is a modest beginning to the classification theory.

There is a small (but insufficient) amount of probability preserving ergodic theory in the book, and I recommend the uninitiated reader to take advantage of the excellent books available on this subject, including [**Cor-Sin-Fom**], [**De-Gr-Sig**], [**Fu**], [**Mañ**], [**Parr2**], [**Pet**], [**Rudo**], [**Wa**].

The reader will no doubt find that many (but hopefully not the reader's favourite) topics are conspicuous by their absence. By way of excuse I can only say that some of these are better covered elsewhere, while others are deemed too advanced for an introduction and yet others are too "fresh" for a book (there being no time to write about them).

Lastly I come to the thanks. I would like to thank the people who worked with me on the topics described in the book (see bibliography). Without them, none of this would have been possible. Also I would like to thank my colleagues Gilat and Lemańczyk; and my student Omri Sarig who found mistakes in early versions (any remaining errors being my sole responsibility having been introduced subsequently while correcting mistakes).

Jon. Aaronson
Tel Aviv, October 1996

CHAPTER 1

Non-singular transformations

§1.0 Standard measure spaces

Apart from the well known, classical theory of abstract measure spaces to be found (for example) in [**Halm1**], we'll also need certain results from the theory of *standard* measure spaces.

This section is a review of that theory. Some (but not all) proofs are supplied here. Complete treatments of the subject can be found in [**Coh**], [**Kec**], [**Kur**] and [**Part**].

DEFINITION: POLISH SPACE, BOREL SETS.

A *Polish space* is a complete, separable metric space. Let X be a Polish space X. The collection of *Borel sets* $\mathcal{B}(X)$ is the σ-algebra of subsets of X generated by the collection of open sets.

DEFINITION: STANDARD MEASURABLE SPACE.

A *standard measurable space* (or standard Borel space) $(X, \mathcal{B})$ is a Polish space X equipped with its collection of Borel sets $\mathcal{B} = \mathcal{B}(X)$.

DEFINITION: MEASURABLE FUNCTION.

Let X, X' be Polish spaces. A function $f : X \to X'$ is called *(Borel) measurable* if $f^{-1}\mathcal{B}(X') \subset \mathcal{B}(X)$.

Given a standard measurable space $(X, \mathcal{B})$, we consider the collection of probability measures defined on $(X, \mathcal{B})$

$$\mathcal{P} = \mathcal{P}(X, \mathcal{B}) := \{p : \mathcal{B} \to [0, 1] : \ p \text{ a probability measure}\}.$$

Let $\mathcal{B}(\mathcal{P})$ be the smallest σ-algebra of subsets of $\mathcal{P}$ such that for each $A \in \mathcal{B}$, the function $\mu \mapsto \mu(A)$ is measurable ($\mathcal{P} \to [0, 1]$). It follows that $(\mathcal{P}, \mathcal{B}(\mathcal{P}))$ is also a standard measurable space.

To see this, choose a compact topology on X generating $\mathcal{B}$, then with respect to the corresponding vague topology (inherited from the weak $*$ topology on $C(X)^*$): $\mathcal{P}$ is compact metric space and $\mathcal{B}(\mathcal{P})$ is its collection of Borel sets.

DEFINITION: STANDARD MEASURE SPACE.

A *standard measure space* is a measure space $(X, \mathcal{B}, m)$ where $(X, \mathcal{B})$ is a standard measurable space.

We sometimes suppress the σ-algebra $\mathcal{B}$ in these notations, denoting the standard measurable space $(X, \mathcal{B}(X))$ by X, and the standard measure space $(X, \mathcal{B}(X), m)$ by (X, m).

A *standard probability space* is a probability space which is a standard measure space. A measure space is called *pure* if it is either non-atomic or purely atomic,

1

and *symmetric* if it is either non-atomic or purely atomic with all atoms having the same measure.

The first result we review simplifies the treatment of measurable functions on standard spaces.

1.0.0 LUSIN'S THEOREM.
Suppose that $(X, \mathcal{B}, m)$ is a standard probability space, that X' is a Polish space, and that $f : X \to X'$ is measurable, then $\forall \ \epsilon > 0$, $\exists$ a compact set $K \subset X$ with $m(K) > 1 - \epsilon$, and such that f is continuous on K.

A Polish space is either finite, countable or has the cardinality of the continuum. The next result shows that there is essentially only one standard measurable space with the cardinality of the continuum.

1.0.1 KURATOWSKI'S ISOMORPHISM THEOREM.
Suppose that $(X, \mathcal{B})$ and $(X', \mathcal{B}')$ are standard measurable spaces with the same cardinality, then $(X, \mathcal{B})$ and $(X', \mathcal{B}')$ are isomorphic in the sense that there is a bijection $\pi : X \to X'$ such that $\pi \mathcal{B} = \mathcal{B}'$.

DEFINITION: ANALYTIC SET. Let $(X, \mathcal{B})$ be a standard measurable space. A subset $A \subset X$ is called *analytic* if $\exists$ another standard measurable space $(X', \mathcal{B}')$, a measurable function $f : X' \to X$ and a set $A' \in \mathcal{B}'$ such that $f A' = A$.

Clearly Borel sets are analytic.

It was shown by Souslin [**So**] that in any uncountable standard measurable space there are analytic sets which are not Borel. This fact (unknown to Lebesgue) contributes some subtlety to the subject and we shall therefore need the following three results.

1.0.2 UNIVERSAL MEASURABILITY THEOREM. *Let $(X, \mathcal{B}, m)$ be a standard measure space.*
If $A \subset X$ is analytic, then $\exists \ B, D \in \mathcal{B}$ such that $A \triangle B \subset D$ and $m(D) = 0$.

1.0.3 MEASURABLE IMAGE THEOREM. *Suppose that $(X, \mathcal{B})$ and $(X', \mathcal{B}')$ are standard measurable spaces and that $f : X \to X'$ is measurable and 1-1, then $f(A) \in \mathcal{B}' \ \forall \ A \in \mathcal{B}$.*

1.0.4 ANALYTIC SECTION THEOREM. *Suppose that $(X, \mathcal{B})$ and $(X', \mathcal{B}')$ are standard measurable spaces and that $f : X \to X'$ is measurable, then $\exists \ g : X' \to X$ which is analytically measurable (in the sense that $g^{-1}A$ is an analytic subset of X' whenever $A \in \mathcal{B}$) such that $f \circ g = Id_{X'}$.*

Let $(X, \mathcal{B})$ be standard.
Given $A \in \mathcal{B}$, let $\mathcal{B} \cap A := \{B \in \mathcal{B} : \ B \subset A\}$. There is a Polish topology on A so that Id: $A \to X$ is continuous. It follows from the measurable image theorem that $\mathcal{B} \cap A = \mathcal{B}(A)$, whence $(A, \mathcal{B} \cap A)$ is a standard measurable space.

Now let $(X, \mathcal{B}, m)$ be a standard measure space and let $A \in \mathcal{B}_+ := \{B \in \mathcal{B} : m(A) > 0\}$. The *induced* measure space is $(A, \mathcal{B} \cap A, m|_A)$ (where $\mathcal{B} \cap A := \{B \in \mathcal{B} : \ B \subset A\}$ and $m|_A(B) := m(B \cap A))$ and this is standard by the above.

Here, and throughout, we'll denote, for a collection $\mathcal{C} \subseteq \mathcal{B}$ of measurable sets, $\mathcal{C}_+ = \{C \in \mathcal{C} : m(C) > 0\}$, and, for $A \subset X$, $\mathcal{C} \cap A = \{C \in \mathcal{C} : C \subset A\}$.

DEFINITION: MEASURABLE MAP, AND INVERTIBLE MAP.

Let $(X, \mathcal{B}, m)$ and $(X', \mathcal{B}', m')$ be measure spaces, and let $A \in \mathcal{B}$, $A' \in \mathcal{B}'$. The map $f : A \to A'$ is *measurable* if $f^{-1}C \in \mathcal{B}$ $\forall C \in \mathcal{B}'$.

The measurable map $f : A \to A'$ is called *invertible* on $B \in \mathcal{B} \cap A$ if f is $1-1$ on B, $fB \in \mathcal{B}'$, and $f^{-1} : fB \to B$ is measurable.

DEFINITION: NON-SINGULAR MAP, AND MEASURE PRESERVING MAP. The measurable map $f : A \to A'$ is called (two-sided) *non-singular* if for $C \in \mathcal{B}' \cap A'$, $m(f^{-1}C) = 0$ iff $m'(C) = 0$; and *measure preserving* if $m(f^{-1}C) = m'(C)$ for $C \in \mathcal{B}' \cap A'$.

If $f : A \to A'$ is measurable, invertible, and non-singular on $B \in \mathcal{B} \cap A$, then by the Radon-Nikodym theorem, $\exists \phi \in L^1(A)_+$ such that $m(fC) = \int_C \phi \, dm$ for $C \in \mathcal{B}' \cap A'$. The function ϕ is called the *Radon-Nikodym derivative* of f on A and is denoted by $f' = \frac{dm \circ f}{dm}$. Evidently $f : A \to A'$ is measurable, invertible, and non-singular is measure preserving iff $f' \equiv 1$.

The chain rule for Radon-Nikodym derivatives applies. If $f : A \to B$ and $g : B \to C$ are measurable, invertible, and non-singular on A and B (respectively), then $g \circ f : A \to C$ is measurable, invertible, and non-singular on A and $(g \circ f)' = g' \circ f f'$.

In case $f : B \to fB$ is measurable, invertible, and non-singular, then $f^{-1} : fB \to B$ is non-singular and $f^{-1\prime} = \frac{1}{f' \circ f^{-1}}$.

EXAMPLE. Let $(X, \mathcal{B}, m)$ be the unit interval equipped with Borel sets and Lebesgue measure, and suppose that $T : X \to X$ is a strictly increasing homeomorphism; then $T : X \to X$ is invertible.

We have that T is nonsingular iff both T and T^{-1} are absolutely continuous functions, and $T' := \frac{dm \circ T}{dm} = |DT|$ where $DT(x) := \lim_{h \to 0} \frac{T(x+h) - T(x)}{h}$.

DEFINITION: FACTOR MEASURE SPACE, AND FACTOR MAP.

The measure space $(X', \mathcal{B}', m')$ is a *factor space* of $(X, \mathcal{B}, m)$ if there are subsets $Y \in \mathcal{B}$, $Y' \in \mathcal{B}'$ such that $m(X \setminus Y) = m'(X' \setminus Y') = 0$, and a measurable, measure preserving map $\pi : Y \to Y'$. This map is called a *factor map* and we sometimes denote it $\pi : X \to X'$.

In the same situation, we sometimes call the measure space $(X, \mathcal{B}, m)$ an *extension* of $(X', \mathcal{B}', m')$.

DEFINITION: CARTESIAN PRODUCT SPACE. If $(\Omega, \mathcal{F}, p)$ is a probability space, then the *Cartesian product space* $(X \times \Omega, \mathcal{B} \otimes \mathcal{F}, m \times p)$ is always an extension of $(X, \mathcal{B}, m)$. Here $\mathcal{B} \otimes \mathcal{F} := \sigma(\mathcal{B} \times \mathcal{F})$ and $(m \times p)(A \times B) := m(A)p(B)$.

DEFINITION: ISOMORPHISM OF MEASURE SPACES.

An *isomorphism* between the measure spaces $(X, \mathcal{B}, m)$ and $(X', \mathcal{B}', m')$ is an invertible factor map $\pi : X \to X'$, and the measure spaces $(X, \mathcal{B}, m)$ and $(X', \mathcal{B}', m')$ are *isomorphic* if there is an isomorphism between them.

REMARK: NON-ATOMIC STANDARD SPACES.

Any non-atomic standard probability space is isomorphic with the unit interval $[0, 1] \subset \mathbb{R}$ equipped with its Borel sets $\mathcal{B}$ and Lebesgue measure λ. This is proven using theorem 1.0.1 to obtain isomorphism with the unit interval $[0, 1] \subset \mathbb{R}$ equipped with its Borel sets and some non-atomic probability p; and then using $\lambda(J) = p(J)$ for intervals $J \subset I$ where $\pi(x) := p([0, x])$ to obtain the final isomorphism with $([0, 1], \mathcal{B}, \lambda)$.

This can be used to show that any non-atomic, σ-finite standard measure space whose total mass is infinite is isomorphic to $\mathbb{R}$ equipped with its Borel sets and Lebesgue measure.

DEFINITION: COMPLETION OF A MEASURE SPACE.
Given a measure space $(X, \mathcal{B}, m)$, the *completion* of $\mathcal{B}$ (with respect to m) is the collection (a σ-algebra)

$$\overline{\mathcal{B}}_m := \{A \subset X : \exists\ B, D \in \mathcal{B} \text{ such that } A \triangle B \subset D, \ m(D) = 0\}.$$

The measure space $(X, \overline{\mathcal{B}}_m, m)$ is also known as the *completion* of $(X, \mathcal{B}, m)$. A measure space $(X, \mathcal{B}, m)$ is *complete* if $\overline{\mathcal{B}}_m = \mathcal{B}$.

DEFINITION: SEPARABLE MEASURE SPACE.
A measure space $(X, \mathcal{B}, m)$ is *separable* if $\exists$ a countable collection $\mathcal{A} \subset \mathcal{B}$ such that $\sigma(\mathcal{A}) = \mathcal{B} \mod m$ (that is: $\overline{\sigma(\mathcal{A})}_m = \overline{\mathcal{B}}_m$) which separates points in the sense that

$$1_A(x) = 1_A(y) \ \forall\ A \in \mathcal{A} \implies x = y.$$

DEFINITION: LEBESGUE SPACE.
A *Lebesgue space* is a complete measure space which is isomorphic to the completion of a standard measure space.

A Lebesgue space $(X, \mathcal{B}, m)$ is evidently separable and complete, and there is a subset $X_0 \in \mathcal{B}$, $m(X \setminus X_0) = 0$ endowed with a Polish topology such that $\mathcal{B} \cap X_0 = \overline{\mathcal{B}(X_0)}_m$.

Let $(X, \mathcal{C}, m)$ be separable and complete with $\mathcal{A} = \{A_n\}_{n \in \mathbb{N}}$ as the countable generating collection. Consider the compact metric space $\Omega := \{0, 1\}^{\mathbb{N}}$, and define $\pi : X \to \Omega$ by $\pi(x)_n := 1_{A_n}(x)$. Evidently, $\pi : X \to \pi(X))$ is 1-1, and measurable. In fact if $\mu := m \circ \pi^{-1} : \mathcal{B}(\Omega) \to [0, \infty)$, then π is an isomorphism of the measure spaces $(X, \mathcal{B}, m)$ and $(\pi(X), \overline{\mathcal{B}(\Omega) \cap \pi(X)}_\mu, \mu)$.

As shown in ([**Ro1**], [**Rudo**]), the space $(X, \mathcal{C}, m)$ is a Lebesgue space if and only if $\pi(X) \in \overline{\mathcal{B}(\Omega)}_\mu$. To see this, if $\pi(X) \in \overline{\mathcal{B}(\Omega)}_\mu$, then $\exists\ \Omega_0 \in \mathcal{B}(\Omega)$ such that $\Omega_0 \subset \pi(X)$ and $\mu(\Omega_0) = 1$ with the consequence that $(X, \mathcal{C}, m)$ is isomorphic with $(\Omega_0, \overline{\mathcal{B}(\Omega_0 m)}_\mu, \mu)$ - a Lebesgue space. Conversely, if $(X, \mathcal{C}, m)$ is a Lebesgue space then $\exists\ X_0 \in \mathcal{B}$, $m(X_0) = 1$ such that X_0 is Polish, $\mathcal{C} \cap X_0 = \overline{\mathcal{B}(X_0)}_m$ and $\pi : X_0 \to \Omega$ is Borel measurable. By the universal measurability theorem (or by Lusin's theorem) $\pi(X_0) \in \overline{\mathcal{B}(\Omega)}_\mu$ and since $\mu(\pi(X_0)) = 1$, $\pi(X_0) \subset \pi(X)$ we have $\pi(X) \in \overline{\mathcal{B}(\Omega)}_\mu$.

This shows that $\exists$ complete, separable measure spaces which are not Lebesgue spaces, for example $(X, \mathcal{C}, m)$ where $X \subset [0, 1]$ has full outer measure and zero inner measure, $\mathcal{C} := \overline{\mathcal{B}([0, 1])}_m \cap X$ and m is outer measure.

DEFINITION: MEASURE ALGEBRA. Let $(X, \mathcal{B}, m)$ be a measure space, and define the relation $\sim$ on $\mathcal{B}$ by $A \sim B$ if $m(A \triangle B) = 0$, then (see [**Halm1**]) $\sim$ is an equivalence relation,
$A_n, A_n' \in \mathcal{B}, \ A_n \sim A_n' \ (n \geq 1) \implies$

$$A_1^c \sim A_1'^c, \ \bigcup_{n=1}^{\infty} A_n \sim \bigcup_{n=1}^{\infty} A_n', \ \& \ \bigcap_{n=1}^{\infty} A_n \sim \bigcap_{n=1}^{\infty} A_n'.$$

The collection of equivalence classes

$$\mathcal{S}(X, \mathcal{B}, m) := \{\{A' \in \mathcal{B} : A' \sim A\} : A \in \mathcal{B}\}$$

is called the *measure algebra* of $(X, \mathcal{B}, m)$.

DEFINITION: MEASURE ALGEBRA CONJUGACY. A *measure algebra conjugacy* between the measure spaces $(X, \mathcal{B}, m)$ and $(X', \mathcal{B}', m')$ is a bijection $\pi :$ $\mathcal{S}(X, \mathcal{B}, m) \to \mathcal{S}(X', \mathcal{B}', m')$ such that $\pi(A \setminus B) = \pi(A) \setminus \pi(B)$, $\pi\left(\bigcup_{n=1}^{\infty} A_n\right) = \bigcup_{n=1}^{\infty} \pi(A_n)$ and $m' \circ \pi = m$.

Isomorphic measure spaces are measure algebra conjugate, a measure algebra conjugacy being induced by an isomorphism.

Conversely, any measure algebra conjugacy between Lebesgue spaces induces an isomorphism ([**Ro1**], [**Rudo**]).

All separable, non-atomic probability spaces are measure algebra conjugate ([**Halm1**]), and every separable measure space is measure algebra conjugate to a standard one ([**Fu**]).

A non-singular- (or measure preserving-) transformation is a non-singular- (or measure preserving-) map mod m. The next definitions make this precise.

DEFINITION: NON-SINGULAR TRANSFORMATION AND MEASURE PRESERVING TRANSFORMATION.

A *non-singular transformation* T of X is a measurable, non-singular map $T :$ $Y \to Y$ where $Y \in \mathcal{B}$ and $m(X \setminus Y) = 0$.

A *measure preserving transformation* T of X is a measurable, measure preserving map $T : Y \to Y$ where $Y \in \mathcal{B}$ and $m(X \setminus Y) = 0$.

1.0.5 PROPOSITION. *Suppose that* $(X, \mathcal{B}, m)$ *is standard*, $Y \in \mathcal{B}$, $m(X \setminus Y) = 0$ *and* $T : Y \to X$ *is a non-singular map, then* T *is a non-singular transformation of* X.

PROOF. By the universal measurability theorem, $TY \in \overline{\mathcal{B}}_m$. Also,

$$0 = m(X \setminus Y) \geq m(X \setminus T^{-1}TY) = m(T^{-1}(X \setminus TY))$$

whence $m(X \setminus TY) = 0$ since $T : Y \to X$ is non-singular.

Choose $Z \in \mathcal{B}$, $Z \subset TY$, $m(X \setminus Z) = 0$ and set $U := \bigcap_{n=1}^{\infty} T^{-n} Z$, then $TU = Z \cap U$ and $TU \in \mathcal{B}$, $TU \subset U$, and $T : U \to TU$ is a nonsingular map.

By non-singularity of $T : Y \to X$, $m(X \setminus T^{-n} Z) = 0 \ \forall \ n \geq 0$ whence $m(X \setminus U) = m(X \setminus TU) = m(U \setminus TU) = 0$;
$T : U \to U$ is a non-singular map,
and T is a non-singular transformation of X. $\qquad\qquad \square$

EXAMPLE 1.0.1, A NON-SINGULAR TRANSFORMATION. Let $(X, \mathcal{B}, m)$ be the unit interval equipped with Borel sets and Lebesgue measure, and let $\{A_j : j \geq 1\}$ be a partition of X into open intervals (i.e. the $\{A_j : j \geq 1\}$ are disjoint open intervals, and $X = \bigcup_{j=1}^{\infty} A_j$ mod m).

Given a collection $\{B_j : j \geq 1\}$ of open intervals, define for $j \geq 1$, $T : A_j \to B_j$ to be an absolutely continuous bijection whose inverse is also absolutely continuous (e.g. the increasing linear bijection).

Note that T is only defined on $U = \bigcup_{j=1}^{\infty} A_j$. Clearly $T : U \to TU = \bigcup_{j=1}^{\infty} B_j$ is measurable, and non-singular. Therefore by proposition 1.0.5, T is a non-singular transformation of X iff

$$TU = \bigcup_{j=1}^{\infty} B_j = X \qquad \text{mod } m.$$

If T is a non-singular transformation of a σ-finite measure space $(X, \mathcal{B}, m)$, and p is another measure on $(X, \mathcal{B})$ *equivalent* to m (denoted $p \sim m$) in the sense that

$$p(A) = 0 \iff m(A) = 0,$$

then T is a non-singular transformation of $(X, \mathcal{B}, p)$.

Thus, a non-singular transformation of a σ-finite measure space is actually a non-singular transformation of a probability space.

EXAMPLE 1.0.2, A PROBABILITY PRESERVING TRANSFORMATION.   Let $X = [0,1]^{\mathbb{N}}$ and let $\mathcal{B}$ be the σ-algebra generated by *cylinder* sets of form $[A_1, \ldots, A_n] := \{\underline{x} \in X : x_j \in A_j, \ 1 \le j \le n\}$, where $A_1, \ldots, A_n \in \mathcal{B}(I)$ (the Borel subsets of $I = [0,1]$), and let the *shift* $T : X \to X$ be defined by $(Tx)_n = x_{n+1}$. Note that X is a compact metric space when equipped with the product topology, and $\mathcal{B}$ is its collection of Borel sets.

Define using Kolmogorov's existence theorem ([**Kol**], [**Part**]) a probability $m : \mathcal{B}(X) \to [0,1]$ by

$$m([A_1, ..., A_n]) := \prod_{k=1}^{n} |A_k| \quad (A_1, \ldots, A_n \in \mathcal{B}(I))$$

where $|A|$ denotes the Lebesgue measure of $A \in \mathcal{B}(I))$. Evidently,

$$m(T^{-1}[A_1, ..., A_n]) = m([I, A_1, ..., A_n]) = m([A_1, ..., A_n]) \quad (A_1, \ldots, A_n \in \mathcal{B}(I))$$

whence

$$m \circ T^{-1} = m.$$

The measure space X represents the set of (possibly random) *"configurations"* of some system, and T represents the change under *"passage of time"*. The non-singularity of T reflects the assumed property of the system that configuration sets that are impossible sometimes are always impossible. A probability preserving transformation would describe a system in a "steady state", where configuration sets occur with the same likelihood at all times.

One might conjecture that each non-singular transformation is obtained by starting with a measure preserving transformation, and then "passing" to some equivalent measure, however we'll see in §1.2 that this is not the case.

DEFINITION: INVERTIBLE, AND LOCALLY INVERTIBLE NON-SINGULAR TRANS-
FORMATIONS.

The non-singular transformation T of X is called *invertible* if T is invertible
on some $Y \in \mathcal{B}$ with $m(X \setminus Y) = 0$, and *locally invertible* if there is are disjoint
measurable sets $\{A_j : j \geq 1\}$ such that $m(X \setminus \bigcup_{j \geq 1} A_j) = 0$, and T is invertible on
each A_j.

The non-singular transformation in example 1.0.1 is locally invertible (being
invertible on each A_j). It is invertible iff $\{B_j : j \geq 1\}$ is a partition of X mod m.

Evidently, if T is a locally invertible, non-singular transformation of X, then T
is *positively non-singular* in the sense that

$$A \in \mathcal{B}, \ m(A) = 0 \implies m(TA) = 0.$$

The probability preserving transformation T in example 1.0.2 does not have this
property. If $C \in \mathcal{B}(I)$ is a non-empty set of Lebesgue measure zero, and $D := [C]$,
then $m(D) = 0$ and $TD = X$.

A non-singular transformation T of a standard probability space $(X, \mathcal{B}, m)$ is
locally invertible iff $T^{-1}\{x\}$ is countable for m-a.e. $x \in X$.

Clearly if T is locally invertible, then $T^{-1}\{x\}$ is countable $\forall \ x \in X$. The
converse follows from the

1.0.6 LOCAL INVERTIBILITY LEMMA.

*Let $(X, \mathcal{B}, m)$ and $(Y, \mathcal{C}, \mu)$ be standard probability spaces, and suppose that $\pi :$
$X \to Y$ is a measurable, measure preserving map with $\pi^{-1}\{x\}$ countable $\forall \ x \in X$,
then $\exists$ a countable, measurable partition α of X such that $\pi : a \to Ta$ is non-
singular and invertible $\forall \ a \in \alpha$.*

We prove the local invertibility lemma using two results which will also be
important in the sequel:
the exhaustion lemma 1.0.7; and the disintegration theorem 1.0.8.

DEFINITION: HEREDITARY COLLECTION, MEASURABLE UNION.

Let $(X, \mathcal{B}, m)$ be a measure space. A collection $\mathfrak{H} \subset \mathcal{B}$ is called *hereditary* if

$$C \in \mathfrak{H}, \ B \subset C, \ B \in \mathcal{B} \implies B \in \mathfrak{H}.$$

A set $U \in \mathcal{B}$ is said to *cover* the hereditary collection $\mathfrak{H}$ if $A \subset U$ mod
$m \ \forall \ A \in \mathfrak{H}$.

A hereditary collection $\mathfrak{H} \subset \mathcal{B}$ is said to *saturate* $A \in \mathcal{B}$ if $\forall \ B \in \mathcal{B}, \ B \subset$
$A, \ m(B) > 0, \ \exists \ C \in \mathfrak{H}, \ m(C) > 0, \ C \subset B$.

The set $U \in \mathcal{B}$ is called a *measurable union* of the hereditary collection $\mathfrak{H} \subset \mathcal{B}$
if it both covers, and is saturated by $\mathfrak{H}$.

There is no more than one measurable union of a hereditary collection. To see
this, let $U, U' \in \mathcal{B}$ be measurable unions of the hereditary collection $\mathfrak{H}$, and suppose
that $m(U \setminus U') > 0$, then (since $\mathfrak{H}$ saturates U) $\exists \ C \in \mathfrak{H}, \ m(C) > 0, \ C \subset U \setminus U'$
whence (since U' covers $\mathfrak{H}$) $C \subset U'$ mod m contradicting $m(C) > 0$. This shows
that $U \subset U'$ mod m and by symmetry, $U = U'$ mod m.

The exhaustion lemma (below) shows existence of measurable unions.

1.0.7 EXHAUSTION LEMMA. *Let $(X, \mathcal{B}, m)$ be a probability space and let $\mathfrak{H} \subset \mathcal{B}$
be hereditary, then $\exists \ A_1, A_2, \cdots \in \mathfrak{H}$ disjoint such that $U(\mathfrak{H}) = \bigcup_{n=1}^{\infty} A_n$ is a
measurable union of $\mathfrak{H}$.*

PROOF.
Let
$$\epsilon_1 := \sup\{m(A): A \in \mathfrak{H}\},$$
choose $A_1 \in \mathfrak{H}$ such that $m(A_1) \geq \frac{\epsilon_1}{2}$, and let
Let
$$\epsilon_2 := \sup\{m(A): A \in \mathfrak{H}, A \cap A_1 = \emptyset\},$$
choose $A_2 \in \mathfrak{H}$ such that $A_2 \cap A_1 = \emptyset$ and $m(A_2) \geq \frac{\epsilon_2}{2}$.

Continuing the process, we obtain a sequence of disjoint $\{A_n\}_{n=1}^{\infty} \subset \mathfrak{H}$ and $\epsilon_n \downarrow$ such that

$$\epsilon_n := \sup\{m(A): A \in \mathfrak{H}, A \cap A_k = \emptyset \ \forall \ k < n\}, \ \ m(A_n) \geq \frac{\epsilon_n}{2}.$$

Clearly $\sum_{n=1}^{\infty} \epsilon_n \leq 2\sum_{n=1}^{\infty} m(A_n) \leq 2$, whence $\epsilon_n \to 0$.

We claim that $U := \bigcup_{n=1}^{\infty} A_n = X$ is a measurable union of $\mathfrak{H}$. Evidently $\mathfrak{H}$ saturates U. To see that U covers $\mathfrak{H}$ assume otherwise, then $\exists A \in \mathfrak{H}$, $m(A) > 0$ such that $A \cap A_n = \emptyset \ \forall n \geq 1$ whence $m(A) \leq \epsilon_n \to 0$ contradicting $m(A) > 0$. $\square$

We denote the measurable union of the hereditary collection $\mathfrak{H}$ by $U(\mathfrak{H})$.

1.0.8 DISINTEGRATION THEOREM. *Let* $(X, \mathcal{B}, m)$, $(Y, \mathcal{C}, \mu)$ *be standard probability spaces and suppose that* $\pi : X \to Y$ *is a measurable, and* $m = \mu \circ \pi^{-1}$, *then* $\exists Y_0 \in \mathcal{C}$ *such that* $\mu(Y_0) = 1$ *and* $\exists$ *a measurable function* $y \mapsto m_y$ $(Y_0 \to \mathcal{P}(X, \mathcal{B}))$ *such that* $m_y(\pi^{-1}\{y\}) = 1 \ \forall \ y \in Y_0$ *and*

$$m(A \cap \pi^{-1}B) = \int_B m_y(A)d\mu(y) \ \forall \ A \in \mathcal{B}, \ B \in \mathcal{C}.$$

The measure m_y is called the *fibre measure* over $\pi^{-1}\{y\}$.

PROOF. For each $f \in L^1(m)_+$, define a measure ν_A on $\mathcal{C}$ by

$$\nu_f(C) := \int_{\pi^{-1}C} f dm.$$

The measure ν_f is μ-absolutely continuous and so, by the Radon-Nikodym theorem there is a measurable function $E(f|\pi) = \frac{d\nu_f}{d\mu} \in L^1(\mu)$, such that $\int_C E(f|\pi)d\mu = \int_{\pi^{-1}C} f dm$.

Set
$$u_y(A) = E(1_A|\pi)(y),$$
then
$$\int_C u_y(A)d\mu(y) = m(A \cap \pi^{-1}C) \ \ \forall \ A \in \mathcal{B}, \ C \in \mathcal{C}.$$

Also, if $A_1, ... \in \mathcal{B}$ are disjoint, then

$$u_y(\bigcup_{k=1}^{\infty} A_k) = \sum_{k=1}^{\infty} u_y(A_k),$$

for m-a.e. $y \in Y$.

Since X is standard and uncountable, we may assume by Kuratowski's isomorphism theorem that $X = \{0,1\}^{\mathbb{N}}$. Let $\mathcal{A}$ denote the algebra of finite unions of cylinder sets in Ω, then $\mathcal{A}$ is countable and generates $\mathcal{B}$, and each set in $\mathcal{A}$ is both open and compact.

Since $\mathcal{A}$ is countable, there is a set $Y_0 = Y \mod m$ such that

$$u_y\left(\bigcup_{k=1}^{n} A_k\right) = \sum_{k=1}^{n} u_y(A_k) \ \forall y \in Y_0$$

whenever $A_1, ..., A_n \in \mathcal{A}$ are disjoint.

For $y \in Y_0$ and $E \subset X$ define

$$m_y(E) = \inf\{\sum_{n=1}^{\infty} u_y(A_n) : \bigcup_{n=1}^{\infty} A_n \supseteq E, \ A_n \in \mathcal{A}\}.$$

By Caratheodory's theorem, $m_y : \mathcal{B} \to [0,1]$ is a measure. Since sets in $\mathcal{A}$ are both compact and open, we have that $u_y(A) = m_y(A) \ \forall \ A \in \mathcal{A}$.

Clearly $y \mapsto m_y(A)$ is measurable $(Y_0 \mapsto [0,1]) \ \forall \ A \in \mathcal{A}$. A monotone class argument shows measurability $\forall \ A \in \mathcal{B}$ and that $m_y(A) = u_y(A)$ for a.e. $y \in Y_0$.

To see that $m_y(\pi^{-1}\{y\}) = 1$ for a.e. $y \in Y$, note first that for $C \in \mathcal{C}$ fixed,

$$\int_C m_y(\pi^{-1}C)d\mu(y) = m(\pi^{-1}C) = \int_C m_y(X)d\mu(y) = \mu(C)$$

whence $m_y(\pi^{-1}C) = 1$ for a.e. $y \in C$.

Now fix a metric on Y and let $\beta_n \uparrow$ be an increasing sequence of countable, measurable partitions on Y such that $\sup_{b \in \beta_n} \operatorname{diam} b \to 0$ as $n \to \infty$. For $y \in Y$, $n \geq 1$ write $y \in b_n(y) \in \beta_n$. Choose $Y_1 \in \mathcal{C} \cap Y_0$, $\mu(Y_1) = 1$ such that

$$m_y(\pi^{-1}b) = 1 \ \forall \ y \in b \cap Y_1, \ b \in \beta_n, \ n \geq 1.$$

We now have for $y \in Y_1$ that

$$m_y(\pi^{-1}\{y\}) \ \leftarrow \ m_y(\pi^{-1}b_n(y)) = 1.$$

$\square$

REMARK: FIBRE EXPECTATIONS AND CONDITIONAL EXPECTATIONS.

Let $(X, \mathcal{B}, m)$, $(Y, \mathcal{C}, \mu)$ be standard probability spaces and suppose that $\pi : X \to Y$ is measurable, and $m = \mu \circ \pi^{-1}$, with fibre measures $y \mapsto m_y \in \mathcal{P}(\pi^{-1}\{y\})$.

If $f : X \to \mathbb{R}$ is bounded and measurable, the function $y \mapsto \int_X f dm_y$ is called the *fibre expectation* of f on $\pi^{-1}\{y\}$. Evidently $\int_X f dm_y = E(f|\pi)(y)$ (as defined above) for μ-a.e. $y \in Y$. It follows that the conditional expectation of f with respect to $\pi^{-1}\mathcal{C}$ is given by

$$E(f|\pi^{-1}\mathcal{C}) = E(f|\pi) \circ \pi \quad m - \text{a.e. on } X.$$

PROOF OF THE LOCAL INVERTIBILITY LEMMA.

Let $y \mapsto m_y \in \mathcal{P}(\pi^{-1}\{y\})$ $(y \in Y_0)$ be the fibre measures on $\pi^{-1}\{y\}$ as in the disintegration theorem.

Since $m_y(\pi^{-1}\{y\}) = 1 \ \forall \ y \in Y_0$, and $\pi^{-1}\{y\}$ is countable $\forall \ y \in Y_0$, the probabilities m_y $(y \in Y_0)$ are purely atomic, and we may assume (possibly discarding a null set) that $m_{\pi x}(\{x\}) > 0 \ \forall \ x \in X$.

Call a set $A \in \mathcal{B}$ a π-*section* if $\pi A \in \mathcal{B}$ and $\pi : A \to \pi A$ is measurable, nonsingular and invertible, and call a section $A \in \mathcal{B}$ *onto* if $\pi A = Y \mod \mu$. Denote the collection of π-sections by $\mathfrak{S}$. It is enough to show that there is a partition of X into π-sections.

We claim that the hereditary collection $\mathfrak{S}$ saturates X. To see this let $B \in \mathcal{B}$, $m(B) > 0$, then by the universal measurability theorem $\exists\ C \in \mathcal{C}$, $C \subset \pi B$, $\mu(C) > 0$ and by the analytic section theorem $\exists\ f : C \to B$ analytically measurable such that $\pi \circ f = \mathrm{Id}|_C$. It follows that $f(C) = B \cap \pi^{-1}C =: A$ and that $A \in \mathfrak{S}$. To see that $m(A) > 0$,

$$m(A) = \int_Y m_y(A)d\mu(y) = \int_C m_y(A)d\mu(y) = \int_C m_y(\{f(y)\})d\mu(y) > 0$$

because $\mu(C) > 0$ and $m_{\pi x}(\{x\}) > 0\ \forall\ x \in X$.

The result now follows from the exhaustion lemma. $\qquad\square$

If T is a non-singular transformation of $(X, \mathcal{B}, m)$, and $A \in \mathcal{B}$, $m(A) > 0$, $T^{-1}A = A$, then T is a non-singular transformation of $(A, \mathcal{B} \cap A, m|_A)$. The concept of irreducibility for non-singular transformations is called *ergodicity*.

DEFINITION: ERGODIC.
A non-singular transformation T is called *ergodic* if $A \in \mathcal{B}$, $T^{-1}A = A$ mod m implies $m(A) = 0$ or $m(A^c) = 0$.

This condition actually implies a stronger condition.

1.0.9 PROPOSITION. *Suppose that T is an ergodic non-singular transformation. If $f : X \to \mathbb{R}$ is measurable, and $f \circ T = f$ a.e., then*

$$\exists c \in \mathbb{R} \text{ such that } f = c \text{ a.e.}$$

PROOF. For $c \in \mathbb{R}$ the set $[f \le c]$ is T-invariant, and hence $[f \le c] = \emptyset$, X mod m. If

$$c_0 = \inf\{c \in \mathbb{R} :\ [f \le c] = X \quad \text{mod } m\},$$

then

$$[f = c_0] = [f \le c_0] \setminus \bigcup_{n \ge 1}[f \le c_0 - \frac{1}{n}] = X \quad \text{mod } m.$$

$\qquad\square$

DEFINITION: FACTOR TRANSFORMATION, ISOMORPHISM.
The non-singular transformation T' of $(X', \mathcal{B}', m')$ is a *factor* of the non-singular transformation T of $(X, \mathcal{B}, m)$ if there are sets $Y \in \mathcal{B}$, $Y' \in \mathcal{B}'$ such that $m(X \setminus Y) = m'(X' \setminus Y') = 0$, $TY \subset Y$, $T'Y' \subset Y'$; and there is a measurable, measure preserving map $\pi : Y \to Y'$ so that $\pi \circ T' = T \circ \pi$ on Y.

The map $\pi : Y \to Y'$ is called a *factor map* and is sometimes denoted $\pi : T \to T'$.

An *isomorphism* between the non-singular transformations T of $(X, \mathcal{B}, m)$ and T' of $(X', \mathcal{B}', m')$ is a factor map $\pi : T \to T'$ which is invertible in the sense that there are sets $Y \in \mathcal{B}$, $Y' \in \mathcal{B}'$ such that $m(X \setminus Y) = m'(X' \setminus Y') = 0$, $TY \subset Y$, $T'Y' \subset Y'$; and such that $\pi : Y \to Y'$ is invertible.

The non-singular transformations T of $(X, \mathcal{B}, m)$ and T' of $(X', \mathcal{B}', m')$ are *isomorphic* if is an isomorphism between them.

1.0.10 FACTOR PROPOSITION.   *Suppose that T is a non-singular transformation of the standard probability space $(X, \mathcal{B}, m)$.*

For every T-invariant, sub-σ-algebra of $\mathcal{F} \subset \mathcal{B}$, there is a factor U, and a factor map $\pi : T \to U$ with $\mathcal{F} = \pi^{-1} \mathcal{B}_U$.

The factor is invertible iff the sub-σ-algebra is strictly T-invariant.

PROOF. Let $\mathcal{F}_0 = \{F_n : n \in \mathbb{N}\}$ be a T-invariant, countable algebra of sets generating $\mathcal{F}$. Define $\pi : X \to \{0,1\}^{\mathbb{N}}$ by $(\pi x)_n = 1_{F_n}(x)$. This map is clearly measurable as

$$\pi^{-1}[a_1, \ldots, a_n] = \bigcap_{k=1}^{n} F_{a_k} \in \mathcal{B},$$

indeed $\pi^{-1} : \mathcal{B}(\{0,1\}^{\mathbb{N}}) \cap \pi(X) \to \mathcal{F}$ is a Boolean isomorphism.

Define a probability $\mu : \mathcal{B}(\{0,1\}^{\mathbb{N}}) \cap \pi(X) \to [0,1]$ by $\mu = m \circ \pi^{-1}$.

Clearly, $\pi(x) = \pi(y)$ if and only if $1_F(x) = 1_F(y) \ \forall \ F \in \mathcal{F}$. By the T-invariance of $\mathcal{F}$, if $\pi(x) = \pi(y)$ then $\pi(Tx) = \pi(Ty)$ and we can define a transformation $U : \pi(X) \to \pi(X)$ by $U\pi(x) := \pi(Tx)$. The transformation U is measurable since for $A \in \mathcal{B}(\{0,1\}^{\mathbb{N}})$,

$$U^{-1}(A \cap \pi(X)) = U^{-1}\pi(\pi^{-1}A) = \pi(T^{-1}\pi^{-1}A) \in \mathcal{B}(\{0,1\}^{\mathbb{N}}) \cap \pi(X).$$

The probability μ is U-non-singular because m is T-non-singular. It remains to represent U as a non-singular transformation of a standard probability space.

By the universal measurability theorem $\pi(X) \in \overline{\mathcal{B}(\{0,1\}^{\mathbb{N}})}_\mu$, hence $\exists \ X_U \in \mathcal{B}(\{0,1\}^{\mathbb{N}})$, $X_U \subset \pi(X)$ such that $\mu(\{0,1\}^{\mathbb{N}} \setminus X_U) = 0$.

To prove the second part of the proposition, we show that if T is a non-singular transformation of a standard probability space, and $T^{-1}\mathcal{B} = \mathcal{B}$, then T is invertible.

Let $\mathcal{A} \subset \mathcal{B}$ be a countable collection generating $\mathcal{B}$ and separating points in the sense that

$$1_A(x) = 1_A(y) \ \forall \ A \in \mathcal{A} \implies x = y.$$

We may assume that $T^{-1}\mathcal{A} = \mathcal{A}$, since if not, we may pass to the larger, countable T-invariant collection

$$\bigcup_{n=0}^{\infty} T^{-n}\mathcal{A} \cup \{A \in \mathcal{B} : \ \exists \ n \geq 1 \text{ such that } T^{-n}A \in \mathcal{A}\}.$$

Define $S : \{0,1\}^{\mathcal{A}} \to \{0,1\}^{\mathcal{A}}$ by $(Sx)_A := x_{T^{-1}A}$. Clearly S is a homeomorphism of $\{0,1\}^{\mathcal{A}}$. Now define $\pi : X \to \{0,1\}^{\mathcal{A}}$ by $\pi(x)_A := 1_A(x)$. Evidently, $\pi : X \to \{0,1\}^{\mathcal{A}}$ is measurable, injective and $\pi \circ T = S \circ \pi$. Let $\mu := m \circ \pi^{-1}$. By the universal measurability theorem, $\pi : (X, m) \to (\{0,1\}^{\mathcal{A}}, \mu)$ is an isomorphism of measure spaces and hence of the non-singular transformations S and T. The invertibility of T now follows from that of S. $\square$

REMARK.

Probability preserving transformations with no (non-trivial) factors are mentioned in [**Rudo**]. Non-singular transformations without absolutely continuous invariant probabilities and without invertible factors are constructed in [**Rudo-Sil**].

We continue with a description of the way factors "sit inside" their extensions.

1.0.11 PROPOSITION: DILATION OF FIBRE MEASURES OVER A FACTOR.

Let $(X, \mathcal{B}, m, T)$, $(Y, \mathcal{C}, \mu, S)$ be invertible non-singular transformations of standard probability spaces and suppose that $\pi : T \to S$ is a factor map with fibre measures $y \mapsto m_y \in \mathcal{P}(\pi^{-1}\{y\})$ measurable $Y_0 \to \mathcal{P}(X, \mathcal{B})$ where $Y_0 \in \mathcal{C}$, $\mu(Y_0) = 1$, then $m_{Sy} \circ T \sim m_y$ and

$$\frac{dm_{Sy} \circ T}{dm_y} = \frac{T'}{S'(y)} \quad m_y - a.e. \quad \forall \, y \in Y_0$$

where $T' := \frac{dm \circ T}{dm}$ and $S' := \frac{d\mu \circ S}{d\mu}$.

PROOF.

To see this let $f \in L^1(m)_+$, $g \in L^1(\mu)_+$, then

$$\int_Y \int_X T' f dm_y g(y) \mu(dy) = \int_X T' f g \circ \pi dm = \int_X f \circ T^{-1} g \circ S^{-1} \circ \pi dm$$

$$= \int_Y \int_X f \circ T^{-1} dm_y g(S^{-1}y) \mu(dy) = \int_Y \int_X f dm_{Sy} \circ T g(y) S'(y) \mu(dy)$$

and for μ-a.e. $y \in Y$

$$\int_X T' f dm_y = S'(y) \int_X f dm_{Sy} \circ T.$$

$\square$

Rokhlin's structure theorem gives more detailed information for ergodic transformations, showing that they are skew products (see chapter 8) over their factors.

1.0.12 ROKHLIN'S STRUCTURE THEOREM [**Ro1**].

Let $(X, \mathcal{B}, m, T)$, $(Y, \mathcal{C}, \mu, S)$ be ergodic, invertible, non-singular transformations and suppose that $\pi : T \to S$ is a factor map;

then $\exists$ a pure probability space $(I, \mathcal{T}, \nu)$ such that, up to isomorphism and equivalence of measures, $X = (Y \times I)$, $\mathcal{B} = \mathcal{C} \otimes \mathcal{T}$, $m = \mu \times \nu$ and $T(y, z) = (Sy, f(y, z))$.

In case $(X, \mathcal{B}, m, T)$, $(Y, \mathcal{C}, \mu, S)$ are measure preserving, then $(I, \mathcal{T}, \nu)$ is symmetric.

PROOF. Let $y \mapsto m_y$ be as in the disintegration theorem.

The function $p : X \to \mathbb{R}$ defined by $p(x) := m_{\pi(x)}(\{x\})$ is measurable. To see this let note that if $\alpha \subset \mathcal{B}$ is a countable partition of X, then $x \mapsto m_{\pi(x)}(\alpha(x))$ is measurable where $x \in \alpha(x) \in \alpha$. If α_n is a refining sequence of countable measurable partitions of X such that $\sup_{a \in \alpha_n} \text{diam.}(a) \to 0$ as $n \to \infty$ then $m_{\pi(x)}(\alpha_n(x)) \to p(x)$ a.e., whence p is measurable.

It follows that the set $[p > 0]$ is T-invariant and hence trivial.

Two cases are possible: $p = 0$, and $p > 0$.

In case $p = 0$, each measure m_y is non-atomic and we assume (without loss of generality) that $X = [0, 1]$ and define $\phi : X \to Y \times [0, 1]$ by

$$\phi(x) := (\pi(x), m_{\pi(x)}([0, x])).$$

Defining $\psi : Y \times I \to X$ by

$$\psi(y, t) := \min \{x \in X : \ m_y([0, x]) \geq t\}$$

we have that $\psi \circ \phi = \text{Id}$, whence

$$m(\phi^{-1}(A \times [0, c])) = \int_A m_y(\psi(A \times [0, c])) d\mu(y)$$

$$= \int_A m_y(\psi(\{y\} \times [0, c])) d\mu(y)$$

$$= \int_A m_y([0, \psi(y, c)]) d\mu(y)$$

$$= c\mu(A)$$

and $m \circ \phi^{-1} = \mu \times \lambda$ where λ is Lebesgue measure on $[0, 1]$.

In case $p > 0$, we have that each measure m_y is purely atomic and we may assume that $\pi^{-1}\{y\}$ is countable $\forall\ y \in Y_0$.

By the local invertibility lemma $\exists$ a partition $\{A_k\ :\ k \in \mathbb{N}\}$ of X into π-sections.

The function $y \mapsto |\pi^{-1}\{y\}| = \sum_{n=1}^{\infty} 1_{\pi A_k}(y)$ is measurable and S-invariant so $\exists\ N \in \mathbb{N} \cup \{\infty\}$ such that $|\pi^{-1}\{y\}| = N$ for a.e. $y \in Y_0$.

We claim that $\exists$ a partition $\{B_k\ :\ 1 \le k < N+1\}$ of X into onto π-sections; indeed supposing $\mu(Y \setminus \pi A_1) > 0$ set

$$C_{n+1} = A_{n+1} \cap \pi^{-1}(Y \setminus \bigcup_{k=1}^{n} \pi A_k)\quad (n \ge 1), \text{ and }\quad B_1 := A_1 \cup \bigcup_{n=2}^{\infty} C_k.$$

It follows that $\{B_1, A_2 \setminus C_2, \dots\}$ is a partition of X into π-sections and B_1 is an onto π-section. In case $N = 1$, $B_1 = X$.

In case $N > 1$, we consider $\pi_1 := \pi|_{X \setminus B_1} : X \setminus B_1 \to Y$. We have $|\pi_1^{-1}\{y\}| = |\pi_1^{-1}\{y\} \cap X \setminus B_1 = N - 1$ a.e., and that $\{A_n' := A_n \setminus C_n\}_{n=2}^{\infty}$ is a partition of $X \setminus B_1$ into π_1-sections. As before, in case $\pi_1 A_2' \ne Y$, we set

$$C_{n+1}' = A_{n+1}' \cap \pi^{-1}(Y \setminus \bigcup_{k=2}^{n} \pi A_k')\quad (n \ge 2), \text{ and }\quad B_2 := A_2' \cup \bigcup_{n=3}^{\infty} C_k'.$$

It follows that $\{B_2, A_3 \setminus C_3, \dots\}$ is a partition of $X \setminus B_1$ into π_1-sections and B_2 is an onto π_1-section. In case $N = 2$, $B_2 = X \setminus B_1$.

In case $N > 2$, we continue this process to obtain disjoint, onto π-sections $\{B_n\ :\ 1 \le n < N+1\}$, which form a partition of X since $\bigcup_{n=1}^{\infty} B_n \subseteq \bigcup_{n=1}^{\infty} A_n = X$ mod μ.

The theorem follows from this with $I = \{n \in \mathbb{N}\ :\ 1 \le n < N+1\}$.

Here, in case $(X, \mathcal{B}, m, T)$, $(Y, \mathcal{C}, \mu, S)$ are measure preserving we have that p is T-invariant, whence $N \in \mathbb{N}$. $\qquad\square$

Topological groups of non-singular transformations.

Given a σ-finite measure space $(Y, \mathcal{C}, \nu)$ let $\mathfrak{A}(Y, \mathcal{C}, \nu)$ denote the group of invertible non-singular transformations of $(Y, \mathcal{C}, \nu)$, and let $\mathfrak{B}(L^2(\nu))$ denote the group of invertible, bounded linear operators on $L^2(Y, \mathcal{C}, \nu)$.

$\mathfrak{B}(L^2(\nu))$ is a Polish space when equipped with the *weak topology*, defined by the metric

$$\rho(Q, R) := \sum_{n=1}^{\infty} \frac{1}{2^n}(\|Qf_n - Rf_n\|_2 + \|Q^{-1}f_n - R^{-1}f_n\|_2)$$

where $\{f_n : n \in \mathbb{N}\}$ is a orthonormal basis in $L^2(\nu)$.

It is also a group under composition, and a Polish (topological) group as $(V, W) \mapsto V \circ W^{-1}$ is continuous $\mathfrak{B}(L^2(\nu)) \times \mathfrak{B}(L^2(\nu)) \to \mathfrak{B}(L^2(\nu))$.

The subgroup of invertible unitary operators (isometries) $\mathcal{U}(L^2(\nu))$ is closed in $\mathfrak{B}(L^2(\nu))$, hence Polish too.

To each invertible non-singular transformation $T \in \mathfrak{A}(Y, \mathcal{C}, \nu)$, there is an associated positive isometry $U_T \in \mathcal{U}(L^2(\nu))$ defined by

$$U_T f := \sqrt{\frac{d\nu \circ T}{d\nu}} \, f \circ T,$$

and in fact, every positive, invertible isometry of $L^2(\nu)$ is obtained in this way, whence $\mathcal{U}_+(L^2(\nu)) := \{U_T \,:\, T \in \mathfrak{A}\}$ is closed in $\mathfrak{B}(L^2(\nu))$, and $\mathfrak{A}(Y, \mathcal{C}, \nu)$ is a Polish group.

We note that $T_n \to \mathrm{Id}$ in $\mathfrak{A}(Y, \mathcal{C}, \nu)$ iff

$$\frac{d\nu \circ T_n^{\pm 1}}{d\nu} \to 1 \text{ in measure, and } m(T_n^{\pm 1} A \triangle A) \to 0 \text{ as } n \to \infty \ \forall \ A \in \mathcal{C}.$$

The collection of invertible measure preserving transformations $\mathfrak{A}_0(Y, \mathcal{C}, \nu)$ is evidently a closed subgroup of $\mathfrak{A}(Y, \mathcal{C}, \nu)$, and hence Polish too.

§1.1 Recurrence and Conservativity

In this section (which is the real beginning of the book) we concern ourselves first with the *recurrence* properties of non-singular transformations.

There are non-singular transformations T of X which are *recurrent* in the sense that

$$\liminf_{n \to \infty} |h \circ T^n - h| = 0 \text{ a.e. } \forall \ h : X \to \mathbb{R} \text{ measurable.}$$

For a trivial example, a permutation S of a finite set F (preserving counting measure) is periodic in the sense that $\exists \ p$ such that $S^p \equiv \mathrm{Id}.$, and hence S is recurrent: $h \circ S^{pn} = h \ \forall \ n \geq 0, \ h : X \to \mathbb{R}$.

> A function $h : X \to \mathbb{R}$ (being a *"measurement"* of configurations of the system) defines a *stochastic process* $\{h \circ T^n : n \geq 0\}$ under the passage of time, (the values of which can be thought of as a sequences of measurements of the system at successive times).
>
> The property of recurrence is that all measurements almost surely, eventually almost recur, that history "repeats itself".
>
> The story is told about a disappointed angler who caught a large dolphin which escaped. The angler comforted himself with the thought that history repeats itself and hence at some time in the future, he would catch the same dolphin again. The dolphin had the same impression and lived in fear but after pondering the matter, realised that if history indeed repeats itself, he would again escape.

We begin by studying the extremely non-recurrent behaviour exhibited by *wandering sets*.

Let T be a non-singular transformation of the standard measure space $(X, \mathcal{B}, m)$.

A set $W \subset X$ is called a *wandering* set (for T) if the sets $\{T^{-n}W\}_{n=0}^\infty$ are disjoint. Let $\mathcal{W} = \mathcal{W}(T)$ denote the collection of measurable wandering sets.

Evidently, $\mathcal{W}$ is a hereditary collection (any subset of a wandering set is also wandering), and $T^{-1}\mathcal{W} \subseteq \mathcal{W}$.

The next results dichotomise, and localise the "recurrence" behaviour of a non-singular transformation.

1.1.1 HALMOS' RECURRENCE THEOREM [**Halm2**].
Suppose that $A \in \mathcal{B}$ $m(A) > 0$, then

$$m(A \cap W) = 0 \; \forall \, W \in \mathcal{W} \quad \Leftrightarrow \quad \sum_{n=1}^{\infty} 1_B \circ T^n = \infty \; a.e. \; on \; B \;\; \forall B \in \mathcal{B}_+ \cap A.$$

PROOF. We show first $\Rightarrow$.
Assume that $B \in \mathcal{B}$ and that $m(B \cap W) = 0 \; \forall \, W \in \mathcal{W}$. Let

$$B^- = \bigcup_{n=1}^{\infty} T^{-n} B.$$

Clearly, $T^{-1}B^- \subseteq B^-$. We claim that $W_B := B \setminus B^-$ is a wandering set for T. To see this, we note that $W_B \subset T^{-n}B^c \subset T^{-n}W_B^c \; \forall \, n \geq 1$, whence $T^{-n}W_B \cap T^{-m}W_B = \emptyset \; \forall \, n \neq m$. Since $W_B \in \mathcal{W}$, $W_B \subset B$, we have (by assumption) that $m(W_B) = 0$. Thus

$$B \subseteq B^- \quad \mathrm{mod} \; m.$$

By the non-singularity of T, $\forall \, n \geq 1$, $m(T^{-n}W_B) = 0$ and so

$$T^{-n}B \subseteq T^{-n}B^- \quad \mathrm{mod} \; m,$$

whence $(\mathrm{mod} \; m)$,

$$B \subseteq B^- = T^{-1}B^- = T^{-2}B^- = \cdots = \bigcap_{k=1}^{\infty} \bigcup_{n=k}^{\infty} T^{-n}B = [\sum_{n=1}^{\infty} 1_B \circ T^n = \infty].$$

Conversely, if $\exists W \in \mathcal{W}$ so that $m(W \cap A) > 0$, then clearly

$$\sum_{n=1}^{\infty} 1_B \circ T^n = 0 \; \text{on} \; B := W \cap A \in \mathcal{B}_+.$$

$\square$

DEFINITION: DISSIPATIVE, DISSIPATIVE PART.
The *dissipative part* of the non-singular transformation T is $\mathfrak{D}(T) := U(\mathcal{W}(T))$ the measurable union of the collection of wandering sets for T.

The non-singular transformation T is called (totally) *dissipative* if $\mathfrak{D}(T) = X$ mod m.

Since $T^{-1}\mathcal{W}(T) \subseteq \mathcal{W}(T)$, we have that $T^{-1}\mathfrak{D} \subseteq \mathfrak{D}$ mod m.

DEFINITION: CONSERVATIVE, CONSERVATIVE PART, HOPF DECOMPOSITION.
The set $\mathfrak{C}(T) := X \setminus \mathfrak{D}(T)$ is called the *conservative part* of T. The non-singular transformation T is called *conservative* if $\mathfrak{C}(T) = X$ mod m.

The *Hopf decomposition* of T is the partition $\{\mathfrak{C}(T), \mathfrak{D}(T)\}$.

1.1.2 PROPOSITION.
If T is invertible, then $\exists\ W \in \mathcal{W}$ such that

$$\mathfrak{D}(T) = \bigcup_{n \in \mathbb{Z}} T^n W \qquad \mod\ m.$$

PROOF.
Let $\mathfrak{I} := \{A \in \mathcal{B}:\ T^{-1}A = A\}$ and consider the collection

$$\mathfrak{w} := \{\bigcup_{n \in \mathbb{Z}} T^n W :\ W \in \mathcal{W}\}.$$

This is a hereditary subcollection of $\mathfrak{I}$. Clearly $U(\mathfrak{w}) \supset U(\mathcal{W})$. By the exhaustion lemma, $\exists\ W_k \in \mathcal{W}\ (k \geq 1)$ such that

$$U(\mathfrak{w}) = \bigcup_{k=1}^{\infty} \bigcup_{n \in \mathbb{Z}} T^n W_k,$$

whence if

$$W := W_1 \cup \bigcup_{k=1}^{\infty} \left(W_{k+1} \setminus \bigcup_{j=1}^{k} \bigcup_{n \in \mathbb{Z}} T^n W_j \right)$$

then $W \in \mathcal{W}$ and

$$U(\mathfrak{w}) = \bigcup_{n \in \mathbb{Z}} T^n W = \mathfrak{D}.$$

$\square$

REMARK. If $A \in \mathcal{B}$, and $T^{-1}A \subset A \subset \mathfrak{C}$, then $A = T^{-1}A \mod m$. This is because $A \setminus T^{-1}A \in \mathcal{W}$, whence $A \setminus T^{-1}A \subset \mathfrak{C} \cap \mathfrak{D} \mod m$.

Thus a conservative, ergodic non-singular transformation $T : X \to X$ has a stronger property that
$A \in \mathcal{B}$, $T^{-1}A \subset A \implies m(A) = 0$ or $m(A^c) = 0$.

A non-singular transformation $T : X \to X$ is called *incompressible* if $A \in \mathcal{B}$ and $T^{-1}A \subset A$ imply $A = T^{-1}A \mod m$.

1.1.3 PROPOSITION. *A non-singular transformation is incompressible iff it is conservative*

PROOF. The last remark shows that conservativity implies incompressibity. Conversely, suppose that a non-singular transformation $T : X \to X$ is not conservative and let $W \in \mathcal{W}_+$. Set $A := \bigcup_{n=0}^{\infty} T^{-n}W$, then $T^{-1}A \subset A$ but $A \setminus T^{-1}A = W$ whence $m(A \setminus T^{-1}A) > 0$ and T is not incompressible. $\square$

1.1.4 COROLLARY [**Halm2**]. *If T is a non-singular transformation, then*

$$\mathfrak{C}(T^n) = \mathfrak{C}(T) \mod m, \quad \forall\ n \geq 1$$

PROOF. We prove that $\mathfrak{D}(T^n) = \mathfrak{D}(T)$.
Since $\mathcal{W}(T) \subset \mathcal{W}(T^n)$ it is clear that

$$\mathfrak{D}(T) \subseteq \mathfrak{D}(T^n).$$

On the other hand, if $W \in \mathcal{W}(T^n)_+$, then $T^{-j}W \in \mathcal{W}(T^n)_+$ and

$$\sum_{k=1}^{\infty} 1_W \circ T^k = \sum_{j=1}^{n} \sum_{k=0}^{\infty} 1_{T^{-j}W} \circ T^{nk} \leq n \quad \text{a.e. on } W.$$

By Halmos' recurrence theorem,

$$\exists W' \in \mathcal{W}(T)_+ \cap W.$$

Thus $m(W \setminus \mathfrak{D}(T)) = 0 \ \forall \ W \in \mathcal{W}(T^n)$, whence

$$\mathfrak{D}(T^n) \subseteq \mathfrak{D}(T) \quad \mod \ m.$$

$\square$

1.1.5 POINCARÉ'S RECURRENCE THEOREM   [**Poi**].
Suppose that $T : X \to X$ *is a conservative, non-singular transformation of* $(X, \mathcal{B}, m)$. *If* (Z, d) *is a separable metric space, and* $f : X \to Z$ *is a measurable map, then*

$$\liminf_{n \to \infty} d(f(x), f(T^n x)) = 0 \text{ for a.e. } x \in X.$$

PROOF. Let $Z_1 = \{z \in Z : m(f^{-1}N(z, \epsilon)) > 0 \ \forall \ \epsilon > 0\}$, a closed subset of Z. Since (Z, d) is separable, $\exists \ N(x_n, \epsilon_n) \subset Z_1^c$ such that $Z_1^c = \bigcup_n N(x_n, \epsilon_n)$, whence $m(X \setminus f^{-1}(Z_1)) = 0$. We show that $\liminf_{n \to \infty} d(f(x), f(T^n x)) = 0$ for a.e. $x \in X_1 := f^{-1}(Z_1)$.
By the Halmos recurrence theorem, for $z \in Z_1$, and $\epsilon > 0$,

$$\sum_{k=1}^{\infty} 1_{f^{-1}N(z,\epsilon)} \circ T^k = \infty \text{ a.e. on } f^{-1}N(z, \epsilon),$$

whence

$$\liminf_{n \to \infty} d(f, f \circ T^n) < 2\epsilon \text{ a.e. on } f^{-1}N(z, \epsilon), \ \forall \ z \in Z_1, \ \epsilon > 0.$$

Fix $\epsilon > 0$. By separability, $\exists \ z_n \in Z_1$ such that $Z_1 = \bigcup_{n=1}^{\infty} N(z_n, \epsilon)$, whence

$$\liminf_{n \to \infty} d(f, f \circ T^n) < 2\epsilon \text{ a.e. on } f^{-1}Z_1 = X_1.$$

$\square$

If $m(\mathfrak{C}) > 0$, then $T^{-1}\mathfrak{C} \supset \mathfrak{C} \mod m$, and $T|_{\mathfrak{C}}$ is a conservative non-singular transformation on $\mathfrak{C}$ with respect to the measure $m|_{\mathfrak{C}}(\cdot) = m(\mathfrak{C} \cap \cdot)$.

EXERCISE 1.1.1. For (Z, d) a separable metric space, call a non-singular transformation T on $(X, \mathcal{B}, m)$ Z-*recurrent* if $\liminf_{n \to \infty} d(f, f \circ T^n) = 0$ a.e. whenever $f : X \to Z$ is a measurable.[1]

Prove that the following conditions on T are equivalent.

1) T is Z-recurrent for some separable metric space (Z, d) containing at least two points.

2) T is Z-recurrent for every separable metric space (Z, d).

3) T is conservative.

Conditions for conservativity.

If there exists a finite, T-invariant measure $q << m$, then clearly there can be no wandering sets with positive q-measure, whence $q(\mathfrak{D}) = 0$ and $[\frac{dq}{dm} > 0] \subseteq \mathfrak{C}$ mod m.

In particular, any probability preserving transformation is conservative. A measure preserving transformation of a σ-finite, infinite measure space may not be conservative. For example $x \mapsto x + 1$ is a measure preserving transformation of $\mathbb{R}$ equipped with Borel sets, and Lebesgue measure, which is totally dissipative. The following propositions help establish conservativity of measure preserving transformations of σ-finite measure spaces.

1.1.6 PROPOSITION   [**Hop1**]. *Suppose that T is a measure preserving transformation of the σ-finite measure space $(X, \mathcal{B}, m)$.*
1) If $f \in L^1(m)$, $f \geq 0$, then

$$\left[\sum_{n=1}^{\infty} f \circ T^n = \infty\right] \subseteq \mathfrak{C}(T) \qquad \text{mod } m.$$

2 If $f \in L^1(m), f > 0$ a.e., then

$$\mathfrak{C}(T) = \left[\sum_{n=1}^{\infty} f \circ T^n = \infty\right] \qquad \text{mod } m.$$

PROOF. We first show 1). Let $f \in L^1(m)_+$, it is sufficient to show that $\sum_{n=1}^{\infty} f \circ T^n < \infty$ a.e. on any $W \in \mathcal{W}$.

To see this let $W \in \mathcal{W}$ and $n \geq 1$, then

$$\int_W \sum_{k=0}^{n} f \circ T^k dm = \sum_{k=0}^{n} \int_X 1_W f \circ T^{n-k} dm$$

$$= \sum_{k=0}^{n} \int_X 1_W \circ T^k f \circ T^n dm$$

$$= \int_X \left(\sum_{k=0}^{n} 1_W \circ T^k\right) f \circ T^n dm$$

$$\leq \int_X f \circ T^n dm$$

$$= \int_X f dm$$

---

[1]Note that "recurrence" is now $\mathbb{R}$-recurrence.

whence $\int_W \sum_{k=0}^{\infty} f \circ T^k dm \leq \int_X f dm < \infty$ and $\sum_{n=1}^{\infty} f \circ T^n < \infty$ a.e. on W.

We now prove 2). In view of 1), for $f \in L^1(m), f > 0$ a.e., it suffices to show that

$$\mathfrak{C}(T) \subset \left[\sum_{n=1}^{\infty} f \circ T^n = \infty \right] \quad \text{mod } m.$$

Indeed, $\forall A \in \mathcal{B}_+$, $A \subset \mathfrak{C}(T)$, $\exists B \in \mathcal{B}_+$, $B \subset A$ and $\epsilon > 0$ such that $f \geq \epsilon 1_B$ on B; whence

$$\sum_{n=1}^{\infty} f \circ T^n \geq \epsilon \sum_{n=1}^{\infty} 1_B \circ T^n = \infty$$

a.e. on B by Halmos's recurrence theorem (1.1.1). $\qquad\square$

1.1.7 MAHARAM'S RECURRENCE THEOREM   [**Mah**]. *Suppose that T is a measure preserving transformation of the σ-finite measure space $(X, \mathcal{B}, m)$.*

If $\exists A \in \mathcal{B}$, $m(A) < \infty$ such that

$$X = \bigcup_{n=0}^{\infty} T^{-n} A \quad \text{mod } m,$$

then T is conservative.

PROOF. Note that

$$X = T^{-k} X = \bigcup_{n=k}^{\infty} T^{-n} A \quad \text{mod } m \ \forall \, k \geq 1,$$

$$\therefore \sum_{n=1}^{\infty} 1_A \circ T^n = \infty \text{ a.e.}$$

and the proposition follows from proposition 1.1.6. $\qquad\square$

EXERCISE 1.1.2.   Let $X = \mathbb{R}$, m be Lebesgue measure, and $T : \mathbb{R} \to \mathbb{R}$ be "Boole's transformation" defined by $Tx = x - \frac{1}{x}$.
(1)   Prove the theorem of G.Boole ([**Boo**]) that $m \circ T^{-1} = m$.
(2)   Show using Maharam's recurrence theorem that T is conservative.

Suppose that T is a measure preserving transformation of the σ-finite measure space $(X, \mathcal{B}, m)$ and let $A \in \mathcal{B}$, $0 < m(A) < \infty$.

If $m(A \cap \mathfrak{C}) > 0$, then

$$\sum_{k=0}^{\infty} m(A \cap T^{-k} A) = \int_A \left(\sum_{k=0}^{\infty} 1_A \circ T^k \right) dm$$

$$\geq \int_{A \cap \mathfrak{C}} \left(\sum_{k=0}^{\infty} 1_{A \cap \mathfrak{C}} \circ T^k \right) dm$$

$$= \infty.$$

Unfortunately, the converse to this is wrong. There are measure preserving transformations T and sets of finite measure $A \in \mathcal{B}$ so that

$$\sum_{k=0}^{\infty} m(A \cap T^{-k}A) = \infty, \text{ but } \sum_{k=0}^{\infty} 1_A \circ T^k < \infty, \text{ a.e.}$$

EXAMPLE.
Let $(X, \mathcal{B}, m)$ be $\mathbb{R}$ equipped with Borel sets and Lebesgue measure, and let $Tx = x + 1$ which is totally dissipative.
Let

$$A := \bigcup_{n=1}^{\infty} \left[n, n + \frac{1}{n(n+1)} \right),$$

then $0 < m(A) < \infty$ whence $\sum_{k=0}^{\infty} 1_A \circ T^k < \infty$ a.e.. However

$$A \cap T^{-k}A = \bigcup_{n=1}^{\infty} \left[n, n + \frac{1}{(n+k)(n+k+1)} \right),$$

whence

$$\sum_{k=0}^{\infty} m(A \cap T^{-k}A) = \sum_{k=0}^{\infty} \frac{1}{(k+1)} = \infty.$$

Nevertheless,

1.1.8 PROPOSITION.
Let T be a measure preserving transformation of the σ-finite measure space $(X, \mathcal{B}, m)$ and let $A \in \mathcal{B}$, $0 < m(A) < \infty$ satisfy

$$\sum_{k=0}^{\infty} m(A \cap T^{-k}A) = \infty.$$

If

(1)
$$\sup_{n \geq 1} \frac{\int_A \left(\sum_{k=0}^{n-1} 1_A \circ T^k \right)^2 dm}{\left(\int_A \sum_{k=0}^{n-1} 1_A \circ T^k dm \right)^2} < \infty,$$

then $m(A \cap \mathfrak{C}) > 0$.

PROOF. (c.f.[**Ren2**]) We'll prove that

(2)
$$m\left(A \cap [\sum_{k=0}^{\infty} 1_A \circ T^k = \infty] \right) > 0.$$

The proposition will then follow from proposition 1.1.6 1).
Let

$$S_n := \sum_{k=0}^{n-1} 1_A \circ T^k, \ a_n = \int_A S_n dm, \ \& \ \phi_n = \frac{S_n}{a_n},$$

then $\int_A \phi_n dm = 1$, and by (1) $\exists\ M > 0$ such that

$$\int_A \phi_n^2 dm \le M\ \forall\ n \ge 1.$$

Set

$$A_\infty := [\sum_{k=0}^\infty 1_A \circ T^k = \infty] = [S_n \to \infty].$$

By assumption $a_n \uparrow \infty$ as $n \uparrow \infty$, whence $\phi_n dm \to 0$ as $n \to \infty$ on $A \setminus A_\infty$.
   We claim that

$$\int_{A \setminus A_\infty} \phi_n dm \to 0.$$

To see this, note that for any $K > 0$, $n \ge 1$,

$$\int_{A \setminus A_\infty} \phi_n dm \le \int_{A \setminus A_\infty} \left(\phi_n 1_{[0,K]}(|\phi_n|) + \phi_n 1_{[K,\infty)}(|\phi_n|) \right) dm$$

$$\le \int_{A \setminus A_\infty} \phi_n 1_{[0,K]}(|\phi_n|) dm + \frac{M}{K}.$$

For any $K > 0$,

$$\int_{A \setminus A_\infty} \phi_n 1_{[0,K]}(|\phi_n|) dm \to 0$$

as $n \to \infty$ by the bounded convergence theorem, whence, given $\epsilon > 0$ we can choose
$K > \frac{M}{\epsilon}$ to obtain

$$\limsup_{n \to \infty} \int_{A \setminus A_\infty} \phi_n dm < \epsilon.$$

Thus

$$1 \leftarrow 1 - \int_{A \setminus A_\infty} \phi_n dm = \int_{A_\infty} \phi_n dm \le \sqrt{m(A_\infty)} \|\phi_n\|_2 \le \sqrt{M m(A_\infty)}.$$

$\square$

DEFINITION: RENYI INEQUALITY. We'll call the inequality (1) a *Renyi in-equality*.

REMARKS.
   1) Proposition 1.1.8 tells us that a measure preserving transformation which has a set satisfying a Renyi inequality and infinite mean number of returns is not totally dissipative. This is useful when it is already known that the Hopf decomposition is trivial (see chapter 7).
   2) Any set in a finite measure preserving transformation satisfies a Renyi inequality. This is not the case when the invariant measure is infinite (as seen above).
   3) For ergodic properties following from Renyi inequalities see §3.3.

§1.2 Ergodicity

Recall that a non-singular transformation T of the measure space $(X, \mathcal{B}, m)$ is called *ergodic* if

$$A \in \mathcal{B},\ T^{-1}A = A \quad \mathrm{mod}\ m \ \Rightarrow\ m(A) = 0,\ \text{or}\ m(A^c) = 0.$$

1.2.1 PROPOSITION. *An invertible ergodic non-singular transformation of a non-atomic measure space is necessarily conservative.*

PROOF. Suppose that T is an invertible ergodic non-singular transformation of the non-atomic measure space $(X, \mathcal{B}, m)$. By the Hopf decomposition for invertible transformations, $T^{-1}\mathfrak{D} = \mathfrak{D}$ mod m, whence if T is not conservative, then T is totally dissipative, and

$$\exists\, W \in \mathcal{W}_+ \ni X = \bigcup_{n \in \mathbb{Z}} T^n W \quad \text{mod } m.$$

Since $(X, \mathcal{B}, m)$ is non-atomic, $\exists\, W_1 \in \mathcal{B}$, $W_1 \subset W$, $m(W_1)$, $m(W \setminus W_1) > 0$, with the consequence that if $A = \bigcup_{n \in \mathbb{Z}} T^n W_1$, then $T^{-1}A = A$, $m(A) > 0$, and $m(A^c) > 0$ contradicting the ergodicity of T. □

EXERCISE 1.2.1. Prove that an invertible, dissipative, ergodic, non-singular transformation of a σ-finite measure space is necessarily isomorphic to $x \mapsto x + 1$ on $\mathbb{Z}$ equipped with an appropriate measure.

REMARK. There are totally dissipative, ergodic measure preserving transformations of non-atomic, σ-finite measure spaces. These transformations, to be seen in the sequel, are not invertible.

1.2.2 PROPOSITION. *Let T be a non-singular transformation, then*

$$T \text{ is conservative and ergodic} \quad \Leftrightarrow \quad \sum_{n=1}^{\infty} 1_A \circ T^n = \infty \text{ a.e. } \forall A \in \mathcal{B}_+.$$

PROOF. Assume first that T is conservative and ergodic. For any $A \in \mathcal{B}_+$, the set

$$[\sum_{n=1}^{\infty} 1_A \circ T^n = \infty] \in \mathcal{B}$$

and is T-invariant. By conservativity, via Halmos' recurrence theorem,

$$A \subset [\sum_{n=1}^{\infty} 1_A \circ T^n = \infty] \quad \text{mod } m,$$

whence by ergodicity of T,

$$[\sum_{n=1}^{\infty} 1_A \circ T^n = \infty] = X \quad \text{mod } m.$$

Conversely, if $A \in \mathcal{B}_+$ is either a T-invariant set with $m(A^c) > 0$, or a wandering set, then

$$[\sum_{n=1}^{\infty} 1_A \circ T^n = \infty] \neq X \quad \text{mod } m.$$

□

In analogue to Poincaré's recurrence theorem, we have

1.2.3 COROLLARY. *Suppose that $T : X \to X$ is a conservative, ergodic non-singular transformation of $(X, \mathcal{B}, m)$. If (Z, d) is a separable metric space, and $f : X \to Z$ is a measurable map, then for a.e. $x \in X$,*

$$\overline{\{f(T^n x) : n \in \mathbb{N}\}} = \operatorname{spt} m \circ f^{-1}.$$

PROOF. Set $Z_1 := \operatorname{spt} m \circ f^{-1} = \{z \in Z : m(f^{-1} N(z, \epsilon)) > 0 \, \forall \, \epsilon > 0\}$.

For $z \notin Z_1$, $\exists \, \epsilon > 0$ such that $m(f^{-1} N(z, \epsilon)) = 0$, whence $\liminf_{n \to \infty} d(z, f \circ T^n) \geq \epsilon$ a.e., and so for a.e. $x \in X$, $\overline{\{f(T^n x) : n \in \mathbb{N}\}} \subseteq Z_1$.

By proposition 1.2.2, for fixed $z \in Z_1$, and $\epsilon > 0$,

$$\sum_{k=1}^{\infty} 1_{f^{-1} N(z, \epsilon)} \circ T^k = \infty \text{ a.e.}$$

whence

$$\liminf_{n \to \infty} d(z, f \circ T^n) < \epsilon \text{ a.e.}$$

This is true $\forall \, \epsilon > 0$, and so

$$\liminf_{n \to \infty} d(z, f \circ T^n) = 0 \text{ a.e.}$$

$\square$

EXERCISE 1.2.2. Prove that a non-singular transformation T is conservative and ergodic if, and only if

$$\sum_{n=1}^{\infty} f \circ T^n = \infty \text{ a.e. } \forall \, f : X \to \mathbb{R}_+, \text{ measurable }, \int_X f \, dm > 0.$$

Cartesian product transformations. Let T_i $(i = 1, 2)$ be non-singular transformations of the standard probability spaces $(X_i, \mathcal{B}_i, m_i)$. The *Cartesian product transformation* $T = T_1 \times T_2$ defined on the product space $(Y, \mathcal{B}, \mu) := (X_1 \times X_2, \mathcal{B}_1 \otimes \mathcal{B}_2, m_1 \times m_2)$ by $T(x_1, x_2) = (T_1 x_1, T_2 x_2)$.

1.2.4 PROPOSITION.

(i) *If T_1 is a probability preserving transformation, and T_2 is conservative, then $T_1 \times T_2$ is conservative.*

(ii) *If T_i $(i = 1, 2)$ are conservative, ergodic then $T_1 \times T_2$ is either conservative, or totally dissipative.*

PROOF.

(i) It's not hard to see that $T_1 \times T_2$ is conservative, for if $W \in \mathcal{W}(T_1 \times T_2)_+$, and $f : X_2 \to \mathbb{R}_+$ is defined by

$$f(y) = \int_{X_1} 1_W(x, y) \, dm_1(x),$$

then

$$\int_{X_2} f \, dm_2 = m_1 \times m_2(W),$$

and

$$\sum_{n=1}^{\infty} f(T_2^n y) = \sum_{n=1}^{\infty} \int_{X_1} 1_W(x, T_2^n y) dm_1(x)$$

$$= \sum_{n=1}^{\infty} \int_{X_1} 1_W(T_1^n x, T_2^n y) dm_1(x) \quad \because m_1 \circ T_1^{-1} = m_1$$

$$\leq 1$$

contradicting conservativity of T_2.

(ii) Note that $R^{-1}\mathfrak{D}(T) \subset \mathfrak{D}(T)$ whenever R is a non-singular transformation of Y commuting with T (i.e. $TR = RT$). In our case, such transformations are given by $R_{k,\ell} := T_1^k \times T_2^\ell$ $(k, \ell \geq 0)$.

Our claim is established by showing that the transformations $\{R_{k,\ell} : k, \ell \geq 0\}$ act ergodically in the sense that

$$A \in \mathcal{B}, \ R_{k,\ell}^{-1} A \subset A \implies m(A) = 0, 1.$$

To see this, for $y \in X_2$ set $A_y := \{x \in X_1 : (x, y) \in A\}$, then $(R_{k,\ell}^{-1} A)_y = T_1^{-k}(A_{T_2^\ell y})$, whence:

$T_1^{-1} A_y = (R_{1,0}^{-1} A)_y \subset A_y$, and $m_1(A_y) = 0, 1$ for $y \in X_2$;

$A = X_1 \times B \bmod \mu$ where $B := \{y \in X_2 : m_1(A_y) = 1\} \in \mathcal{B}_2$,

whence $X_1 \times B \supset R_{0,1}^{-1} X_1 \times B = X_1 \times T_2^{-1} B; T_2^{-1} B \subset B$ and $m_2(B) = 0, 1.$ $\square$

One of the basic skills of ergodic theory is knowing how to prove ergodicity. There are many methods of proving ergodicity of non-singular transformations. We'll illustrate some by example.

One of our examples will be a conservative, ergodic, invertible, non-singular transformation without absolutely continuous invariant measure (see theorem 1.2.9 below). We'll consider conditions for existence of invariant measures later.

DEFINITION: ROTATIONS OF THE CIRCLE.

Let X be the circle $\mathbb{T} = \mathbb{R}/\mathbb{Z} = [0, 1)$, $\mathcal{B}$ be its Borel sets, and m be Lebesgue measure. The *rotation* (or translation) of the circle by $x \in X$ is the transformation $r_x : X \to X$ defined by $r_x(y) = x + y \mod 1$.

Evidently $m \circ r_x = m$ for every $x \in X$ and each r_x is an invertible measure preserving transformation of $(X, , \mathcal{B}, m)$.

1.2.5 PROPOSITION.
r_x is ergodic if and only if x is irrational.

PROOF.
Clearly if $x = \frac{k}{n}$, and

$$A = \bigcup_{j=0}^{n-1} [\frac{j}{n}, \frac{j + 1/2}{n}),$$

then $r_x A = A$, and $m(A) = 1/2$, whence r_x is not ergodic.

If $x \in X$ is irrational, one way to prove ergodicity is to use harmonic analysis. Suppose that $f : X \to \mathbb{R}$ is bounded and measurable, and that $f \circ r_x = f$, then

$$\hat{f}(n) = \int_{[0,1)} f(y)e^{-2\pi iny}dy$$

$$= \int_{[0,1)} f(x+y)e^{-2\pi iny}dy$$

$$= \int_{[0,1)} f(y)e^{-2\pi in(y-x)}dy$$

$$= e^{2\pi inx}\hat{f}(n).$$

It follows that

$$e^{2\pi inx} = 1 \text{ whenever } \hat{f}(n) \neq 0,$$

whence, since $x \neq \frac{k}{n}$, $\hat{f}(n) = 0$ whenever $n \neq 0$ and f is constant. $\qquad \square$

REMARK. This method of proof works, to show that if $(X, \mathcal{B}, m)$ is a compact, Abelian topological group considered with respect to Haar measure (see [**Hew-Ros**]), and $x \in X$ is such that $\overline{\{nx : n \in \mathbb{Z}\}} = X$ then r_x is ergodic on X.

Another way to prove ergodicity is to establish a limit theorem.

DEFINITION: THE ADDING MACHINE.
Let $\Omega = \{0,1\}^{\mathbb{N}}$, and $\mathcal{B}$ be the σ-algebra generated by cylinders. Define the *adding machine* $\tau : \Omega \to \Omega$ by

$$\tau(1, ..., 1, 0, \epsilon_{n+1}, \epsilon_{n+2}, ...) = (0, ..., 0, 1, \epsilon_{n+1}, \epsilon_{n+2}, ...).$$

The reason for the name "adding machine" is that

$$\sum_{k=1}^{\infty} 2^{k-1}(\tau^n \underline{0})_k = n \quad \forall \, n \geq 1$$

where $(\underline{0})_k = 0 \, \forall \, k \geq 1$. More generally, we have the so called *odometer property* that for any $x \in \Omega$, and $n \geq 1$,

$$\{\sum_{j=1}^{n} 2^{j-1}(\tau^k x)_j :: 0 \leq k \leq 2^n - 1\} = \{1, 2, \ldots, 2^n\}$$

and

$$\{((\tau^k x)_1, ..., (\tau^k x)_n) : 0 \leq k \leq 2^n - 1\} = \{0,1\}^n.$$

Define a probability m on Ω by

$$m([\epsilon_1, ..., \epsilon_n]) = \left(\frac{1}{2}\right)^n.$$

1.2.6 PROPOSITION.
τ *is an ergodic measure preserving transformation of* $(\Omega, \mathcal{B}, m)$.

PROOF.

Evidently $m \circ \tau^{-1} = m$.

If $n \in \mathbb{N}$ is fixed, and $g : \{0,1\}^n \to \mathbb{R}$, and $f : \Omega \to \mathbb{R}$ is defined by $f(x) = g(x_1, ..., x_n)$, then by the odometer property,

$$\frac{1}{2^n} \sum_{k=0}^{2^n-1} f \circ \tau^k \equiv \int_\Omega f dm,$$

whence

$$\frac{1}{N} \sum_{k=0}^{N-1} f \circ \tau^k \to \int_\Omega f dm \text{ as } N \to \infty \text{ uniformly on } \Omega.$$

Since functions of this form are dense in $L^1(m)$, it follows that

$$\frac{1}{N} \sum_{k=0}^{N-1} F \circ \tau^k \xrightarrow{L^1(m)} \int_\Omega F dm \text{ as } N \to \infty,$$

whence, if $F \in L^1(m)$, is τ-invariant, then

$$F = \int_\Omega F dm \text{ a.e.}$$

$\square$

EXERCISE 1.2.3. When considered with the product discrete topology, Ω is a compact metric space (with metric for example $d(x, x') = 2^{-\min\{n \geq 1: \ x_n \neq x'_n\}}$) and τ is a homeomorphism. Continuous functions on Ω are uniform limits of functions f of form $f(x) = g(x_1, ..., x_N)$.

Prove that if $f : \Omega \to \mathbb{R}$ is continuous, then

$$\frac{1}{n} \sum_{k=0}^{n-1} f \circ \tau^k \to \int_\Omega f dm \text{ as } n \to \infty \text{ uniformly on } \Omega,$$

whence the only τ-invariant probability on $(\Omega, \mathcal{B})$ is m.

DEFINITION: DYADIC INTEGERS [**Hew-Ros**].

For $x = (x_1, x_2, ...)$, $y = (y_1, y_2, ...) \in \Omega$, define $x + y \in \Omega$ by

$$(x + y)_n = x_n + y_n + \epsilon_n \quad \mod 2$$

where $\epsilon_1 = 0$ and

$$\epsilon_{n+1} = \begin{cases} 0 & x_n + y_n + \epsilon_n \leq 1, \\ 1 & x_n + y_n + \epsilon_n \geq 2. \end{cases}$$

$(\Omega, +)$ is a compact topological group and is called the group of *dyadic integers*.

REMARK.

Normalised Haar measure on Ω is m. The ergodicity of τ could also be established using harmonic analysis (as in proposition 1.2.5) since $\tau(x) \equiv x + \underline{1}$ where $\underline{1} = (1, 0, ...)$, and $\{n\underline{1}\}_{n \in \mathbb{Z}} = \{x \in \Omega : \exists \lim_{n \to \infty} x_n\}$ which is dense in Ω.

DEFINITION: KRONECKER TRANSFORMATION.

A *Kronecker transformation* is an ergodic translation of a compact metric group considered with respect to normalised Haar measure.

REMARKS.

1) Suppose that X is a compact metric group and that $T : X \to X$ is defined by $Tx = \eta x$ (some $\eta \in X$). As above, it can be shown that T is ergodic (with respect to normalised Haar measure) if and only if $\overline{\{n\eta : n \in \mathbb{Z}\}} = X$. Consequently, a compact metric group admits a Kronecker transformation if and only if it is monothetic (and hence Abelian).

2) Suppose that T is a Kronecker transformation of the compact metric group X, and suppose that $Tx = \eta x$. If p is another T-invariant probability on $(X, \mathcal{B})$, then using the continuity of $x \mapsto f \circ r_x$ $(X \to C(X))$ $\forall\, f : X \to \mathbb{R}$ continuous, and the density of $\{n\eta : n \in \mathbb{Z}\}$, we see that $p \circ r_x = p\ \forall\, x \in X$ whence by unicity of Haar measure, $p = m$.

It follows from this that if $f : X \to \mathbb{R}$ is continuous, then

$$\frac{1}{n} \sum_{k=0}^{n-1} f \circ \eta^k \to \int_X f\, dm \text{ as } n \to \infty \text{ uniformly on } X.$$

Sometimes it's easier to prove more than ergodicity.

DEFINITION: ONE-SIDED BERNOULLI SHIFT.

Let $X = \mathbb{R}^{\mathbb{N}}$ and let $\mathcal{B}(X)$ be the σ-algebra generated by *cylinder* sets of form $[A_1, \ldots, A_n] := \{\underline{x} \in X : x_j \in A_j,\ 1 \leq j \leq n\}$, where $A_1, \ldots, A_n \in \mathcal{B}(\mathbb{R})$ (the Borel subsets of $\mathbb{R}$), and let the *shift* $S : X \to X$ be defined by

$$(Sx)_n = x_{n+1}.$$

For $p : \mathcal{B}(\mathbb{R}) \to [0,1]$ a probability, let $\mu_p : \mathcal{B}(X) \to [0,1]$ be the probability satisfying

$$\mu_p([A_1, ..., A_n]) = \prod_{k=1}^{n} p(A_k)\ \ (A_1, \ldots, A_n \in \mathcal{B}(\mathbb{R})).$$

Evidently $S^{-1}[A_1, ..., A_n] = [\mathbb{R}, A_1, ..., A_n]$ whence $\mu_p \circ S^{-1} = \mu_p$.

The *one-sided Bernoulli shift* with marginal distribution p is the measure preserving transformation S of $(X, \mathcal{B}, \mu_p)$.

DEFINITION: TAIL, EXACTNESS.

Let T be a non-singular transformation of $(X, \mathcal{B}, m)$. The *tail* σ-algebra of T is

$$\mathfrak{T}(T) := \bigcap_{n=1}^{\infty} T^{-n} \mathcal{B}.$$

The transformation T is called *exact* if $\mathfrak{T}(T) = \{\emptyset, X\}$ mod m.

Evidently $\mathfrak{I}(T) \subset \mathfrak{T}(T)$ mod m and so exact transformations are ergodic.

1.2.7 KOLMOGOROV'S ZERO-ONE LAW   [Kol].
The one-sided Bernoulli shift is exact.

PROOF.

Suppose that $B \in \mathcal{B}$ is a finite union of cylinders. If the length of the longest cylinder in the union is n, then

$$\mu_p(B \cap S^{-n}C) = \mu_p(B)\mu_p(C) \ \ \forall \, C \in \mathcal{B}.$$

Now suppose $A \in \mathfrak{T}$. Since, for each $n \in \mathbb{N}$,

$$A = S^{-n}A_n \text{ where } A_n \in \mathcal{B}, \ \mu_p(A_n) = \mu_p(A),$$

we have that

$$\mu_p(B \cap A) = \mu_p(B)\mu_p(A)$$

for $B \in \mathcal{B}$ a finite union of cylinders, and hence $\forall \, B \in \mathcal{B}$. This implies that

$$0 = \mu_p(A \cap A^c) = \mu_p(A)(1 - \mu_p(A))$$

demonstrating that $\mathfrak{T}$ is trivial mod μ_p.   □

Note that no invertible non-singular transformation can be exact. Hence an irrational rotation of $\mathbb{T}$ is ergodic, but not exact.

DEFINITION: TWO SIDED BERNOULLI SHIFT.

The *two sided* Bernoulli shift is defined with $X = \mathbb{R}^{\mathbb{Z}}$, $\mathcal{B}(X)$ the σ-algebra generated by cylinder sets of form $[A_1, \ldots, A_n]_k := \{\underline{x} \in X : x_{j+k} \in A_j, \ 1 \leq j \leq n\}$, where $A_1, \ldots, A_n \in \mathcal{B}(\mathbb{R})$. The shift $S : X \to X$ is defined as before by $(Sx)_n = x_{n+1}$, and the S-invariant probability $\mu_p : \mathcal{B}(X) \to [0,1]$ is defined (for $p : \mathcal{B}(\mathbb{R}) \to [0,1]$ a probability) by

$$\mu_p([A_1, ..., A_n]) = \prod_{k=1}^{n} p(A_k) \ \ (A_1, \ldots, A_n \in \mathcal{B}(\mathbb{R})).$$

The two sided Bernoulli shift is an invertible measure preserving transformation (and is hence not exact).

EXERCISE 1.2.4.   Show that the two sided Bernoulli shift is *mixing* in the sense that

$$\mu_p(A \cap T^{-n}B) \to \mu_p(A)\mu_p(B) \text{ as } n \to \infty \ \ \forall \, A, B \in \mathcal{B}(X),$$

and hence ergodic.

More generally, if Y is a Polish space, and for $n \geq 1$, $P_n \in \mathcal{P}(Y^n)$ are such that

$$P_{n+1}(A_1 \times \cdots \times A_n \times Y) = P_{n+1}(Y \times A_1 \times \cdots \times A_n) = P_n(A_1 \times \cdots \times A_n),$$

it can be shown using Kolmogorov's existence theorem ([**Kol**], see also [**Part**]) that there are shift invariant measures

$$P_1 \in \mathcal{P}(Y^{\mathbb{N}}) \ni \ P_n([A_1, \cdots, A_n]) = P_n(A_1 \times \cdots \times A_n),$$

$$P_2 \in \mathcal{P}(Y^{\mathbb{Z}}) \ni \ P_n([A_1, \cdots, A_n]_k) = P_n(A_1 \times \cdots \times A_n),$$

the *one sided-* and *two sided* shifts of $\{P_n\}_{n\in\mathbb{N}}$ (which may, or may not be ergodic).

A non-singular adding machine.

Let Ω be the group of dyadic integers, and let $\tau x = x + \underline{1}$.

For $p \in (0,1)$, define a probability μ_p on Ω by

$$\mu_p([\epsilon_1, ..., \epsilon_n]) = \prod_{k=1}^{n} p(\epsilon_k)$$

where $p(0) = 1 - p$ and $p(1) = p$.

Recall that $\mu_{\frac{1}{2}} = m$ is Haar measure on Ω, whence $\mu_{\frac{1}{2}} \circ \tau = \mu_{\frac{1}{2}}$. It is no longer true that τ preserves μ_p if $p \neq 1/2$.

1.2.8 PROPOSITION.

τ *is an invertible, conservative, ergodic non-singular transformation of* $(\Omega, \mathcal{B}, \mu_p)$.

PROOF.

We show first that $\mu_p \circ \tau \sim \mu_p$. More specifically,

$$\frac{d\mu_p \circ \tau}{d\mu_p} = \left(\frac{1-p}{p}\right)^{\phi}$$

where

$$\phi(x) = \min\{n \in \mathbb{N} : x_n = 0\} - 2.$$

To see this we show that for any set $A \in \mathcal{B}$,

$$\mu_p(\tau A) = \int_A \left(\frac{1-p}{p}\right)^{\phi} d\mu_p.$$

Consider first a cylinder set $A \subset [\phi = k - 2]$ $(k \geq 1)$

$$A = [\underbrace{1, \ldots, 1}_{k-1 \text{ times}}, 0, a_1, \ldots, a_k],$$

then

$$\tau A = [\underbrace{0, \ldots, 0}_{k-1 \text{ times}}, 1, a_1, \ldots, a_k],$$

and

$$\mu_p(\tau A) = \mu_p([\underbrace{0, \ldots, 0}_{k-1 \text{ times}}, 1]) \mu_p([a_1, \ldots, a_k])$$

$$= \left(\frac{1-p}{p}\right)^{k-2} \mu_p(A)$$

$$= \int_A \left(\frac{1-p}{p}\right)^{\phi} d\mu_p,$$

so $\mu_p(\tau A) = \int_A \left(\frac{1-p}{p}\right)^{\phi} d\mu_p$.

Now consider the collection $\mathcal{C}$ of sets $A \in \mathcal{B}$ satisfying $\mu_p(\tau A) = \int_A \left(\frac{1-p}{p}\right)^{\phi} d\mu_p$.

By the above, $\mathcal{C}$ contains all cylinder sets, whence the algebra generated by cylinder sets. It is evidently a monotone class, and hence the σ-algebra generated by cylinder sets. Thus $\mathcal{C} = \mathcal{B}$ and our claim is established.

We show that τ is ergodic on $(\Omega, \mathcal{B}, \mu_p)$. Suppose $x \in \Omega$, then using the odometer property,

$$\tau^{-\nu_n} x = (0, ..., 0, x_{n+1}, x_{n+2}, ...)$$

where

$$\nu_n = \sum_{k=1}^{n} 2^{k-1} x_k.$$

Thus, if $f : \Omega \to \mathbb{R}$, then, for every $n \in \mathbb{N}$, there exists $g_n : \Omega \to \mathbb{R}$ such that

$$\sum_{k=0}^{2^n - 1} f(\tau^{k-\nu_n} x) = \sum_{\underline{\epsilon} \in \{0,1\}^n} f(\underline{\epsilon}, x_{n+1}, x_{n+2}, ...) := g_n(S^n x).$$

In particular, if f is τ-invariant, and $\mathcal{B}$-measurable, then f is $S^{-n}\mathcal{B}$-measurable for $n \in \mathbb{N}$, and, by Kolmogorov's $0 - 1$ law (proposition 1.2.7), constant.

By proposition 1.2.1, τ is conservative on (the non-atomic) $(\Omega, \mathcal{B}, \mu_p)$. □

1.2.9 THEOREM [**Arn**].   *There is no τ-invariant μ_p-absolutely continuous measure on Ω when $p \neq 1/2$.*

1.2.10 LEMMA. *If $f : \Omega \to \mathbb{R}$ is measurable, then $\forall \, \epsilon > 0$,*

$$\mu_p([|f \circ \tau^{2^n} - f| \geq \epsilon]) \to 0 \text{ as } n \to \infty.$$

PROOF. Firstly, note that if $f : \Omega \to \mathbb{R}$ and f is defined by $f(x) = g(x_1, \ldots, x_n)$ for some $n \in \mathbb{N}$, then $f \circ \tau^{2^k} \equiv f$ for every $k \geq n$. To enable approximation, we show that

$$\mu_p(\tau^{-2^n} A) \to 0 \text{ as } \mu_p(A) \to 0, A \in \mathcal{B},$$

uniformly in $n \in \mathbb{N}$. Note that

$$\frac{d\mu_p \circ \tau^{-1}}{d\mu_p} = \left(\frac{p}{1-p}\right)^{\psi}$$

where

$$\psi(x) = \min\{n \in \mathbb{N} : x_n = 1\} - 2.$$

Calculations show that
$\exists \, q > 1$ such that

$$\left\|\left(\frac{p}{1-p}\right)^{\psi}\right\|_{L^q(\mu_p)} := M < \infty,$$

and that
for $n \in \mathbb{N}$, and $x \in \Omega$,

$$\sum_{k=0}^{2^n - 1} \psi(\tau^{-k} x) = \psi(S^n x)$$

(where S is the shift as before). Therefore

$$\frac{d\mu_p \circ \tau^{-2^n}}{d\mu_p} = \prod_{k=0}^{2^n-1} \left(\frac{d\mu_p \circ \tau^{-1}}{d\mu_p}\right) \circ \tau^{-k}$$

$$= \prod_{k=0}^{2^n-1} \left(\frac{p}{1-p}\right)^{\psi \circ \tau^{-k}}$$

$$= \left(\frac{p}{1-p}\right)^{\psi \circ S^n}.$$

Thus, for $A \in \mathcal{B}$,

$$\mu_p(\tau^{-2^n} A) = \int_A \left(\frac{p}{1-p}\right)^{\psi \circ S^n} d\mu_p$$

$$\leq \left\| \left(\frac{p}{1-p}\right)^\psi \right\|_q \mu_p(A)^{\frac{1}{q'}}$$

$$= M\mu_p(A)^{\frac{1}{q'}}$$

by Hölder's inequality where $q' := \frac{q}{q-1}$.

Now, suppose that $F : \Omega \to \mathbb{R}$ is measurable, and let $\epsilon > 0$ be given. There exist $n \in \mathbb{N}$, and $f : \Omega \to \mathbb{R}$ and f defined by $f(x) = g(x_1, \ldots, x_n)$ for some $g : \{0,1\}^n \to \mathbb{R}$ such that $\mu_p([|F - f| \geq \epsilon/2]) < \epsilon$. For $k \geq n$, we have $f \circ \tau^{2^k} \equiv f$, whence

$$\mu_p([|F \circ \tau^{2^k} - F| \geq \epsilon]) \leq \mu_p([|F \circ \tau^{2^k} - f \circ \tau^{2^k}| \geq \epsilon/2]) + \mu_p([|F - f| \geq \epsilon/2])$$

$$\leq \epsilon + M\epsilon^{\frac{1}{q'}},$$

establishing that indeed

$$F \circ \tau^{2^n} \xrightarrow{\mu_p} F.$$

$\square$

PROOF OF ARNOLD'S THEOREM. Suppose that $\nu << \mu_p$ is τ-invariant. We'll show that $p = \frac{1}{2}$.

Set

$$\frac{d\nu}{d\mu_p} = h,$$

then, $\nu(A) = 0$ iff $A \subset [h = 0]$ mod μ_p. By nonsingularity, $\nu(\tau[h = 0]) = 0$ hence $\tau[h = 0] \subset [h = 0]$. By conservativity and ergodicity, $h > 0$ μ_p-a.e., and $\mu \sim m_p$.

By the chain rule for Radon Nikodym derivatives for $k \geq 1$,

$$\frac{d\mu_p \circ \tau^k}{d\mu_p} = \prod_{j=0}^{k-1} \frac{d\mu_p \circ \tau}{d\mu_p} \circ \tau^j = \left(\frac{1-p}{p}\right)^{\phi_k}$$

where

$$\phi_k = \sum_{j=0}^{k-1} \phi \circ \tau^j.$$

Thus, for $k \geq 1$, $A \in \mathcal{B}$,

$$\int_A \left(\frac{1-p}{p}\right)^{\phi_k} d\mu_p = \mu_p(\tau^k A)$$

$$= \int_\Omega \frac{1}{h} 1_A \circ \tau^{-k} d\nu$$

$$= \int_\Omega \frac{1}{h \circ \tau^k} 1_A d\nu$$

$$= \int_A \frac{h}{h \circ \tau^k} d\mu_p$$

whence

$$\left(\frac{1-p}{p}\right)^{\phi_k} = \frac{h}{h \circ \tau^k} \text{ a.e.}$$

Again, calculations show that

$$\phi_{2^n} = \phi \circ S^n \ \forall \, n \geq 1,$$

whence

$$\left(\frac{1-p}{p}\right)^{\phi \circ S^n} = \left(\frac{1-p}{p}\right)^{\phi_{2^n}} = \frac{h}{h \circ \tau^{2^n}} \text{ a.e. } \forall \, n \geq 1.$$

By lemma 1.2.10, $\exists \, n_k \to \infty$ such that

$$h \circ \tau^{2^{n_k}} \to h \text{ a.e. as } k \to \infty.$$

The events

$$A_n = [\phi \circ S^n = -1] = \{x \in \Omega : x_{n+1} = 0\}$$

are independent, and $\mu_p(A_n) = 1 - p$. By the Borel-Cantelli lemma,

$$\phi \circ S^{n_k} = -1 \text{ i.o. a.e.}$$

whence

$$\frac{p}{1-p} = \left(\frac{1-p}{p}\right)^{\phi \circ S^{n_k}} \text{ i.o. a.e.}$$

$$= \left(\frac{1-p}{p}\right)^{\phi_{2^{n_k}}}$$

$$= \frac{h}{h \circ \tau^{2^{n_k}}}$$

$$\to 1 \text{ a.e.}$$

and $p = \frac{1}{2}$. $\qquad\qquad\qquad\qquad\qquad\qquad\qquad\qquad\qquad\qquad$ $\square$

§1.3 The dual operator

If $T : X \to X$ is non-singular then $f \to f \circ T$ defines a linear isometry of $L^\infty(m)$. The *dual*, or *Frobenius-Perron* or *transfer* operator, $\widehat{T} : L^1(m) \to L^1(m)$, is defined by

$$f \mapsto \nu_f(\cdot) = \int f dm \mapsto \widehat{T}f = \frac{d\nu_f \circ T^{-1}}{dm}$$

and satisfies

$$\int_X \widehat{T}f.gdm = \int_X f.g \circ Tdm \quad f \in L^1(m), \ g \in L^\infty(m).$$

Note that the domain of definition of $\widehat{T}$ can be extended to all non-negative measurable functions. This definition can be made when m is infinite, but σ-finite. Clearly, if T is invertible, then

$$\widehat{T}f = \frac{dm \circ T^{-1}}{dm}f \circ T^{-1}.$$

The following proposition is a "dual" version of proposition 1.1.6.

1.3.1 PROPOSITION.

$$[\sum_{n=1}^\infty \widehat{T}^k f = \infty] = \mathfrak{C} \quad \mod \ m \ \forall \ f \in L^1(m), f > 0.$$

PROOF. Fix $f \in L^1(m)$, $f > 0$ a.e. If $A \in \mathcal{B}_+$ and

$$\sum_{n=1}^\infty \widehat{T}^k f < \infty$$

on A, then there exists $B \in (\mathcal{B} \cap A)_+$ such that

$$\int_B \left(\sum_{n=1}^\infty \widehat{T}^k f \right) dm < \infty.$$

It follows that

$$\int_X f \left(\sum_{n=1}^\infty 1_B \circ T^k \right) dm < \infty,$$

whence, since $f > 0$ a.e.,

$$\sum_{n=1}^\infty 1_B \circ T^k dm < \infty \ \text{a.e.}$$

and by the Halmos recurrence theorem, $B \subset \mathfrak{D}$. This proves that

$$[\sum_{n=1}^\infty \widehat{T}^k f = \infty] \supset \mathfrak{C} \quad \mod \ m.$$

Conversely, if $W \in \mathcal{W}_+$, and $f \in L^1(m)_+$, then

$$\sum_{n=1}^\infty \widehat{T}^n f < \infty \ \text{a.e. on } W,$$

since

$$\int_W \left(\sum_{n=1}^{\infty} \widehat{T}^k f \right) dm = \int_X f \left(\sum_{n=1}^{\infty} 1_W \circ T^n \right) dm$$

$$\leq \|f\|_1 < \infty.$$

Thus

$$[\sum_{n=1}^{\infty} \widehat{T}^k f < \infty] \supset \mathfrak{D} \qquad \text{mod} \ \ m.$$

$\square$

EXERCISE 1.3.1. Call a non-singular transformation T *dual incompressible* if

$$f : X \to [0, \infty) \ \text{measurable}, \ \widehat{T} f \leq f \ \implies \ \widehat{T} f = f \ \text{a.e.}.$$

1) A non-singular transformation is dual incompressible if it is conservative.

2) A non-singular, dual incompressible transformation is conservative if it is either invertible, or measure preserving.

3) There are non-singular, dual incompressible transformations which are not conservative.

1.3.2 PROPOSITION.

T *is conservative and ergodic* $\Leftrightarrow$

$$\sum_{n=0}^{\infty} \widehat{T}^n f = \infty \ \ a.e. \ \forall f \in L^1(m), \ \ f \geq 0 \ a.e. \ , \ \int_X f dm > 0.$$

PROOF. This is a dual version of proposition 1.2.2, and follows from that proposition via the identity

$$\int_A \left(\sum_{n=0}^{\infty} \widehat{T}^n f \right) dm = \int_X f \left(\sum_{n=0}^{\infty} 1_A \circ T^n \right) dm$$

where $f \in L^1(m)$ and $A \in \mathcal{B}$. $\square$

The last result of this section is a characterisation of exactness in terms of the dual operator. Recall that a non-singular transformation $T : X \to X$ is called *exact* if

$$A \in \bigcap_{n=1}^{\infty} T^{-n} \mathcal{B} \ \Rightarrow \ m(A)m(A^c) = 0.$$

1.3.3 THEOREM [Lin].
Let T be a non-singular transformation of $(X, \mathcal{B}, m)$, then

$$T \ \text{is exact} \ \Leftrightarrow \ \|\widehat{T}^n f\|_1 \to_{n \to \infty} 0 \ \forall f \in L^1, \int_X f dm = 0.$$

PROOF. Suppose first that T is exact, and let $f \in L^1$, $\int_X f dm = 0$. There are functions $g_n \in L^\infty$, such that $\|g_n\|_\infty = 1$, and $\int_X f g_n \circ T^n dm = \|\widehat{T}^n f\|_1$. Any weak-$*$-limit of $\{g_n \circ T^n\}$ is measurable with respect to $\bigcap_{n \geq 1} T^{-n} \mathcal{B}$, and is hence constant by exactness. Whence,

$$\lim_{n \to \infty} \|\widehat{T}^n f\|_1 = \lim_{n \to \infty} \int_X f g_n \circ T^n dm = 0.$$

Conversely, suppose that T is not exact, and let $A \in \bigcap_{n \geq 1} T^{-n} \mathcal{B}$ satisfy $m(A), m(A^c) > 0$. There exists $f \in L^1$ such that $\int_X f dm = 0$, but $\int_A f dm > 0$. If $A = T^{-n} A_n$ for $n \geq 1$ where $A_n \in \mathcal{B}$, then

$$\|\widehat{T}^n f\|_1 \geq \int_{A_n} \widehat{T}^n f dm = \int_A f dm > 0.$$

$\square$

EXERCISE 1.3.2. Let $X = [0,1)$, m be Lebesgue measure, and $T : X \to X$ be defined by $Tx = 2x \bmod 1$.
(1) Compute $\widehat{T}$ and show that if $f : X \to \mathbb{R}$ is continuous, then

$$\widehat{T}^n f \to \int_X f dm \text{ uniformly on } X,$$

whence T is exact.
(2) Show that $\forall \lambda \in \mathbb{C}$, $\exists f : X \to \mathbb{C}$ measurable such that $\widehat{T} f = \lambda f$.

EXERCISE 1.3.3. Let $X = [0,1)$, m be Lebesgue measure, and $T : X \to X$ be defined by

$$Tx = \begin{cases} 2x & x \in [0, 1/4) \\ 2x - 1/2 & x \in [1/4, 1/2) \\ 2x - 1 & x \in [1/2, 1) \end{cases}$$

(1) Show that $\mathfrak{C} = [0, 1/2] \bmod m$.
(2) Show that T is ergodic. Is it exact?

EXERCISE 1.3.4. Let $X = \mathbb{R}$, m be Lebesgue measure, and $T = T_{\alpha,\beta} : X \to X$ be defined by

$$Tx = \alpha x - \frac{\beta}{x}$$

where α, $\beta > 0$. Note that $T_{1,1}$ is "Boole's transformation" of exercise 1.1.2.
(1) Show that

$$\widehat{T} f(y) = f(x_+(y)) x'_+(y) + f(x_-(y)) x'_-(y)$$

where $x_\pm(y)$ are the solutions to $Tx = y$.
(2) For $\omega = a + ib \in \mathbb{C}$, $b > 0$ let $\varphi_\omega : \mathbb{R} \to \mathbb{R}_+$ be defined by

$$\varphi_\omega(x) = \operatorname{Im} \frac{1}{\pi(x - \omega)} = \frac{b}{\pi((x-a)^2 + b^2)}.$$

Show that $\widehat{T} \varphi_\omega = \varphi_{T\omega}$.
(3) Show that if $\alpha < 1$, then the probability p on $\mathbb{R}$ defined by $dp(x) = \varphi_{i\sqrt{\frac{\beta}{1-\alpha}}}$ is $T_{\alpha,\beta}$-invariant.
(4) Show that Lebesgue measure is $T_{1,\beta}$-invariant.

(5) Show that $T_{1,\beta}^n \omega = a_n + ib_n$ where $\sup_n |a_n| < \infty$ and $b_n \sim \sqrt{2n}$, whence

$$\frac{1}{a(n)} \sum_{k=0}^{n-1} \widehat{T}_{1,\beta}^k \varphi_\omega \to 1$$

uniformly on compact subsets, where $a(n) := \frac{\sqrt{2n}}{\pi}$.

(6) Show that

$$\|\widehat{T}^n(\varphi_{ib} * f)\|_1 \to 0 \ \forall \ b > 0, \ f \in L^1(m)_0,$$

whence $T_{1,\beta}$ is exact.

For more details, see chapter 6.

§1.4 Invariant measures

We've seen that there are conservative, ergodic, non-singular transformations without absolutely continuous invariant measures.

In this section, we consider conditions for existence of an invariant measure. The connection with the dual operator is given by

1.4.1 PROPOSITION. *Let T be a non-singular transformation.*
1) If $\mu << m$ and $\frac{d\mu}{dm} = h$, then

$$\mu \circ T^{-1} = \mu \ \Leftrightarrow \ \widehat{T}h = h.$$

2) If T is conservative, ergodic, and $\mu << m$ satisfies $\mu \circ T^{-1} \sim \mu$ then $\mu \sim m$.

PROOF. To see 1), for $f \in L^1(m)_+$

$$\int_X f \circ T d\mu = \int_X f \circ T h dm = \int_X f \widehat{T} h dm,$$

whence

$$\int_X f \circ T d\mu = \int_X f d\mu \ \forall \ f \in L^1(m)$$

iff $\widehat{T}h = h$ mod m.

To prove 2), note first that $\mu(A) = \int_A h dm = 0$ iff $A \subset [h = 0]$ mod m. Next,

$$0 = \mu([h = 0]) = \mu(T^{-1}[h = 0])$$

whence $T^{-1}[h = 0] \subset [h = 0]$ mod m. By incompressibility, $T^{-1}[h = 0] = [h = 0]$ mod m, whence by ergodicity either $h = 0$ a.e., or $h > 0$ a.e.. $\qquad \square$

Let $(X, \mathcal{B}, m)$ be a probability space. Recall that a collection $\mathcal{F} \subset L^1(m)$ is called *uniformly integrable* if $\forall \ \epsilon > 0$, $\exists \ M > 1$ such that $\int_{[|f| \geq M]} |f| dm \leq \epsilon \ \forall \ f \in \mathcal{F}$.

A collection $\mathcal{F}$ is uniformly integrable iff it is weakly sequentially precompact in $L^1(m)$ – every sequence in $\mathcal{F}$ has a subsequence $\{f_n : n \geq 1\}$ which converges weakly to some $f \in L^1(m)$ (written $f_n \rightharpoonup f$ and meaning $\int_A f_n dm \to \int_A f dm \ \forall \ A \in \mathcal{B}$).

As is well known, for each $p > 1$, $K > 0$ the collection $\{f \in L^p(m) : \|f\|_p \leq K\}$ is uniformly integrable (but not when $p = 1$).

1.4.2 PROPOSITION   (EXISTENCE OF INVARIANT PROBABILITY).   *Let T be a non-singular transformation of $(X, \mathcal{B}, m)$.*

If $\exists \ f \in L^1(m)_+$ such that $\{\hat{A}_n f : \ n \geq 1\}$ is a uniformly integrable family (here $\hat{A}_n f := \frac{1}{n} \sum_{k=1}^{n} \hat{T}^k f$), then $\exists$ a T-invariant probability $P << m$.

PROOF. It follows from uniform integrability that $\exists \ n_k \to \infty$ and $h \in L^1(m)$ such that $\hat{A}_{n_k} f \rightharpoonup h$. Evidently, $h \geq 0$ and $\hat{T} h = h$. Furthermore,

$$\int_X h \, dm \ \leftarrow \ \int_X \hat{A}_{n_k} f \, dm = \int_X f \, dm > 0$$

and we have that $\frac{h}{\int_X h \, dm}$ is the density of a T-invariant probability $P << m$. $\square$

1.4.3 PROPOSITION   (C.F. [**Ren1**]).   *Let T be a non-singular transformation of $(X, \mathcal{B}, m)$.*

If $\exists \ M > 0$ and $1 < p \leq \infty$ such that $\|\hat{T} f\|_p \leq M\|f\|_p$, then $\exists$ a T-invariant probability $P << m$ with $\frac{dP}{dm} \in L^p(m)$.

PROOF. This follows from proposition 1.4.2 as for any $f \in L^p(m)_+$, $\hat{A}_n f \in L^p(m)$ and $\|\hat{A}_n f\|_p \leq M\|f\|_p \ \forall \ n \geq 1$ whence $\{\hat{A}_n f : \ n \geq 1\}$ is a uniformly integrable family.

If $\hat{A}_{n_k} f \rightharpoonup h$ then $\liminf_{k \to \infty} \|\hat{A}_{n_k} f\|_p \geq \|h\|_p$. $\square$

1.4.4 KRENGEL'S THEOREM [**Kre**].
Let T be a non-singular transformation of $(X, \mathcal{B}, m)$.
Either $\exists$ an m-absolutely continuous T-invariant probability, or

$$\frac{1}{n} \sum_{k=0}^{n-1} \hat{T}^k h \xrightarrow{\ m \ } 0 \ as \ n \to \infty \ \forall \ h \in L^1(m).$$

For the proof of Krengel's theorem, we'll need the following lemma.

1.4.5 LEMMA.
Let $(X, \mathcal{B}, m)$ be a probability space and suppose that
$f_n \in L^2(m) \ (n \geq 1), \quad \sup_{n \geq 1} \|f_n\|_2 < \infty,$
then $\exists \ f \in L^2(m)$ and $n_k \to \infty$ such that whenever $\{n'_j : \ j \geq 1\}$ is a subsequence of $\{n_k : \ k \geq 1\}$,

$$\frac{1}{K^2} \sum_{j=1}^{K^2} f_{n'_j} \to f \ a.e. \ as \ K \to \infty.$$

PROOF.
$\exists \ f \in L^2(m)$ and $n_k \to \infty$ such that $f_{n_k} \rightharpoonup f$. By possibly choosing a further subsequence (also denoted $n_k \to \infty$) we can ensure that

$$\left| \int_X (f_{n_j} - f)(f_{n_k} - f) \, dm \right| < \frac{1}{2^k} \ \forall \ 1 \leq j < k.$$

Evidently this property passes to subsequences, and it follows that for any subsequence $\{n'_j : j \geq 1\}$ of $\{n_k : k \geq 1\}$,

$$\int_X \left(\sum_{j=1}^{K} (f_{n'_j} - f) \right)^2 dm = O(K) \text{ as } K \to \infty$$

whence by Chebyshev's inequality, $\forall \, \epsilon > 0$,

$$m([|\sum_{j=1}^{K} (f_{n'_j} - f)| \geq K\epsilon]) = O\left(\frac{1}{K}\right) \text{ as } K \to \infty.$$

The lemma follows from this. $\square$

PROOF OF KRENGEL'S THEOREM.

There is no loss of generality in assuming that
T conservative as $\frac{1}{n} \sum_{k=0}^{n-1} \widehat{T}^k h \to 0$ a.e. on $\mathfrak{D}$ as $n \to \infty$ $\forall \, h \in L^1(m)$;
and that
$h \geq 0$ since $|\frac{1}{n} \sum_{k=0}^{n-1} \widehat{T}^k h| \leq \frac{1}{n} \sum_{k=0}^{n-1} \widehat{T}^k |h|$ $\forall \, h \in L^1(m)$.

Suppose that $\nexists$ an m-absolutely continuous T-invariant probability, fix $h \in L^1(m)_+$ and set for $n \geq 1$ $\hat{A}_n := \hat{A}_n h = \frac{1}{n} \sum_{k=0}^{n-1} \widehat{T}^k h$.

We must show that $\hat{A}_n \xrightarrow{m} 0$ and to do this we show that for any subsequence $n_k \to \infty$, $\exists$ a sub-subsequence $\nu_r = n_{k_r} \to \infty$ such that $\hat{A}_{\nu_r} \to 0$ a.e. as $r \to \infty$.

First (using lemma 1.4.5 successively) choose a subsequence $n_k \to \infty$ of the given subsequence such that $\frac{\widehat{T}^{n_k} h}{n_k} \to 0$ a.e. as $k \to \infty$ and such that

$$\frac{1}{N^2} \sum_{k=1}^{N^2} \hat{A}_{n_k} \wedge M \to g_M$$

a.e. $\forall \, M \in \mathbb{N}$ where $g_M \in L_+^\infty$ and $g_M \leq g_{M+1}$.

Next, since $\int_X \hat{A}_n dm = \int_X h dm := c$ $\forall \, n \geq 1$, we have that $\int_X g_M dm \leq c$ $\forall \, M \in \mathbb{N}$ and it follows that $g_M \uparrow g \in L_+^1$, $\int_X g dm \leq c$. We show that $\widehat{T} g = g$.

To see this,

$$\widehat{T} g_M \quad \leftarrow \quad \frac{1}{N^2} \sum_{k=1}^{N^2} \widehat{T} (\hat{A}_{n_k} \wedge M)$$

$$\leq \frac{1}{N^2} \sum_{k=1}^{N^2} (\widehat{T} \hat{A}_{n_k}) \wedge (M \widehat{T} 1)$$

$$= \frac{1}{N^2} \sum_{k=1}^{N^2} \hat{A}_{n_k} \wedge (M \widehat{T} 1) + o(1)$$

a.e., and it follows that $\forall \, K \geq 1$,

$$\widehat{T} g_M 1_{[\widehat{T} 1 \leq K]} \leq 1_{[\widehat{T} 1 \leq K]} \liminf_{N \to \infty} \frac{1}{N^2} \sum_{k=1}^{N^2} \hat{A}_{n_k} \wedge (MK) = 1_{[\widehat{T} 1 \leq K]} g_{MK} \leq 1_{[\widehat{T} 1 \leq K]} g,$$

whence (sending $K \to \infty$) $\widehat{T} g_M \leq g$ and (sending $M \to \infty$) $\widehat{T} g \leq g$. Since $\int_X \widehat{T} g dm = \int_X g dm$ we have $\widehat{T} g = g$.

It follows from our assumption that $g = 0$ (else $g/\int_X g\,dm$ would be the density of an m-absolutely continuous T-invariant probability) whence

$$\frac{1}{N^2}\sum_{k=1}^{N^2} \hat{A}_{n_k} \wedge M \to 0 \text{ a.e. } \forall\, M \in \mathbb{N}.$$

By Egorov's theorem, $\exists\, B_n \in \mathcal{B}$ $(n \geq 1)$, $m(X \setminus B_n) < \frac{1}{2^n}$ such that

$$\frac{1}{N^2}\sum_{k=1}^{N^2} \int_{B_n} \hat{A}_{n_k} \wedge M\,dm \to 0 \text{ as } N \to \infty\ \forall\, M, n \geq 1.$$

Choosing $L_r \uparrow$ such that $\frac{1}{L_r}\sum_{k=1}^{L_r} \int_{B_r} \hat{A}_{n_k} \wedge 2^r dm < \frac{1}{2^r}$, it follows that $\exists\, \nu_r = m_{\ell_r} \to \infty$ such that $\int_{B_r} \hat{A}_{\nu_r} \wedge 2^r dm < \frac{1}{2^r}\ \forall\, r \geq 1$.
Evidently $\forall\, \epsilon > 0$,

$$m([\hat{A}_{\nu_r} \geq \epsilon]) \leq m(B_r^c) + m([\hat{A}_{\nu_r} \geq 2^r]) + \frac{1}{\epsilon}\int_{B_r} \hat{A}_{\nu_r} \wedge 2^r dm \leq \frac{2 + 1/\epsilon}{2^r},$$

whence $\hat{A}_{\nu_r} \to 0$ a.e. as $r \to \infty$. $\qquad\square$

EXERCISE 1.4.1.
1) Show that if $\psi_n \in L^2(m)$, $(n \geq 1)$ satisfy $\int_X \psi_k\psi_\ell dm = 0$ $(k \neq \ell)$ and $\sup_{n \geq 1}\|\psi_n\|_2 < \infty$, then $\frac{1}{n}\sum_{k=1}^n \psi_k \to 0$ a.e. as $n \to \infty$.
2) Show that if n_k is as in the proof of lemma 1.4.5, then $\frac{1}{K}\sum_{j=1}^K f_{n_j} \to f$ a.e. as $K \to \infty$.

A more advanced form of the subsequence phenomenon exhibited in lemma 1.4.5 is seen in

KOMLOS' THEOREM   [**Kom**].
Let $(X, \mathcal{B}, m)$ be a probability space and suppose that
$f_n \in L^1(m)$ $(n \geq 1)$, $\sup_{n \geq 1}\|f_n\|_1 < \infty$,
then $\exists\, f \in L^1(m)$ and $n_k \to \infty$ such that whenever $\{n'_j : j \geq 1\}$ is a subsequence of $\{n_k : k \geq 1\}$,

$$\frac{1}{K}\sum_{j=1}^K f_{n'_j} \to f \text{ a.e. as } K \to \infty.$$

1.4.6 THEOREM   (POSITIVE-NULL DECOMPOSITION).
Let T be a non-singular transformation of $(X, \mathcal{B}, m)$, then $\exists\, h \in L^1(m)$, $h \geq 0$ such that $\hat{T}h = h$ and

$$\hat{A}_n f \overset{m}{\to} 0 \text{ on } [h = 0]\ \forall\, f \in L^1(m).$$

PROOF. Since $\hat{A}_n f \to 0$ a.e. on $\mathfrak{D}$, there is no loss of generality in assuming that T is conservative. Let

$$\mathfrak{H} := \{A \in \mathcal{B} : \exists\, h \in L^1(m)_+,\ \hat{T}h = h,\ A \subset [h > 0]\}.$$

Evidently $\mathfrak{H}$ is a hereditary collection and so by the exhaustion lemma, $\exists\ A_n \in \mathfrak{H}$ $(n \geq 1)$ such that the measurable union is given by $U(\mathfrak{H}) = \bigcup_{n=1}^{\infty} A_n$. By definition, $\exists\ h_n \in L^1(m)_+$ such that $\widehat{T}h_n = h_n$, $\int_X h_n dm = 1$ and $A_n \subset [h_n > 0]$.

It follows that $U(\mathfrak{H}) = [h > 0]$ where $h := \sum_{n=1}^{\infty} h_n \in L^1(m)_+$.

The set $[h = 0]$ is T-invariant and since $[h = 0] = U(\mathfrak{H})^c$, $\nexists$ an m-absolutely continuous T-invariant probability on $[h = 0]$. The result now follows from theorem 1.4.4. $\qquad\square$

Let $h \in L^1(m)_+$ be as in theorem 1.4.6. We call $[h > 0]$ the *positive part* of T and denote it by $\mathfrak{P}(T)$. The *null part* of T is $\mathfrak{N}(T) := [h = 0] = X \setminus \mathfrak{P}$.

Let T be a non-singular transformation of $(X, \mathcal{B}, m)$. A measurable set $W \in \mathcal{B}$ is called *weakly wandering* if there is a sequence $n_k \to \infty$ such that $\{T^{-n_k}W : k \geq 1\}$ are disjoint.

Evidently, if $\exists$ a T-invariant probability $\mu \sim m$ then there can be no weakly wandering set with positive measure (else $1 \geq \mu(X) = \sum_{k=1}^{\infty} \mu(T^{-n_k}W) = \infty$). This shows that all weakly wandering sets for a non-singular transformation are contained in its null part.

1.4.7 PROPOSITION [**Haj-Kak**].   *The null part of a non-singular transformation is a union of weakly wandering sets.*

PROOF   [**E-Fr**]. Let T be a non-singular transformation of $(X, \mathcal{B}, m)$ and let $\mathfrak{W}$ denote the collection of weakly wandering sets for T. This collection is clearly hereditary and by the exhaustion lemma, it suffices to show that it saturates $\mathfrak{N}$.

Fix $0 < \epsilon < m(\mathfrak{N})$. It follows from Krengel's theorem that $\forall\ f \in L^1(m)_+$, $\exists\ B \in \mathcal{B}$ and $n \geq 1$ such that $m(B) > m(\mathfrak{N}) - \epsilon$ and $\int_X f 1_B \circ T^n dm < \epsilon$.

Accordingly, choose $\exists\ B_1 \in \mathcal{B}$ and $n_1 \geq 1$ such that $m(B_1) > m(\mathfrak{N}) - \frac{\epsilon}{2}$ and $m(T^{-n_1}B_1) < \frac{\epsilon}{2}$. Next choose $B_2 \in \mathcal{B}$, $B_2 \subset B_1$ such that $m(B_2) > m(B_1) - \frac{\epsilon}{2^2}$ and $\int_X 1_{B_1} \circ T^{n_1} 1_{B_2} \circ T^{n_2} dm < \frac{\epsilon}{2^2}$.

Continue the process, obtaining $B_k \in \mathcal{B}$ $(k \geq 1)$ satisfying

$$B_k \supset B_{k+1},\quad m(B_{k+1}) > m(B_k) - \frac{\epsilon}{2^{k+1}},$$

and $n_k \uparrow$ such that

$$\int_X \sum_{j=1}^{k-1} 1_{B_j} \circ T^{n_j} \cdot 1_{B_k} \circ T^{n_k} dm < \frac{\epsilon}{2^k} \qquad (k \geq 1).$$

It follows that $W := \bigcap_{k=1}^{\infty} B_k \in \mathfrak{W}$ and $m(W) > m(\mathfrak{N}) - \epsilon$. $\qquad\square$

One of our goals will be prove that a conservative, ergodic, non-singular transformation has at most one absolutely continuous invariant measure (up to a multiplicative factor).

If T is an invertible, conservative, ergodic, measure preserving transformation of $(X, \mathcal{B}, m)$, and μ is a T-invariant, m-absolutely continuous measure, then $\frac{d\mu}{dm} = \widehat{T}\frac{d\mu}{dm} = \frac{d\mu}{dm} \circ T^{-1}$ is constant, and μ is a constant multiple of m.

This shows using proposition 1.4.1, that up to multiplication by a constant, there is at most one T-invariant, m-absolutely continuous measure. The same result remains true without the assumption of invertibility (theorem 1.5.6 below), but it is more difficult to prove. We begin the proof here with a special case.

1.4.8 THEOREM (UNICITY OF INVARIANT PROBABILITY). *Let T be a conservative, ergodic non-singular transformation of $(X, \mathcal{B}, m)$. There is at most one T-invariant, m-absolutely continuous probability.*

PROOF. Let $\mu << m$ be a T-invariant, m-absolutely continuous probability. By proposition 1.4.1, $\mu \sim m$, and we may suppose that m itself is a T-invariant probability. We'll prove the theorem by showing that

$$h \in L^1(m), \ \widehat{T}h = h \ \Rightarrow \ h \text{ is constant.}$$

Suppose that $\widehat{T}h = h \in L^1(m)$. Let $A \in \mathcal{B}$, then $\|\frac{1}{n} \sum_{j=0}^{n-1} 1_A \circ T^j\|_\infty \leq 1 \ \forall \ n \geq 1$, and there is a subsequence $n_k \to \infty$, and $g_A \in L^\infty(m)$ such that

$$\frac{1}{n_k} \sum_{j=0}^{n_k-1} 1_A \circ T^k \overset{\text{weak-* in } L^\infty(m)}{\longrightarrow} g_A.$$

Clearly, $g_A \circ T = g_A$, whence by ergodicity g_A is constant. It follows from the T-invariance of m that $\int_X g_A dm = m(A)$, whence $g_A = m(A)$ a.e., and

$$\int_A h \, dm = \int_A \left(\frac{1}{n_k} \sum_{j=0}^{n_k-1} \widehat{T}^k h \right) dm$$

$$= \int_X h \left(\frac{1}{n_k} \sum_{j=0}^{n_k-1} 1_A \circ T^k \right) dm$$

$$\to m(A) \int_X h \, dm \text{ as } k \to \infty.$$

Thus $h = \int_X h \, dm$ a.e.   □

§1.5 Induced transformations and applications

Suppose T is conservative and non–singular, and let $A \in \mathcal{B}_+$, then m–a.e. point of A returns infinitely often to A under iterations of T, and in particular the *return time* function, defined for $x \in A$ by $\varphi_A(x) := \min\{n \geq 1 : T^n x \in A\}$ is finite m–a.e. on A.

The *induced transformation* on A is defined by $T_A x = T^{\varphi_A(x)} x$. The first key observation is that $m|_A \circ T_A^{-1} << m|_A$. This is because

$$T_A^{-1} B = \bigcup_{n=1}^{\infty} [\varphi = n] \cap T^{-n} B.$$

It follows that $\varphi_A \circ T_A$ is defined a.e. on A and an induction now shows that all powers $\{T_A^k\}_{k \in \mathbb{N}}$ are defined a.e. on A, and satisfy

$$T_A^k x = T^{(\varphi_A)_k(x)} x \text{ where } (\varphi_A)_1 = \varphi_A, \ (\varphi_A)_k = \sum_{j=0}^{k-1} \varphi_A \circ T_A^j.$$

We first establish the basic properties of the induced transformation.

1.5.1 PROPOSITION. *The induced transformation T_A is a conservative, non-singular transformation of $(A, \mathcal{B} \cap A, m|_A)$.*

PROOF. As shown above, $m|_A \circ T_A^{-1} \ll m|_A$. Suppose that $B \in \mathcal{B} \cap A$, and that $m(B) > 0$, then $\exists\ n \geq 1$ such that $m(A \cap T^{-n}B) > 0$. Let $n_0 \geq 1$ be the smallest such $n \in \mathbb{N}$, then $A \cap T^{-n_0}B \subset T_A^{-1}B$ mod m, and $m(T_A^{-1}B) > 0$. This shows that T_A is a non-singular transformation of $(A, \mathcal{B} \cap A, m|_A)$. To prove conservativity of T_A, note that, because φ_A is defined as a minimum, we have

$$x, T^n x \in A \ \Rightarrow\ \exists k \leq n \ni T^n x = T_A^k x.$$

We obtain, using conservativity of T, that for $A \in \mathcal{B}, B \in \mathcal{B} \cap A$,

$$\sum_{n=1}^{\infty} 1_B \circ T_A^n = \sum_{n=1}^{\infty} 1_B \circ T^n = \infty$$

a.e. on B, proving the conservativity of T_A. $\square$

1.5.2 PROPOSITION.   *Suppose that T is conservative, and $A \in \mathcal{B}_+$, then*

(1) T *ergodic* $\Rightarrow$ T_A *ergodic.*

(2) T_A *is ergodic, and* $\displaystyle\bigcup_{n=1}^{\infty} T^{-n}A = X$ *mod m* $\Rightarrow$ T *is ergodic.*

PROOF. Suppose that T is ergodic, and that $B \in \mathcal{B}$, $B \subset A$, $m(B) > 0$ is T_A-invariant. We claim that $B = A$ mod m, else for a.e. $x \in A \setminus B$,

$$0 = \sum_{n=1}^{\infty} 1_B \circ T_A^n(x) = \sum_{n=1}^{\infty} 1_B \circ T^n(x) = \infty.$$

Whence $B = A$ mod m and T_A is ergodic.

To prove (2), let $C \in \mathcal{B}$ satisfy $m(C) > 0, T^{-1}C = C$. Since $\bigcup_{n=1}^{\infty} T^{-n}A = X$ mod m, for some $n \in \mathbb{N}$,

$$C \cap T^{-n}A = T^{-n}(C \cap A)$$

has positive measure, whence so does $A \cap C$. This forces

$$C = \bigcup_{n=1}^{\infty} T^{-n}C \supseteq \bigcup_{n=1}^{\infty} T_A^{-n}(A \cap C) = A \quad \text{mod } m$$

by conservativity and ergodicity of T_A, whence

$$C = \bigcup_{n=1}^{\infty} T^{-n}C \supseteq \bigcup_{n=1}^{\infty} T^{-n}A = X \quad \text{mod } m,$$

establishing (2). $\square$

Invariant measures.

One of the uses of the induced transformation is in the study of σ-finite, absolutely continuous, invariant measures.

The next few results are established in the general (not necessarily invertible) case. They have easy proofs in the invertible case.

1.5.3 PROPOSITION.   *If T is a conservative measure preserving transformation of $(X, \mathcal{B}, m)$, and $A \in \mathcal{B}$, $0 < m(A) < \infty$, then $m|_A \circ T_A^{-1} = m|_A$.*

PROOF. We have, for $B \in \mathcal{B} \cap A$,

$$m(T_A^{-1}B) = \sum_{n=1}^{\infty} m([\varphi_A = n] \cap T^{-n}B)$$

$$= \sum_{n=1}^{\infty} m(A \cap T^{-n}B \setminus \bigcup_{k=1}^{n-1} T^{-k}A)$$

$$= \sum_{n=1}^{\infty} m(A \cap T^{-1}B_{n-1}),$$

where, $B_0 = B$, and for $n \geq 1$, $B_n = T^{-n}B \setminus \bigcup_{k=0}^{n-1} T^{-k}A$.

Now $T^{-1}B_n = (A \cap T^{-1}B_n) \cup B_{n+1}$, whence $m(A \cap T^{-1}B_n) = m(B_n) - m(B_{n+1})$ and

$$m(T_A^{-1}B) = \sum_{n=1}^{\infty} m(A \cap T^{-1}B_{n-1}) = m(B) - \lim_{n \to \infty} m(B_n) \leq m(B).$$

Also

$$m(T_A^{-1}B) = m(A) - m(T_A^{-1}A \setminus B) \geq m(A) - m(A \setminus B) = m(B),$$

thus $m(T_A^{-1}B) = m(B)$ and $m(B_n) \to 0$ as $n \to \infty \ \forall \ B \in \mathcal{B} \cap A$. $\square$

REMARK. It follows from the proof of proposition 1.5.3 that

$$m(T^{-n}A \setminus \bigcup_{k=0}^{n-1} T^{-k}A) \to 0 \text{ as } n \to \infty \ \ \forall \ A \in \mathcal{B}, \ m(A) < \infty.$$

In case T is invertible, this can be seen more directly since

$$m(T^{-n}A \setminus \bigcup_{k=0}^{n-1} T^{-k}A) = m(A \setminus \bigcup_{k=1}^{n} T^k A) \to 0 \text{ as } n \to \infty$$

by conservativity of T^{-1}. A similar argument also works generally as

$$m(T^{-n}A \setminus \bigcup_{k=0}^{n-1} T^{-k}A) = m(A \setminus \bigcup_{k=1'}^{n} T^{-k}A).$$

1.5.4 LEMMA.
If T is a conservative measure preserving transformation of $(X, \mathcal{B}, m)$, $A \in \mathcal{B}_+$, $0 < m(A) < \infty$, and $\bigcup_{n=1}^{\infty} T^{-n}A = X \mod m$; then

$$m(B \setminus A) = \sum_{k=1}^{\infty} m(A \cap T^{-k}B \setminus \bigcup_{j=1}^{k} T^{-j}A) \ \ \forall \ B \in \mathcal{B}.$$

PROOF. For $B \in \mathcal{B}$, define for $n \geq 0$

$$B_n = T^{-n}B \setminus \bigcup_{k=0}^{n} T^{-k}A,$$

then

$$A \cap T^{-k}B \setminus \bigcup_{j=1}^{k} T^{-j}A = T^{-1}B_{k-1} \setminus B_k \text{ for } k \geq 1,$$

and it follows that for $N \geq 1$,

$$\sum_{k=1}^{N} m(A \cap T^{-k}B \setminus \bigcup_{j=1}^{k} T^{-j}A) = \sum_{k=1}^{N}(m(T^{-1}B_{k-1}) - m(B_k))$$

$$= m(B \setminus A) - m(B_N)$$

and to complete the proof of the lemma, we must show that $m(B_N) \to 0$ as $N \to \infty$.

By assumption,

$$B = B \cap A \cup \bigcup_{N=1}^{\infty} B^{(N)} \text{ where } B^{(N)} \subset T^{-N}A \setminus \bigcup_{k=0}^{N-1} T^{-k}A.$$

For each $N, n \geq 1$, we have

$$B_n^{(N)} := T^{-n}B^{(N)} \setminus \bigcup_{k=0}^{n} T^{-k}A \subset T^{-(n+N)}A \setminus \bigcup_{k=0}^{N+n-1} T^{-k}A$$

and by the remark $m(B_n^{(N)}) \to 0$ as $n \to \infty \; \forall \, N \geq 1$.

To see that $m(B_n) \to 0$, let $\epsilon > 0$ and choose $N_\epsilon \geq 1$ such that $\sum_{k=N_\epsilon+1}^{\infty} m(B^{(k)}) < \frac{\epsilon}{2}$. Now choose n_ϵ such that $m(B_n^{(k)}) < \frac{\epsilon}{2N_\epsilon} \; \forall \, n \geq n_\epsilon$, $1 \leq k \leq N_\epsilon$. It follows that for $n \geq n_\epsilon$,

$$m(B_n) = \sum_{k=1}^{N_\epsilon} m(B_n^{(k)}) + \sum_{k=N_\epsilon+1}^{\infty} m(B_n^{(k)})$$

$$\leq \sum_{k=1}^{N_\epsilon} m(B_n^{(k)}) + \sum_{k=N_\epsilon+1}^{\infty} m(B^{(k)}) < \epsilon.$$

$\square$

1.5.5 KAC'S FORMULA [**Kac**].   *If T is a conservative, ergodic measure preserving transformation of $(X, \mathcal{B}, m)$, and $A \in \mathcal{B}$, $0 < m(A) < \infty$, then*

$$\int_A \varphi_A \, dm = m(X).$$

PROOF. By conservativity and ergodicity of T, $\bigcup_{k=0}^{\infty} T^{-k}A = X$ mod m, and so by lemma 1.5.4, $\forall\, B \in \mathcal{B}$,

$$m(B) = m(A \cap B) + \sum_{k=1}^{\infty} m(A \cap T^{-k}B \setminus \bigcup_{j=1}^{k} T^{-j}A).$$

A calculation shows that

$$m(A \cap B) + \sum_{k=1}^{\infty} m(A \cap T^{-k}B \setminus \bigcup_{j=1}^{k} T^{-j}A) = \int_A \sum_{k=0}^{\varphi_A - 1} 1_B \circ T^k \, dm,$$

whence

$$m(B) = \int_A \sum_{k=0}^{\varphi_A - 1} 1_B \circ T^k \, dm \ \forall\, B \in \mathcal{B}$$

and in particular (choosing $B = X$)

$$m(X) = \int_A \varphi_A \, dm.$$

$\square$

Inducing helps to reduce the study of infinite measure preserving transformations to the easier study of finite measure preserving transformations.

1.5.6 THEOREM (UNICITY OF INVARIANT MEASURE).
Let T be a conservative, ergodic, non-singular transformation of $(X, \mathcal{B}, m)$, then, up to multiplication by constants, there is at most one m-absolutely continuous, σ-finite T-invariant measure.

PROOF. Suppose that $\mu << m$ is a σ-finite T-invariant measure, then $\mu \sim m$, and to prove the theorem we must show that if T is a conservative, ergodic, measure preserving transformation of $(X, \mathcal{B}, \mu)$, and $\nu << \mu$ is a σ-finite T-invariant measure, then $\nu = c\mu$ for some $c \in \mathbb{R}_+$. Under these conditions $\nu \sim \mu$. There is a set $A \in \mathcal{B}$ such that $0 < \mu(A), \nu(A) < \infty$. We'll assume that in addition, $\mu(A) = \nu(A) = 1$. By proposition 1.5.3, $\mu|_A$ and $\nu|_A$ are T_A-invariant probabilities, and so by theorem 1.4.2 $\mu|_A \equiv \nu|_A$. By lemma 1.5.4, for any $B \in \mathcal{B}$,

$$\nu(B) = \sum_{k=0}^{\infty} \nu(A \cap T^{-k}B \setminus \bigcup_{j=1}^{k} T^{-j}A)$$

$$= \sum_{k=0}^{\infty} \mu(A \cap T^{-k}B \setminus \bigcup_{j=1}^{k} T^{-j}A)$$

$$= \mu(B).$$

$\square$

The next proposition shows how to construct an invariant measure for a non-singular transformation, given one for an induced transformation. This allows one to exploit the relative ease of finding an invariant probability for a chosen induced transformation, than an invariant measure for a given transformation.

1.5.7 PROPOSITION.
Let T be a conservative, non-singular transformation of $(X, \mathcal{B}, m)$.

Let $A \in \mathcal{B}_+$, and suppose that $q << m|_A$ is a T_A-invariant measure. Set, for $B \in \mathcal{B}$,

$$\mu(B) = \sum_{k=0}^{\infty} q(A \cap T^{-k}B \setminus \bigcup_{j=1}^{k} T^{-j}A),$$

then $\mu << m$ is a T-invariant measure.

PROOF. Clearly $\mu << m$. Moreover

$$\mu(B) = \sum_{k=0}^{\infty} q(A \cap T^{-k}B \setminus \bigcup_{j=1}^{k} T^{-j}A)$$

$$= \int_A \sum_{k=0}^{\varphi-1} 1_B \circ T^k dq,$$

and therefore

$$\mu(T^{-1}B) = \int_A \sum_{k=0}^{\varphi-1} 1_B \circ T^{k+1} dq$$

$$= \int_A \sum_{k=1}^{\varphi-1} 1_B \circ T^k dq + \int_A 1_B \circ T_A dq$$

$$= \mu(B),$$

since

$$\int_A 1_B \circ T_A dq = q(T_A^{-1}(A \cap B)) = q(A \cap B).$$

$\square$

1.5.8 THEOREM (MAXIMALLY SUPPORTED INVARIANT MEASURE).
Let T be a conservative, non-singular transformation of $(X, \mathcal{B}, m)$, and suppose that $\exists$ a σ-finite, T-invariant measure; then $\exists$ σ-finite, T-invariant measure μ which is maximally supported in the sense that $T|_{[\frac{d\mu}{dm}=0]}$ has no σ-finite, T-invariant measure.

PROOF.
Consider the hereditary collection

$$\mathfrak{S} :=$$

$$\left\{ A \in \mathcal{B} : \exists \ \sigma\text{-finite}, T\text{-invariant measure } \nu \text{ such that } A \subset [\frac{d\nu}{dm} > 0] \mod m \right\}.$$

By the exhaustion lemma, $\exists \ A_1, A_2, \cdots \in \mathfrak{S}$ such that $U(\mathfrak{S}) = \bigcup_{n=1}^{\infty} A_n$. Suppose that $\nu_n << m$ is a σ-finite, T-invariant measure and $A_n \subset [\frac{d\nu_n}{dm} > 0] \mod m$. A suitable measure is given by

$$\mu := \sum_{n=1}^{\infty} \frac{1}{2^n} \nu_n$$

because $[\frac{d\mu}{dm} > 0] = U(\mathfrak{S}) \mod m$.

$\square$

REMARK: DECOMPOSITION.

We now know that if T is a conservative, non-singular transformation of $(X, \mathcal{B}, m)$, then

$$X = \mathfrak{P}(T) \cup \Sigma(T) \cup \mathfrak{N}_{III}(T) \qquad \text{mod } m$$

where $\mathfrak{P}(T)$, $\Sigma(T)$, $\mathfrak{N}_{III}(T) \in \mathcal{B}$ are disjoint, and

$\mathfrak{P}(T)$ is the support of a maximally supported absolutely continuous T-invariant probability;

and $\mathfrak{P}(T) \cup \Sigma(T)$ is the support of a maximally supported absolutely continuous T-invariant σ-finite measure.

As shown in theorem 1.2.9, it may be that $\mathfrak{N}_{III}(T) := X \setminus (\mathfrak{P}(T) \cup \Sigma(T))$ is non-null.

Tower constructions. We conclude this section with two structural representations of a conservative non-singular transformation, Rokhlin towers, and Kakutani towers (or skyscrapers).

1.5.9 ROKHLIN'S TOWER THEOREM [**Ro2**]. *Let T be a conservative, ergodic non-singular transformation of the σ-finite measure space $(X, \mathcal{B}, m)$. For $N \geq 1$, and $\epsilon > 0$, $\exists\, E \in \mathcal{B}$ such that $\{T^{-j}E\}_{j=0}^{N-1}$ are disjoint, and $m(X \setminus \bigcup_{j=0}^{N-1} T^{-j}E) < \epsilon$.*

PROOF. Choose $A \in \mathcal{B}$ such that $m(\bigcup_{k=0}^{N-1} T^{-k}A) < \epsilon$, and set $A_0 := A$, $A_n := T^{-n}A \setminus \bigcup_{j=0}^{n-1} T^{-j}A$, $(n \geq 1)$, then $\bigcup_{n=0}^{\infty} A_n = \bigcup_{n=0}^{\infty} T^{-n}A = X$.

Set $E := \bigcup_{p=1}^{\infty} A_{pN}$, then $\{T^{-j}E\}_{j=0}^{N-1}$ are disjoint, and for $0 \leq k \leq N - 1$, $T^{-k}E \supset \bigcup_{p=1}^{\infty} A_{pN+k}$, whence $\bigcup_{k=0}^{N-1} T^{-k}E \supset \bigcup_{n=N}^{\infty} A_n$, and

$$m(X \setminus \bigcup_{j=0}^{N-1} T^{-j}E) \leq m(\bigcup_{n=0}^{N-1} A_n) = m(\bigcup_{k=0}^{N-1} T^{-k}A) < \epsilon.$$

$\square$

1.5.10 COROLLARY. *Let T be a conservative, ergodic non-singular transformation of the σ-finite measure space $(X, \mathcal{B}, m)$. Let $N \geq 1$, and $A \in \mathcal{B}$, $m(A) > 0$, then*

$$\exists\, B \in \mathcal{B}_+ \cap A \ni \varphi_B \geq N \text{ on } B.$$

PROOF. If E is as in theorem 1.5.9, then $\varphi_E \geq N$. $\square$

Suppose that S is a conservative, non-singular transformation of the σ-finite measure space $(X_S, \mathcal{B}_S, m_S)$ and that $\varphi : X_S \to \mathbb{N}$ is measurable. The (Kakutani) *tower*, (or skyscraper) over S with *height function* φ is the transformation T of the σ-finite measure space $(X_T, \mathcal{B}_T, m_T)$ defined as follows.

$$X_T = \{(x, n) : x \in X_S,\ 1 \leq n \leq \varphi(x)\},$$

$$\mathcal{B}_T = \sigma\{A \times \{n\} : n \in \mathbb{N},\ A \in \mathcal{B}_S \cap [\varphi \geq n]\},\ m_T(A \times \{n\}) = m_S(A),$$

and

$$T(x, n) = \begin{cases} (Sx, \varphi(Sx)) & \text{if } n = 1, \\ (x, n - 1) & \text{if } n \geq 2. \end{cases}$$

EXERCISE 1.5.1   [**Kak**].
Show that

(1) T is a conservative, non-singular transformation.

(2) $T_{X_S \times \{1\}}(x, 1) \equiv (Sx, 1)$, & $\varphi_{X_S \times \{1\}}(x, 1) \equiv \varphi(x)$.

(3) $m_S \circ S^{-1} = m_S \Rightarrow m_T \circ T^{-1} = m_T$.

(4) S is ergodic $\Rightarrow$ T is ergodic.

EXERCISE 1.5.2.   Let T be an invertible, conservative, measure preserving transformation of the σ-finite measure space $(X, \mathcal{B}, m)$, and let $A \in \mathcal{B}$. Prove that T is isomorphic with the tower over T_A with height function φ_A.

§1.6 Group actions and flows

In this section, we consider actions of groups other than $\mathbb{Z}$. Further details on the subject can be found in [**Te**] and [**Zi**].

Let G be a locally compact, second countable topological group, and suppose that $\{T_s : s \in G\}$ is a collection of invertible nonsingular transformations of $(X, \mathcal{B}, m)$.

DEFINITION: NON-SINGULAR ACTION.
We say that $T := \{T_s : s \in G\}$ is a nonsingular *action* of G on $(X, \mathcal{B}, m)$ if
(i) $\exists Y \in \mathcal{B}$ such that $m(X \setminus Y) = 0$ and $T_s : Y \to Y \quad \forall s \in G$,
(ii) $(s, x) \mapsto T_s(x)$ is measurable ($G \times Y \to Y$),
(iii) $T_t \circ T_s(x) = T_{st}(x) \quad \forall x \in Y$, $s, t \in G$.

REMARK. It follows from Banach's theorem for Polish groups (see [**Ban**] p. 20) that T is a nonsingular action of G on $(X, \mathcal{B}, m)$ iff $T : G \to \mathfrak{A}(X, \mathcal{B}, m)$ is a continuous homomorphism of Polish groups.

A nonsingular action of G on $(X, \mathcal{B}, m)$ is called *measure preserving* if $m \circ T_s = m \ \forall s \in G$.

A nonsingular action of $\mathbb{R}$ is called a *flow*.

Actions of countable groups. Suppose that G is countable and that T is a nonsingular action of G on $(X, \mathcal{B}, m)$.

DEFINITION: WANDERING SET, DISSIPATIVE PART.
A *wandering set for* T is a set $A \in \mathcal{B}$ so that $\{T_s A : s \in G\}$ are disjoint.
Let $\mathcal{W} = \mathcal{W}(T)$ denote the collection of wandering sets for T.
The measurable union $\mathfrak{D}(T)$ of the hereditary collection of wandering sets for T is called the *dissipative part* of T, and T is called *dissipative* if $\mathfrak{D}(T) = X$ mod m.

1.6.1 PROPOSITION.
$$\exists \ W_* \in \mathcal{W} \ such \ that \ \mathfrak{D}(T) = \bigcup_{s \in G} T_s W_* \quad mod \ m.$$

PROOF. Assume m is a probability. Set

$$\epsilon_1 := \sup\{m(W): W \in \mathcal{W}\}$$

and choose $W_1 \in \mathcal{W}$ with $m(W_1) > \frac{\epsilon_1}{2}$; set

$$\mathcal{W}_2 := \{W \in \mathcal{W}: W \cap T_s W_1 = \emptyset \ \forall \ s \in G\},$$

$$\epsilon_2 := \sup_{W \in \mathcal{W}_2} m(W) \text{ and choose } W_2 \in \mathcal{W}_2 \text{ with } m(W_2) > \frac{\epsilon_2}{2}.$$

Continuing, we obtain $\mathcal{W}_n$, ϵ_n, W_n such that

$$\mathcal{W}_n = \{W \in \mathcal{W}: W \cap T_s W_k = \emptyset \ \forall \ s \in G, \ 1 \le k \le n-1\},$$

$$\epsilon_n := \sup_{W \in \mathcal{W}_n} m(W),$$

and

$$W_n \in \mathcal{W}_n \text{ with } m(W_n) > \frac{\epsilon_n}{2}.$$

The disjointness of $\{W_n: n \ge 1$ implies that

$$\sum_{n=1}^{\infty} \epsilon_n \le 2 \sum_{n=1}^{\infty} m(W_n) \le m(X) = 1.$$

Let

$$W_* := \bigcup_{n=1}^{\infty} W_n,$$

then $W_* \in \mathcal{W}$ since $T_s W_k \cap T_t W_\ell = \emptyset$ for $(s,k) \ne (t,\ell)$.

The required set is

$$\mathfrak{D} := \bigcup_{s \in G} T_s W_*.$$

Since otherwise, $\exists \ W \in \mathcal{W}$, $W \cap \mathfrak{D} = \emptyset$, whence
$\forall \ n \ge 1: W \in \mathcal{W}_n, \ \rightarrow \ m(W) \le \epsilon_n \rightarrow 0.$ $\qquad\square$

Set $\mathfrak{C}(T) := \mathfrak{D}(T)^c$. Note that $T_s \mathfrak{C} = \mathfrak{C} \mod m \ \forall \ s \in G$.

An action T of G is *free* if $m(\{x \in X: T_s x = x\}) = 0 \ \forall \ s \in G \setminus \{e\}$. In order for this definition to make sense, we need that $\{x \in X: T_s x = x\} \in \mathcal{B}$ which follows from the standardness of $(X, \mathcal{B}, m)$.

An action T of G is called *conservative* if $\mathfrak{C} = X \mod m$, and *totally dissipative* if $\mathfrak{D} = X \mod m$.

1.6.2 PROPOSITION. *Let T be a nonsingular, free action of G. If $A \in \mathcal{B}$ and $A \subset \mathfrak{C}$, then*

$$\sum_{s \in G} 1_A \circ T_s = \infty \text{ a.e. on } A.$$

PROOF. If not, then $\exists \ A \in \mathcal{B}_+$, $A \subset \mathfrak{C}$, such that

$$\sum_{s \in G} 1_A \circ T_s < \infty \text{ a.e. on } A.$$

Note that $\forall \ B \in \mathcal{B}_+ \cap A$, $\exists \ B_1 \in \mathcal{B}_+ \cap B$ and $F \subset G$ finite such that

$$B_1 \cap T_s B_1 = \emptyset \ \forall \ s \in G \setminus F.$$

Let F_* be a minimal finite subset obtained in this way, and suppose that $B_* \in \mathcal{B}_+ \cap A$ satisfies

$$B_* \cap T_s B_* = \emptyset \ \forall \ s \in G \setminus F_*.$$

We claim $T_s x = x$ a.e. on $B_* \ \forall \ s \in F_*$. If this is not the case, $\exists \ A \in \mathcal{B}_+ \cap B_*$ and $s \in F_*$ with $m(A \setminus T_s A) > 0$, whence, if $C = A \setminus T_s A$, then $C \cap T_s C = \emptyset$ and so

$$C \cap T_s C = \emptyset \ \forall \ s \in G \setminus F \setminus \{s\}$$

contradicting the minimality of F_*. Thus $T_s x = x$ a.e. on $B_* \ \forall \ s \in F_*$ and the assumption of freeness of T forces $F_* = \{e\}$ and $B_* \in \mathcal{W}$ contradicting $B_* \subset A \subset \mathfrak{C}(T)$. $\qquad \square$

1.6.3 THEOREM. *Let T be a free, measure preserving action of G on $(X, \mathcal{B}, m)$, and let $f \in L^1(m)$, $f > 0$; then*

$$\mathfrak{D}(T) = \left[\sum_{s \in G} f \circ T_s < \infty \right] \qquad \text{mod } m.$$

PROOF. Let $A \in \mathcal{B}_+ \cap \left[\sum_{s \in G} f \circ T_s < \infty \right]$, then $\exists \ B \in \mathcal{B}_+ \cap A$ such that

$$\int_B \left(\sum_{s \in G} f \circ T_s \right) dm < \infty \qquad \text{mod } m,$$

whence $\sum_{s \in G} 1_B \circ T_s < \infty$ a.e., and by the previous $B \subset \mathfrak{D}(T)$. This shows

$$\mathfrak{D}(T) \supset \left[\sum_{s \in G} f \circ T_s < \infty \right] \qquad \text{mod } m.$$

The other inclusion follows from

$$\int_W \left(\sum_{s \in G} f \circ T_s \right) dm \leq \int_X f \, dm < \infty \ \forall \ f \in L^1(m)_+.$$

$\qquad \square$

Actions of uncountable groups. Let G be a locally compact, second countable topological group, and let $dm_G(g) = dg$ denote the left Haar measure.

Recall that a *lattice* in G is a subgroup $\Gamma \subset G$ with the property that $\exists \ F \in \mathcal{B}_G$, $F \subset G$ with $m_G(F) < \infty$ such that $\{\gamma F : \gamma \in \Gamma\}$ are disjoint, and $\bigcup_{\gamma \in \Gamma} \gamma F = G$.

1.6.4 THEOREM. *Let T be a free, measure preserving action of G on $(X, \mathcal{B}, m)$, and let $f \in L^1(m)$, $f > 0$; then for any lattice $\Gamma \subset G$,*

$$\mathfrak{D}(T|_\Gamma) = \left[\int_G f \circ T_s \, ds < \infty \right] \qquad \text{mod } m$$

where $T|_\Gamma$ denotes the subaction of Γ.

PROOF.

$$\int_G f \circ T_g dg = \sum_{\gamma \in \Gamma} \int_{F\gamma} f \circ T_g dg$$

$$= \sum_{\gamma \in \Gamma} \left(\int_F f \circ T_g dg \right) \circ T_\gamma dg$$

$\square$

1.6.5 COROLLARY.
Let T be a non-singular flow on the σ-finite measure space $(X, \mathcal{B}, m)$, then

$$\mathfrak{C}(T_s) = \mathfrak{C}(T_1) \ \forall \ s \neq 0.$$

PROOF. For each $s \neq 0$, $s\mathbb{Z}$ is a lattice in $\mathbb{Z}$ and $\mathfrak{C}(T|_{s\mathbb{Z}}) = \mathfrak{C}(T_s)$. $\square$

DEFINITION: ERGODIC.
The nonsingular action T of G on $(X, \mathcal{B}, m)$ is *ergodic* if $A \in \mathcal{B}$, $m(A \triangle T_s A) = 0 \ \forall \ s \in G$ implies $m(A) = 0$ or $m(A^c) = 0$.

1.6.6 PROPOSITION. *An ergodic, non-singular, free action of a countable group on a nonatomic space is conservative.*

PROOF. If T is an action of G on X which is ergodic and not conservative, then T is totally dissipative, and $\mathfrak{D} = X = \bigcup_{s \in G} T_s W$ for some $W \in \mathcal{W}$. Ergodicity forces W to be an atom, whence up to isomorphism, $X = G$ and $T_s x = sx$, contradicting nonatomicity of X. $\square$

1.6.7 PROPOSITION. *Suppose that T is an ergodic, nonsingular action of G on X, and let $\Gamma \subset G$ be a dense subgroup;*
then $T|_\Gamma$ (the subaction of Γ on X) is also ergodic.

PROOF. Suppose that $A \in \mathcal{B}$ and $m(A \triangle T_s A) = 0 \ \forall \ s \in \Gamma$. By denseness of Γ in G and continuity of $T : G \to \mathfrak{A}(X, \mathcal{B}, m)$, $m(A \triangle T_s A) = 0 \ \forall \ s \in G$. $\square$

EXAMPLE. Let $(X, \mathcal{B}, m)$ be $\mathbb{R}$ equipped with Borel sets and a probability $m \sim$ Lebesgue measure. Let T be the flow defined by $T_s x := x + s$, $x, s \in X = \mathbb{R}$. Clearly T is ergodic on X.
Each element T_s $(s \neq 0)$ is totally dissipative. However, the subaction $T|_\mathbb{Q}$, being ergodic by proposition 1.6.7, is conservative by proposition 1.6.6.

Let G be a locally compact, second countable topological group, then G is σ-compact, and Polish. Let m_G be left Haar measure on G.

1.6.8 PROPOSITION.
The action of G on $(G, \mathcal{B}(G), m_G)$ by left translation is ergodic.

PROOF. Suppose $A \in \mathcal{B}(G)_+$ and $m_G(gA \triangle A) = 0 \ \forall \ g \in G$. The measure m' defined by $dm' = 1_A dm_G$ is a left Haar measure on G, and by unicity of such, $m' = m_G$ whence $A = G$ mod m_G. $\square$

The ergodicity of the action of G by right translation is obtained in a similar manner, as right Haar measure is equivalent to m_G.
Define maps $L, R : G \to \mathcal{M}(G)$ by $L_g(x) := gx$, $R_g(x) := xg$.

1.6.9 COROLLARY.

The maps L, $R : G \to \mathcal{M}(G)$ are continuous, and their ranges are closed in $\mathcal{M}(G)$.

PROOF. This follows from the ergodicity of the actions of G by translation; indeed, if $R = \lim_{n \to \infty} R_{g_n}$, set $f(x) = x^{-1}R(x)$, which is L_g-invariant $\forall\, g \in G$, hence constant, and $R = R_h$ for some $h \in G$. Let the ranges of these maps in $\mathcal{M}(G)$ be $\tilde{G}_L$ and $\tilde{G}_R$, considered with their inherited (Polish) topologies. By the measurable image theorem the inverse maps $L^{-1} : \tilde{G}_L \to G$ and $R^{-1} : \tilde{G}_R \to G$ are both measurable, and (being group isomorphisms) are continuous by Banach's theorem ([**Ban**], p.20). $\square$

GROUP NORM.

A *(group) norm* on the Abelian group G is a function $y \mapsto \|y\|$ $(G \to \mathbb{R})$ such that
$$\|y\| = \|-y\| \geq 0 \ \forall\, y \in G$$
with equality if and only if $y = 0$, and such that
$$\|x + y\| \leq \|x\| + \|y\| \ \forall\, x, y \in G.$$

Any group norm on G is canonically associated to an invariant metric: if $\rho(x, y) := \|x - y\|$ is a metric on G which is *invariant* in the sense that
$$\rho(x + z, y + z) = \rho(x, y).$$
Also if ρ is an invariant metric on G then $y \mapsto \rho(0, y)$ defines a norm on G.

G is a topological group under the topology generated by a norm, indeed $\|(x - y) - (x' - y')\| \leq \|x - x'\| + \|y - y'\|$ showing the continuity of $(x, y) \mapsto x - y$.

Even if $G = \mathbb{R}^d$, there are more group norms than vector space norms on G. Indeed if $\|\cdot\|$ is a group norm on G, then so is $\|\cdot\|_* := \|\cdot\| \wedge 1$.

1.6.10 COROLLARY.

If G is Abelian, then $\exists$ a topology generating norm $\|\cdot\|_G$ on G, and $\exists\, \kappa_G > 0$ such that
$$B_G(y, \epsilon) := \{z \in G : \ \|z - y\|_G \leq \epsilon$$
is compact $\forall\, \epsilon < \kappa_G$, $y \in G$.

PROOF. It follows from corollary 1.6.9 that the metric $d : G \times G \to \mathbb{R}$ defined by
$$d(x, y) = \rho(L_x, L_y) := \sum_{n=1}^{\infty} \frac{1}{2^n} (\|L_x f_n - L_y f_n\|_2 + \|L_{x^{-1}} f_n - L_{y^{-1}} f_n\|_2)$$
where $\{f_n : n \in \mathbb{N}\}$ is a complete, orthonormal system in $L^2(m_G)$. generates the topology of G. Since G is Abelian, this metric is invariant $(d(a + x, a + y) = d(x, y))$ and the norm $\|x\|_G := d(x, \text{Id})$ is a topology generating norm on G. Because of this $\exists\, \kappa_G > 0$ such that $B_G(y, \epsilon) := \{z \in G : \ \|z - y\|_G \leq \epsilon\}$ is compact $\forall\, \epsilon < \kappa_G$ when $y = 0$. The compactness $\forall\, y \in G$ follows from $B_G(y, \epsilon) = B_G(0, \epsilon) + y$. $\square$

CHAPTER 2

General Ergodic and Spectral Theorems

Let $(X, \mathcal{B}, m)$ be a σ-finite measure space, and suppose that for some $p \in [1, \infty)$, $P : L^p(m) \to L^p(m)$ is a continuous linear operator.

The *ergodic sums* of $f \in L^p(m)$ (under P) are the sums $\sum_{k=0}^{n-1} P^k f$, $(n \in \mathbb{N})$.

An ergodic theorem (usually, and here) establishes the convergence of normalised ergodic sums, or ratios thereof. For example if T is an ergodic probability preserving transformation of $(X, \mathcal{B}, m)$, then :

von Neumann's ergodic theorem [**Neu-v2**] proved in §2.1 says that

$$\left\| \frac{1}{n} \sum_{k=0}^{n-1} f \circ T^k - \int_X f \, dm \right\|_2 \longrightarrow 0 \quad \forall \, f \in L^2(m),$$

and Birkhoff's ergodic theorem [**Bir**] proved in §2.2 says that

$$\frac{1}{n} \sum_{k=0}^{n-1} f \circ T^k \to \int_X f \, dm \text{ a.e. } \forall \, f \in L^1(m).$$

and if T is a conservative ergodic measure preserving transformation of $(X, \mathcal{B}, m)$, then Hopf's ergodic theorem [**Hop1**] says that for $f, g \in L^1(m)$, $\int_X g \, dm \neq 0$, for a.e. $x \in X$:

$$\frac{\sum_{k=0}^{n-1} f(T^k x)}{\sum_{k=0}^{n-1} g(T^k x)} \longrightarrow \frac{\int_X f \, dm}{\int_X g \, dm}$$

We begin in §2.1 with that of von Neumann (historically the first) which deals with convergence in Hilbert space of the averages of a unitary operator.

In §2.2, we prove Hurewicz's ergodic theorem [**Hur**] establishing almost everywhere convergence of the ratios of ergodic sums under the dual operator of a conservative non-singular transformation, deducing the ergodic theorems of Birkhoff and Hopf from it.

The proof of Hurewicz's theorem has the same template as von Neumann's proof. The maximal inequality used is Hopf's version of Riesz's maximal inequality. We give here the proof of Garsia [**Ga**]. The interested reader may find a similar proof of the more general Chacon-Ornstein theorem [**Cha-Orn**] in [**Ga**] (see also [**Fo**]).

In §2.3, we prove a converse to Birkhoff's theorem, and in §2.4 we show that there is no absolutely normalised pointwise convergence for ergodic, infinite measure preserving transformations.

The rest of the chapter is devoted to spectral theory including unitary operators (§2.5), eigenvalues (§2.6) and ergodicity of Cartesian products (§2.7).

§2.1 von Neumann's mean ergodic theorem.

2.1.1 MEAN ERGODIC THEOREM [**Neu-v2**].
Let H be a Hilbert space, let $U : H \to H$ be a unitary operator, and let

$$H_U = \{f \in H : Uf = f\},$$

then

$$\frac{1}{n} \sum_{k=0}^{n-1} U^k f \xrightarrow{H} Pf \text{ as } n \to \infty, \ \ \forall f \in H,$$

where $P : H \to H_U$ is linear, $P|_{H_U} = Id$, and

$$\langle f, h \rangle = \langle Pf, h \rangle \ \ \forall f \in H, \ h \in H_U.$$

PROOF. Let

$$H_0 = \{g - Ug + h : g \in H, h \in H_U\}.$$

Clearly, for $f = g - Ug + h \in H_0$,

$$\frac{1}{n} \sum_{k=0}^{n-1} U^k f = \frac{1}{n}(g - U^n g) + h \xrightarrow{H} h := Pf.$$

This defines $P : H_0 \to H_U$ by

$$P(g - Ug + h) = h,$$

which satisfies $\|Pf\| \leq \|f\|$, $(f \in H_0)$, $P|_{H_U} = Id$, and

$$\frac{1}{n} \sum_{k=0}^{n-1} U^k f \xrightarrow{H} Pf \ \forall f \in H_0.$$

Next, the operator P can clearly be extended to $\overline{H_0} \in H$, and the extension clearly satisfies $\|Pf\| \leq \|f\|$, $(f \in \overline{H_0})$, $P|_{H_U} = Id$.

Suppose $f \in \overline{H_0}$, then there is a sequence $f_n \in H_0$,

$$f_n \xrightarrow{H} f,$$

and, for every $n, \nu \in \mathbb{N}$,

$$\|\frac{1}{n} \sum_{k=0}^{n-1} U^k f - Pf\| \leq \|\frac{1}{n} \sum_{k=0}^{n-1} U^k (f - f_\nu)\| + \|\frac{1}{n} \sum_{k=0}^{n-1} U^k f_\nu - Pf_\nu\| + \|Pf_\nu - Pf\|$$

$$\leq 2\|f - f_\nu\| + \|\frac{1}{n} \sum_{k=0}^{n-1} U^k f_\nu - Pf_\nu\|,$$

whence

$$\frac{1}{n} \sum_{k=0}^{n-1} U^k f \xrightarrow{H} Pf.$$

We show that $\overline{H_0} = H$. To see this, let $f \in H_0^\perp$, then

$$f \perp g - Ug \ \forall g \in H,$$

whence $U^*f = f = Uf$ and $f \in H_0 \cap H_0^\perp = \{0\}$. Thus $H_0^\perp = \{0\}$ and $\overline{H_0} = H$.

Lastly, we show that

$$\langle f, h \rangle = \langle Pf, h \rangle \quad \forall \, h \in H_U.$$

Let $h \in H_U$, and $f \in H$, then

$$\langle Pf, h \rangle = \lim_{n \to \infty} \frac{1}{n} \sum_{k=0}^{n-1} \langle U^k f, h \rangle$$
$$= \langle f, h \rangle.$$

$\square$

2.1.2 COROLLARY [Neu-v2]. *Let T be an invertible, measure preserving transformation of the probability space $(X, \mathcal{B}, m)$, then*

$$\frac{1}{n} \sum_{k=0}^{n-1} f \circ T^k \xrightarrow{L^2(m)} E(f|\mathfrak{I}) \quad \forall \, f \in L^2(m)$$

where $E(\cdot|\mathfrak{I})$ denotes conditional expectation with respect to the σ-algebra $\mathfrak{I}$ of T-invariant, $\mathcal{B}$-measurable sets.

PROOF. Let $H = L^2(m)$, and define $U : H \to H$ by $Uf := f \circ T$. Clearly, U is a unitary operator.

We must show that $H_U = L^2(\mathfrak{I})$, and $Pf = E(f|\mathfrak{I})$.

Clearly $L^2(\mathfrak{I}) \subset H_U$. Conversely let $f \in H_U$, then $f \circ T = f \mod m$ and for $A \in \mathcal{B}(\mathbb{R})$, $T^{-1}f^{-1}A = f^{-1}A \mod m$, whence $f^{-1}A \in \mathfrak{I}$. This shows that f is $\mathfrak{I}$-measurable, hence $f \in L^2(\mathfrak{I})$. Thus $H_U = L^2(\mathfrak{I})$.

It follows that $P : H \to L^2(\mathfrak{I})$ satisfies

$$\int_I Pf dm = \int_I f dm \quad \forall \, f \in L^2(m), \ I \in \mathfrak{I},$$

whence $Pf = E(f|\mathfrak{I})$.

$\square$

An invertible, non-singular transformation T of the standard σ-finite measure space $(X, \mathcal{B}, m)$ also induces a surjective isometry $U_T : L^2(m) \to L^2(m)$ defined by

$$U_T f := \sqrt{\frac{dm \circ T}{dm}} f \circ T.$$

If $f \in H_{U_T}$, then

$$\widehat{T}|f|^2 = (U_T|f|)^2 = |U_T f|^2 = |f|^2$$

and $dp := \frac{|f|^2 dm}{\|f\|_2^2}$ is a T-invariant probability.

We thus obtain the following information from von Neumann's theorem:

2.1.3 PROPOSITION.   *Let T be an invertible, non-singular transformation the standard σ-finite measure space $(X, \mathcal{B}, m)$ without an m-absolutely continuous, T-invariant probability, then*

$$\frac{1}{n} \sum_{k=0}^{n-1} U_T^k f \xrightarrow{L^2(m)} 0 \quad \forall f \in L^2(m).$$

§2.2 Pointwise ergodic theorems

Pointwise convergence of

$$\frac{1}{n} \sum_{k=0}^{n-1} f \circ T^k$$

for T a measure preserving transformation was proved by Birkhoff [**Bir**], but again the limit here is only non-trivial for finite measure spaces. Generalisation's of Birkhoff's theorem were proved by E.Hopf [**Hop1**] (for infinite measure preserving transformations), W.Hurewicz [**Hur**] (for non-singular transformations), and R.Chacon, D.Ornstein [**Cha-Orn**] (for Markov operators).

We prove Hurewicz's theorem. The proof we give is a development of the proof of von Neumann's theorem, and can easily be adapted to prove the Chacon-Ornstein theorem (see [**Ga**]).

Suppose that T is a conservative, non-singular transformation of the σ-finite measure space $(X, \mathcal{B}, m)$. In this section, we prove

2.2.1 HUREWICZ'S ERGODIC THEOREM   [**Hur**].

$$\frac{\sum_{k=1}^{n} \widehat{T}^k f(x)}{\sum_{k=1}^{n} \widehat{T}^k p(x)} \to_{n \to \infty} h(f, p)(x) \ \textit{for a.e.} \ x \in X, \ \forall f, p \in L^1(m), \quad p > 0,$$

where

$$h(f, p) \circ T = h \ \textit{and} \ \int_X h(f, p) k p \, dm = \int_X k f \, dm \ \forall k \in \ L^\infty(m), \ k \circ T = k.$$

Note that, when T is ergodic, $h(f, p)$ is constant, whence

$$h(f, p) = \frac{\int_X f \, dm}{\int_X p \, dm}.$$

In general,

$$h(f, p) = E_{m_p}(\frac{f}{p} | \mathfrak{J}),$$

where $dm_p = p \, dm$, and $\mathfrak{J}$ is the σ-algebra of T-invariant sets in $\mathcal{B}$.

Set, for $f, p \in L^1(m)$, $p > 0$, $\widehat{T}_0 f = 0$, and $n \in \mathbb{N}$,

$$\widehat{T}_n f = \sum_{k=0}^{n-1} \widehat{T}^k f, \quad R_n(f, p) = \frac{\widehat{T}_n f}{\widehat{T}_n p}.$$

The plan of proof is, to fix $p \in L^1(m)$, $p > 0$, and show, as in the proof of von Neumann's theorem, convergence of the ratios $R_n(f, p)$ for a dense class of functions

$f \in L^1(m)$, consider the limit as a bounded linear operator, and then extend the definition of this operator, and the ratio convergence to all functions $f \in L^1(m)$. The approximation argument extending the ratio convergence needs the *maximal inequality*.

First, suppose that $f \in L^1(m)$ has the form

$$f = hp, \text{ where } h \circ T = h.$$

We have

$$\int_X \widehat{T}^n f \cdot g \, dm = \int_X phg \circ T^n dm$$

$$= \int_X ph \circ T^n g \circ T^n dm = \int_X h\widehat{T}^n p \cdot g \, dm$$

for every $g \in L^\infty(m)$, $n \in \mathbb{N}$, whence

$$\widehat{T}^n f = h\widehat{T}^n p,$$

and

$$R_n(f, p) = h.$$

The class of functions for which we shall prove convergence, and density is

$$\mathcal{H}_p := \{f = hp + g - \widehat{T}g \in L^1(m) : h \circ T = h \in L^\infty(m), g \in L^1(m)\}.$$

For $f = hp + g - \widehat{T}g$, we have

$$R_n(f, p) = h + \frac{g - \widehat{T}^n g}{\widehat{T}_n p},$$

and, the convergence

$$R_n(f, p) \to_{n \to \infty} h, \text{ a.e.,}$$

follows immediately from the

2.2.2 CHACON-ORNSTEIN LEMMA [**Cha-Orn**].

$$\frac{\widehat{T}^n g}{\widehat{T}_n p} \to_{n \to \infty} 0, \text{ a.e. } \forall g \in L^1(m).$$

PROOF. Choose $\epsilon > 0$, and let

$$\eta_n = 1_{[\widehat{T}^n g > \epsilon \widehat{T}_n p]}.$$

We must show that $\sum_{n=1}^{\infty} \eta_n < \infty$ a.e. $\forall \epsilon > 0$.

We have

$$\epsilon p + \widehat{T}^{n+1} g - \epsilon \widehat{T}_{n+1} p = \widehat{T}(\widehat{T}^n g - \epsilon \widehat{T}_n p),$$

whence

$$\epsilon p + \widehat{T}^{n+1} g - \epsilon \widehat{T}_{n+1} p \leq \widehat{T}(\widehat{T}^n g - \epsilon \widehat{T}_n p)_+,$$

where g_+ denotes $g \vee 0$, $f \vee g = \max\{f, g\}$, and

$$\eta_n \epsilon p + (\widehat{T}^{n+1} g - \epsilon \widehat{T}_{n+1} p)_+ \leq \widehat{T}(\widehat{T}^n g - \epsilon \widehat{T}_n p)_+.$$

Integrating, we get

$$\epsilon \int_X p\eta_n dm \leq \int_X (J_n - J_{n+1})dm$$

where

$$J_n = (\widehat{T}^n g - \epsilon \widehat{T}_n p)_+,$$

and, summing over n, we get

$$\epsilon \int_X p \sum_{n=1}^{N} \eta_n dm \leq \int_X J_1 dm < \infty.$$

This shows that indeed

$$\sum_{n=1}^{\infty} \eta_n < \infty \text{ a.e.}$$

and thereby proves the lemma.   □

To establish that

$$\overline{\mathcal{H}_p} = L^1(m),$$

we show that

$$k \in L^\infty(m), \int_X kf dm = 0 \ \forall f \in \mathcal{H}_p \ \Rightarrow \ k = 0 \ \text{a.e.}$$

To see this, let

$$k \in L^\infty(m) \ni \int_X kf dm = 0 \ \forall f \in \mathcal{H}_p,$$

then, in particular

$$\int_X gk \circ T dm = \int_X \widehat{T}g \cdot k dm = \int_X gk dm \ \forall g \in L^1(m),$$

whence $k \circ T = k$ a.e., and $kp \in \mathcal{H}_p$.

Hence,

$$\int_X k^2 p dm = 0 \ \Rightarrow \ k = 0 \ \text{a.e.}$$

The convergence of the ratios $R_n(f, g)$ for $f \in \mathcal{H}_p$ shows that if $f = hp+g-\widehat{T}g \in \mathcal{H}_p$, then h is uniquely determined by f. Moreover, if $k \circ T = k \in L^\infty(m)$, then

$$\int_X kf dm = \int_X k(hp + g - \widehat{T}g)dm$$

$$= \int_X khp dm + \int_X k(g - \widehat{T}g)dm$$

$$= \int_X khp dm.$$

If $k =$sign$h$, then $k \circ T = k$, and

$$\|hp\|_1 = \int_X hkp dm = \int_X kf dm \leq \|f\|_1.$$

What all this means is that we can define $\Phi_p : \mathcal{H}_p \to L^1(m)$ by

$$\Phi_p(hp + g - \widehat{T}g) = hp,$$

and then Φ_p can be extended to all $L^1(m)$, having the properties

$$\|\Phi_p(f)\|_1 \leq \|f\|_1 \ \forall f \in L^1(m),$$

$$\int_X k\Phi_p(f)dm = \int_X kfdm, \ \forall k \circ T = k \in L^\infty(m).$$

We have shown that

$$R_n(f,g) \to_{n\to\infty} \frac{\Phi_p(f)}{p}$$

for $f \in \mathcal{H}_p$, and we extend this convergence to all $f \in L^1(m)$, by an approximation argument which uses the

2.2.3 MAXIMAL INEQUALITY   [**Wi**].
For $f,p \in L^1$, such that $p > 0$ a.e., and $t \in \mathbb{R}_+$,

$$m_p([\sup_{n\in\mathbb{N}} R_n(f,p) > t]) \leq \frac{\|f\|_1}{t},$$

where $dm_p = pdm$.

Before proving the maximal inequality, we show how to complete the proof of the ergodic theorem using it.

Let $f \in L^1(m)$. Fix $\epsilon > 0$. We can write $f = g + k$, where $g \in \mathcal{H}_p$, $\Phi_p(g) = \Phi_p(f)$, and $\|k\|_1 < \epsilon^2$. It follows that

$$\limsup_{n\to\infty} |R_n(f,p) - h| \leq \limsup_{n\to\infty} |R_n(k,p)| \leq \sup_{n\in\mathbb{N}} |R_n(k,p)|,$$

whence, by the maximal inequality,

$$m_p([\limsup_{n\to\infty} |R_n(f,p) - h| > \epsilon]) \leq \frac{\|k\|_1}{\epsilon} \leq \epsilon.$$

This last inequality holds for arbitrary $\epsilon > 0$, whence

$$\limsup_{n\to\infty} |R_n(f,p) - h| = 0 \ \text{ a.e.},$$

and the ergodic theorem is almost established, it remaining only to prove the maximal inequality.

To do this, we begin by establishing

2.2.4 MAXIMAL ERGODIC THEOREM   [**Yo-Kak**], (SEE ALSO [**Hop2**]).

$$\int_{[M_n f > 0]} fdm \geq 0, \ \forall f \in L^1(m), n \in \mathbb{N},$$

where

$$M_n f = \left(\bigvee_{k=1}^n \widehat{T}_k f \right)_+ = \left(\bigvee_{k=0}^n \widehat{T}_k f \right).$$

PROOF [**Ga**]. Note first that if $M_n f(x) > 0$, then

$$M_n f(x) \leq M_{n+1} f(x)$$

$$= \bigvee_{k=1}^{n+1} \widehat{T}_k f(x)$$

$$= f(x) + \bigvee_{k=0}^{n} \widehat{T}_k \widehat{T} f(x)$$

$$= f(x) + M_n \widehat{T} f(x).$$

Also, since

$$\widehat{T} f \vee \widehat{T} g \leq \widehat{T}(f \vee g),$$

we have that

$$M_n \widehat{T} f \leq \widehat{T} M_n f,$$

whence

$$M_n f > 0 \;\Rightarrow\; f \geq M_n f - M_n \widehat{T} f,$$

and

$$\int_{[M_n f > 0]} f \, dm \geq \int_{[M_n f > 0]} (M_n f - \widehat{T} M_n f) dm.$$

Since $\widehat{T} M_n f \geq 0$ a.e., and $M_n f = 0$ on $[M_n f > 0]^c$, we get

$$\int_{[M_n f > 0]} f \, dm \geq \int_{[M_n f > 0]} (M_n f - \widehat{T} M_n f) dm$$

$$= \int_{[M_n f > 0]} M_n f \, dm - \int_{[M_n f > 0]} \widehat{T} M_n f \, dm$$

$$\geq \int_X M_n f \, dm - \int_X \widehat{T} M_n f \, dm$$

$$= 0,$$

whence the theorem. □

PROOF OF THE MAXIMAL INEQUALITY. Suppose f, p, t are as in the maximal inequality, then

$$M_n(f - tp) > 0 \;\Leftrightarrow\; \max_{1 \leq k \leq n} R_k(f, p) > t.$$

Thus, using Hopf's maximal ergodic theorem, we obtain

$$\int_{[M_n(f-tp)>0]} (f - tp) dm \geq 0,$$

whence

$$tm_p([\max_{1 \leq k \leq n} R_k(f, p) > t]) \leq \int_{[\max_{1 \leq k \leq n} R_k(f,p)>t]} f \, dm$$

$$\leq \|f\|_1.$$

The maximal inequality follows from this as $n \to \infty$. □

Hurewicz's ergodic theorem is now established.

Hurewicz's theorem for a conservative, ergodic non-singular transformation T, states that

$$\frac{\sum_{k=0}^{n-1}\widehat{T}^k f(x)}{\sum_{k=0}^{n-1}\widehat{T}^k g(x)} \to \frac{\int_X f\,dm}{\int_X g\,dm}\text{ for a.e. } x \in X$$

whenever $f, g \in L^1(m)$, $\int_X g\,dm \neq 0$.

2.2.5 HOPF'S ERGODIC THEOREM   [**Hop1**].
Suppose that T is a conservative measure preserving transformation of the σ-finite measure space
$(X, \mathcal{B}, m)$, then

$$\frac{\sum_{k=1}^{n} f(T^k x)}{\sum_{k=1}^{n} p(T^k x)} \to_{n\to\infty} E_{m_p}(f|\mathfrak{I})(x)\text{for a.e. } x \in X, \ \forall f, p \in L^1(m), \ \ p > 0.$$

Hopf's ergodic theorem is a special case of Hurewicz's theorem in case T is invertible. It can be proved analogously for T non-invertible.

As a corollary, setting $p \equiv 1$, we obtain

2.2.6 BIRKHOFF'S ERGODIC THEOREM   [**Bir**].
Suppose that T is a probability preserving transformation of $(X, \mathcal{B}, m)$, then

$$\frac{1}{n}\sum_{k=1}^{n} f(T^k x) \to_{n\to\infty} E(f|\mathfrak{I})(x)\text{ for a.e. } x \in X, \ \forall f \in L^1(m).$$

EXERCISE 2.2.1.   Suppose that T is a conservative, ergodic, measure preserving transformation of the σ-finite, infinite measure space $(X, \mathcal{B}, m)$, then

$$\frac{1}{n}\sum_{k=1}^{n} f(T^k x) \to_{n\to\infty} 0\text{ for a.e. } x \in X, \ \forall f \in L^1(m).$$

2.2.7 STOCHASTIC ERGODIC THEOREM   [**Kre**].
Suppose that T is a conservative, non-singular transformation of the probability space $(X, \mathcal{B}, m)$, then

$$\frac{1}{n}\sum_{k=0}^{n-1}\widehat{T}^k f \overset{m}{\to} h(f) \ \ \forall \, f \in L^1(m)$$

where $h(f) \in L^1(m)$ and $h(f) \circ T = h(f)$.

PROOF.
This follows from Hurewicz's theorem, and theorem 1.4.4. Note that $h(f) = 0$ on $X \setminus \mathfrak{P}(T)$. □

EXERCISE 2.2.2.   Let $(X, \mathcal{B}, m)$ be $\mathbb{R}$ equipped with Borel sets and Lebesgue measure, and let $Tx = x - \frac{1}{x}$ be Boole's transformation.

Use Hurewicz's theorem and exercise 1.3.4 show that

$$\frac{1}{a(n)} \sum_{k=0}^{n-1} \widehat{T}^k f \to \int_X f dm \text{ a.e. as } n \to \infty \; \forall \; f \in L^1(m)$$

where $a(n) = \sqrt{2n}/\pi$.

Ergodic Decomposition.

DEFINITION: INVARIANT FACTOR.

Let T be a non-singular transformation of the standard probability space $(X, \mathcal{B}, m)$, and let $\mathfrak{I} := \{A \in \mathcal{B} : T^{-1}A = A\}$.

By the factor proposition (§1.0) $\exists$ a non-singular transformation S of a standard probability space $(Y, \mathcal{C}, \mu)$ and a factor map $\pi : X \to Y$ such that $\pi^{-1}\mathcal{C} = \mathfrak{I}$. Evidently $S = $ Id and $\pi \circ T = \pi$.

The factor $(Y, \mathcal{C}, \mu, \text{Id})$ is the maximal identity factor of T and is called the *invariant factor* of T and $\pi : X \to Y$ is called the *invariant factor map*.

We'll see that the invariant factor is connected with the *ergodic decomposition* of T.

The ergodic decompositions considered here are all descended from the von Neumann's original ergodic decomposition [**Neu-v3**] and their proofs are based on the ergodic theorems. An ergodic decomposition (not based on an ergodic theorem) for actions of countable groups is given in [**Var**].

2.2.8 NON-SINGULAR ERGODIC DECOMPOSITION.

Suppose that T is an invertible, conservative, non-singular transformation of a standard probability space $(X, \mathcal{B}, m)$, then there is a probability space $(Y, \mathcal{C}, \mu)$, and a collection of probabilities

$$\{m_y : y \in Y\}$$

on $(X, \mathcal{B})$ such that

For $y \in Y$, T is an invertible, conservative ergodic non-singular transformation of $(X, \mathcal{B}, m_y)$, and

$$\frac{dm_y \circ T}{dm_y} = \frac{dm \circ T}{dm} \quad m_y\text{-a.e.}$$

For $A \in \mathcal{B}$, the map $y \mapsto m_y(A)$ is measurable, and

$$m(A) = \int_Y m_y(A) d\mu(y).$$

PROOF.

Let $(Y, \mathcal{C}, \mu, \text{Id})$ be the invariant factor of T and let $\pi : X \to Y$ be the invariant factor map.

By the disintegration theorem (§1.0), $\exists \; Y_0 \in \mathcal{C}$, $\mu(Y_0^c) = 0$ and $y \mapsto m_y$ measurable $Y_0 \to \mathcal{P}(X, \mathcal{B})$ such that

$$m_y(\pi^{-1}\{y\}) = 1 \; \forall \; y \in Y_0, \quad m(A \cap \pi^{-1}B) = \int_B m_y(A) d\mu(y) \; \forall \; A \in \mathcal{B}, \; B \in \mathcal{C},$$

and
$$m_y \circ T \sim m_y, \quad \frac{dm_y \circ T}{dm_y} = \frac{dm \circ T}{dm} \quad m_y - \text{ a.e. } \forall \, y \in Y_0.$$

As T is conservative on $(X, \mathcal{B}, m)$,

$$\sum_{n=1}^{\infty} \widehat{T}^n 1 = \sum_{n=1}^{\infty} (T^n)' = \infty \text{ a.e. },$$

with the consequence that

$$\sum_{n=1}^{\infty} (T^n)' = \infty \; m_x\text{-a.e., for a.e. } x \in Y.$$

Again, we may assume that for every $x \in Y$,

$$\sum_{n=1}^{\infty} (T^n)' = \infty \; m_x\text{-a.e.,}$$

with the consequence that T is conservative on $(X, \mathcal{B}, m_x)$ for $x \in Y$.

By the Hurewicz ergodic theorem for T acting on $(X, \mathcal{B}, m)$,

$$\lim_{n \to \infty} \frac{\sum_{k=0}^{n-1} \widehat{T}^k 1_A(x)}{\sum_{k=0}^{n-1} \widehat{T}^k 1(x)} = E(1_A | \mathfrak{I})(x) = m_{\pi x}(A)$$

for a.e. $x \in X$, and $A \in \mathcal{B}$, whence, for a.e. $y \in Y$,

$$\lim_{n \to \infty} \frac{\sum_{k=0}^{n-1} \widehat{T}^k 1_A}{\sum_{k=0}^{n-1} \widehat{T}^k 1} = m_y(A) \quad m_y - \text{ a.e. } \forall \, A \in \mathcal{A}.$$

On the other hand, for $y \in Y$, by the Hurewicz ergodic theorem, for T acting on $(X, \mathcal{B}, m_y)$,

$$\lim_{n \to \infty} \frac{\sum_{k=0}^{n-1} \widehat{T}^k 1_A}{\sum_{k=0}^{n-1} \widehat{T}^k 1} = E_{m_y}(1_A | \mathfrak{I}) \quad m_y - \text{ a.e. } \forall \, A \in \mathcal{A}.$$

It follows that for $A \in \mathcal{A}$, and a.e. $y \in Y$,

$$E_{m_y}(1_A | \mathfrak{I}) = m_y(A) \quad m_y \text{a.e.}$$

whence
$$\mathfrak{I} = \{\emptyset, X\} \quad \text{mod } m_y$$
and T is ergodic on $(X, \mathcal{B}, m_x)$. $\qquad \square$

2.2.9 MEASURE PRESERVING ERGODIC DECOMPOSITION.

Suppose that T is a conservative, invertible, measure preserving transformation of a standard σ-finite measure space $(X, \mathcal{B}, \mu)$, then there is a probability space $(\Omega, \mathfrak{I}, \lambda)$, and a collection of measures

$$\{\mu_\omega : \omega \in \Omega\}$$

on $(X, \mathcal{B})$ such that

1 For $\omega \in \Omega$, T is an invertible, conservative ergodic measure-preserving transformation of $(X, \mathcal{B}, \mu_\omega)$.

2   *For $A \in \mathcal{B}$, the map $\omega \mapsto \mu_\omega(A)$ is measurable, and*

$$\mu(A) = \int_\Omega \mu_\omega(A) d\lambda(\omega).$$

PROOF.   Choose a probability $m \sim \mu$, then $(X, \mathcal{B}, m)$ is a standard probability space, and T is a conservative, invertible, non-singular transformation of $(X, \mathcal{B}, m)$, and, indeed

$$\frac{dm \circ T}{dm} = \frac{f}{f \circ T}$$

where

$$f = \frac{d\mu}{dm}.$$

Applying the previous theorem, we obtain a probability space $(\Omega, \mathfrak{I}, \lambda)$, and a collection of probabilities

$$\{m_\omega : \omega \in \Omega\}$$

on $(X, \mathcal{B})$ such that
 1   For $\omega \in \Omega$, T is an invertible, conservative ergodic non-singular transformation of $(X, \mathcal{B}, m_\omega)$, and

$$\frac{dm_\omega \circ T}{dm_\omega} = \frac{f}{f \circ T} \quad m_\omega\text{-a.e.}$$

2   For $A \in \mathcal{B}$, the map $\omega \mapsto m_\omega(A)$ is measurable, and

$$m(A) = \int_\Omega m_\omega(A) d\lambda(\omega).$$

It is not hard to check that the measures

$$\{\mu_\omega : \omega \in \Omega\}$$

defined by

$$\mu_\omega(A) = \int_A f d\mu_\omega \text{ for } A \in \mathcal{B},$$

satisfy the conclusions of the theorem.   □

EXERCISE 2.2.3.   State and prove a version of the ergodic decomposition for T a conservative, non-singular, non-invertible transformation of the standard probability space $(X, \mathcal{B}, m)$.

§2.3 Converses to Birkhoff's theorem

Let T be an ergodic measure preserving transformation of the probability space $(X, \mathcal{B}, m)$, then, for $f : X \to \mathbb{R}$, $f \geq 0$ measurable,

$$\limsup_{n\to\infty} \frac{1}{n} \sum_{k=1}^{n-1} f \circ T^k < \infty \text{ a.e.} \Rightarrow \int_X f dm < \infty$$

by Birkhoff's ergodic theorem. This can be thought of as a primitive converse to Birkhoff's ergodic theorem. However, it does not eliminate the possibility of convergence of form

$$\frac{1}{b_n} \sum_{k=1}^{n-1} f \circ T^k \to 1 \text{ a.e.}$$

for certain non-integrable functions $f \geq 0$, and constants b_n.

We'll see here that no such ergodic convergence exists. This was shown in [**Cho-Ro**] for $\{f \circ T^n : n \geq 0\}$ independent, and in general in [**A2**].

We begin with a proposition.

2.3.1 PROPOSITION [**A8**].
Suppose that $(X, \mathcal{B}, m, T)$ is an ergodic, probability preserving transformation of a Lebesgue probability space, and suppose that $f : X \to \mathbb{R}$ is measurable.

Let $a : [0, \infty) \to [0, \infty)$ be continuous, strictly increasing, and satisfy $\frac{a(x)}{x} \downarrow 0$ as $x \uparrow \infty$.

If $\int_X a(|f|) dm < \infty$, then,

$$\frac{a(|S_n f|)}{n} \to 0 \ a.e. \ as \ n \to \infty$$

where $S_n f := \sum_{k=0}^{n-1} f \circ T^k$.

PROOF [**A-W2**].
We first establish the proposition under the additional assumption that $a(0) = 0$.

Given $\epsilon > 0$, we prove that

$$\limsup_{n \to \infty} \frac{a(|S_n f|)}{n} \leq \epsilon \text{ a.e.}$$

We claim that there are measurable functions $g, h : X \to \mathbb{R}_+$ such that

$$|f| = g + h, \ \sup g < \infty, \ \& \ \int_X a(h) dm < \epsilon.$$

This is because

$$a\left(|f| 1_{[|f| \geq M]}\right) \to 0 \text{ a.e. as } M \to \infty,$$

whence by the dominated convergence theorem (as $a(|f|)$ is integrable)

$$\int_X a\left(|f| 1_{[|f| \geq M]}\right) dm \to 0 \text{ as } M \to \infty,$$

and for suitably large M we can set

$$g = |f| 1_{[|f| < M]}, \ h = |f| 1_{[|f| \geq M]}.$$

Using that $a(x + y) \leq a(x) + a(y)$, we have that

$$
\begin{aligned}
\frac{a(|S_n f|)}{n} &\leq \frac{a(S_n g)}{n} + \frac{a(S_n h)}{n} \\
&\leq \frac{a(Mn)}{n} + \frac{S_n a(h)}{n} \\
&\longrightarrow \int_X a(h) dm < \epsilon.
\end{aligned}
$$

We now establish the proposition without the additional assumption that $a(0) = 0$.
To do this, choose $m > 0$ such that $a(m) = \alpha m$ where $0 < \alpha < 1$, and define

$$
\tilde{a}(x) = \begin{cases} \alpha x & 0 \leq x \leq m, \\ a(x) & x \geq m. \end{cases}
$$

It is straightforward to verify that $\tilde{a}$ is continuous, increasing, $a \equiv \tilde{a}$ on $[m, \infty)$ and $\frac{\tilde{a}(x)}{x} \downarrow 0$ as $x \uparrow \infty$.

Therefore, by the above

$$
\frac{\tilde{a}(|S_n f|)}{n} \to 0 \text{ a.e. as } n \to \infty,
$$

whence

$$
\frac{a(|S_n f|)}{n} \leq \frac{1}{n}(a(m) + \tilde{a}(|S_n f|)) \text{ a.e. as } n \to \infty.
$$

$\square$

REMARKS.

1) As in lemma 1.2 of [A-W2], the condition $\frac{a(x)}{x} \downarrow 0$ in proposition 2.3.1 can be replaced by the weaker $a(x) = o(x)$, $a(x + y) \leq a(x) + a(y)$. This because for any such $a(x) \uparrow$, $\exists A(x) \uparrow$ such that $\frac{A(x)}{x} \downarrow$ and $a(x) \leq A(x) \leq 2a(x)$.

The required function $A(x)$ is $A(x) := \sup\{\frac{a(xt)}{t} : t \geq 1\}$.

It follows that $A(x) \uparrow$ (because $a(x) \uparrow$), $\frac{A(x)}{x} \downarrow$ (as $\frac{A(x)}{x} = \sup\{\frac{a(y)}{y} : y \geq x\}$), $a(x) \leq A(x)$ (as $a(x)$ participates in the sup) and $A(x) \leq 2a(x)$ (since if $A(x) = \frac{a(tx)}{t}$, set $n = [t] + 1$ and then $A(x) \leq a(nx)/t \leq na(x)/t \leq 2a(x)$).

2) The condition $\int_X a(|f|) dm < \infty$ in proposition 2.3.1 is not necessary.

Let T be an invertible, ergodic, probability preserving transformation on a standard probability space X and suppose $a(x) \uparrow \infty$ and $a(x)/x \downarrow 0$ as $x \uparrow \infty$. There is a measurable function $F : X \to [0, \infty)$ such that $E(a(F)) = \infty$, but $a(S_n F) = o(n)$ a.e. as $n \to \infty$.

To see this, we argue as in [A-W2].

Start with $\epsilon, \delta > 0, K > 0$, and choose $M \in \mathbb{N}$ such that $a(M)\epsilon > K$. Next, choose $N \in \mathbb{N}$ such that $\frac{a(M(2n))}{n} < \delta$ for every $n \geq N$.

By Rokhlin's tower theorem 1.5.9 $\exists A \in \mathcal{B}(X)$ such that $\{T^a A : -2N \leq a \leq 2N\}$ are disjoint, and $m\left(\bigcup_{-2N \leq a \leq 2N} T^a A\right) = 4\epsilon$.

The function $f : X \to \mathbb{R}$ defined by $f = M \sum_{0 \leq a < N} 1_A \circ T^a$ satisfies $\int_X a(f) dm = a(M)\epsilon > K$.

Moreover, $\frac{a(S_n f(x))}{n} < \delta \; \forall \; n \geq N$, $x \in X$ and $\forall \; n \geq 1$, $x \notin B := \bigcup_{-2N \leq a \leq 2N} T^a A$.

To see this, note that for $x \notin B$, $S_n f(x) = 0 \ \forall n \leq N$, and for $n \geq N$, $\forall \, x \in X$, $S_n f(x) \leq (2n)M$.

To construct the measurable function $F : X \to \mathbb{R}$ fix $\epsilon_k = \delta_k = \frac{1}{2^k}$, $K_k = k$, construct $f_k : X \to [0, \infty)$, $A_k, B_k \in \mathcal{B}$, $M_k, N_k \in \mathbb{N}$ accordingly, and set $F = \sum_{k=1}^{\infty} f_k$.

Clearly $\int_X a(F) dm \geq \int_X a(f_k) dm \geq k \to \infty$ as $k \to \infty$.

Let $\kappa(x) = \min \{k \geq 1 : x \notin B_r \ \forall \, r \geq k\}$ which is finite a.e. because $\sum_{k=1}^{\infty} m(B_k) < \infty$.

It follows that for $\ell \geq \kappa(x)$, $n \geq N_\ell$,

$$a(S_n F) \leq \sum_{k=1}^{\infty} a(S_n f_k) = \sum_{k=1}^{\ell} a(S_n f_k) + \sum_{k=\ell+1}^{\infty} a(S_n f_k)$$

$$\leq \ell a(S_n f_\ell) + \sum_{k=\ell+1}^{\infty} \frac{n}{2^k} \leq \frac{\ell+1}{2^\ell} n.$$

The following generalises [**Fe1**].

2.3.2 THEOREM.

Suppose that T is an ergodic measure preserving transformation of the probability space $(X, \mathcal{B}, m)$, and suppose that $f : X \to \mathbb{R}$ is measurable, and

$$b(n) > 0, \ (n \in \mathbb{N}), \quad \frac{b(n)}{n} \uparrow \infty \text{ as } n \uparrow \infty.$$

(1) If $\exists A \in \mathcal{B}_+$ such that

$$\int_A b^{-1} \left(\max_{1 \leq n \leq \varphi_A} \Big| \sum_{k=0}^{n-1} f \circ T^k \Big| \right) dm < \infty,$$

then

$$\frac{1}{b(n)} \sum_{k=0}^{n-1} f \circ T^k \to 0 \text{ a.e. as } n \to \infty.$$

(2) Otherwise,

$$\limsup_{n \to \infty} \frac{1}{b(n)} \Big| \sum_{k=0}^{n-1} f \circ T^k \Big| = \infty \text{ a.e.}$$

PROOF.

Let $a = b^{-1}$, and note that

$$\frac{b(n)}{n} \uparrow \infty \text{ as } n \uparrow \infty \quad \Rightarrow \quad \frac{a(n)}{n} \downarrow 0 \text{ as } n \uparrow \infty.$$

Next, suppose that $A \in \mathcal{B}_+$ and

$$\int_A a \left(\max_{1 \leq n \leq \varphi_A} \Big| \sum_{k=0}^{n-1} f \circ T^k \Big| \right) dm < \infty,$$

then, in particular, by the proposition, $\frac{a(g_n)}{n} \to 0$ a.e. on A as $n \to \infty$, where $g = |\sum_{k=0}^{\varphi_A - 1} f \circ T^k|$, and $g_n = \sum_{k=0}^{n-1} g \circ T_A^k$.

Setting $f_n = \sum_{k=0}^{n-1} f \circ T^k$, and $h = \max_{1 \le n \le \varphi_A} |f_n|$, we obtain

$$\frac{a(|f_{(\varphi_A)_n}|)}{n} \le \frac{a(g_n)}{n} \to 0 \text{ a.e. on } A \text{ as } n \to \infty.$$

Now $\max_{(\varphi_A)_n \le k \le (\varphi_A)_{n+1}} |f_k| \le g_n + h \circ T_A^n$, whence $\max_{(\varphi_A)_n \le k \le (\varphi_A)_{n+1}} a(|f_k|) \le a(g_n) + a(h) \circ T_A^n$.

By assumption, $\int_A a(h) dm < \infty$ and it follows that

$$\frac{a(h) \circ T_A^n}{n} \to 0 \text{ a.e. on } A \text{ as } n \to \infty,$$

whence

$$\frac{\max_{(\varphi_A)_n \le k \le (\varphi_A)_{n+1}} a(|f_k|)}{n} \to 0 \text{ a.e. on } A \text{ as } n \to \infty,$$

and

$$\frac{a(|f_n|)}{n} \to 0 \text{ a.e. on } A \text{ as } n \to \infty.$$

It follows that

$$\frac{|f_n|}{b(n)} \to 0 \text{ a.e. on } A \text{ as } n \to \infty.$$

The set on which this convergence takes place is T-superinvariant, hence, containing A, is $X \bmod m$. This proves (1).

Now suppose that the conclusion of (2) does not hold. By an Egorov type argument, we obtain $M > 1$, $A \in \mathcal{B}_+$ such that

$$|f_n(x)| \le Mb(n) \quad \forall n \in \mathbb{N}, \ x \in A,$$

whence

$$\max_{1 \le n \le \varphi_A} |f_n| \le Mb(\varphi_A),$$

and

$$\int_A b^{-1} \left(\max_{1 \le n \le \varphi_A} |f_n| \right) dm < \infty.$$

$\square$

2.3.3 PROPOSITION. *If $\{f \circ T^k\}_{k \in \mathbb{N}}$ are independent in the theorem, and*

$$\frac{1}{b(n)} \sum_{k=0}^{n-1} f \circ T^k \to 0 \text{ a.e. as } n \to \infty,$$

then

$$\int_X a(|f|) dm < \infty.$$

PROOF. If $\{f \circ T^k\}_{k \in \mathbb{N}}$ are independent, and $\limsup_{n \to \infty} \frac{|f \circ T^n|}{b(n)} < \infty$, a.e., then, by the Borel-Cantelli lemma,

$$\int_X a(|f|) dm \le \sum_{n=0}^{\infty} m([|f \circ T^n| \ge b(n)]) < \infty.$$

$\square$

EXERCISE 2.3.1.   Show that if T is an ergodic, measure preserving transformation of the probability space $(X, \mathcal{B}, m)$, $f : X \to \mathbb{R}$ is measurable, and

$$m\left(\left[\limsup_{n \to \infty} \frac{1}{n} |\sum_{k=0}^{n-1} f \circ T^k| < \infty\right]\right) > 0,$$

then $\frac{1}{n} \sum_{k=0}^{n-1} f \circ T^k$ converges a.e..

Deduce from this that for $f : X \to \mathbb{R}$ is measurable, either $\limsup_{n \to \infty} \frac{|f \circ T^n|}{n} = \infty$ a.e., or $|f \circ T^n| = o(n)$ as $n \to \infty$ a.e. (see [**Ta**]).

2.3.4 COROLLARY [**A8**].

Suppose that T is an ergodic measure preserving transformation of the probability space $(X, \mathcal{B}, m)$, and suppose that $f : X \to \mathbb{R}$ is measurable, and

$$b_n > 0, \ \limsup_{n \to \infty} b_n/n = \infty,$$

then, either

$$(1) \qquad\qquad \limsup_{n \to \infty} \frac{1}{b_n} |\sum_{k=0}^{n-1} f \circ T^k| = \infty \ \text{a.e.}$$

or, there is a sequence $n_k \to \infty$ such that

$$(2) \qquad\qquad \frac{1}{b_{n_k}} \sum_{j=0}^{n_k-1} f \circ T^j \to 0 \ \text{a.e. as } k \to \infty.$$

PROOF.

Let

$$b(n) = n \max_{1 \le k \le n} \frac{b_k}{k},$$

then

$$b(n) \ge b_n, \ (n \in \mathbb{N}), \ \frac{b(n)}{n} \uparrow \infty \ \text{as } n \uparrow \infty, \ \text{and } \exists n_k \to \infty \ni b(n_k) = b_{n_k}.$$

If (1) does not hold, then

$$\limsup_{n \to \infty} \frac{1}{b(n)} |\sum_{k=0}^{n-1} f \circ T^k| < \infty$$

on some set of positive measure, and it follows from the theorem that

$$\frac{1}{b(n)} \sum_{k=0}^{n-1} f \circ T^k \to 0 \ \text{as } n \to \infty,$$

whence (2).   □

We can now see that no convergence

$$\frac{1}{b_n} \sum_{k=0}^{n-1} f \circ T^k \to 1 \ \text{as } n \to \infty$$

is possible for $f \geq 0$ measurable and non-integrable, and $b_n > 0$ constants, since if such convergence did hold, then by the primitive converse to Birkhoff's theorem mentioned above, necessarily

$$\frac{b_n}{n} \to \infty \text{ as } n \to \infty$$

and the corollary would apply.

§2.4 Transformations with infinite invariant measures

We show that when $m(X) = \infty$, there are no constants $a_n > 0$ such that

$$\frac{S_n(f)}{a_n} \to_{n \to \infty} \int_X f dm \text{ a.e. } \forall f \in L^1(m)_+.$$

This follows from theorem 2.4.2 (below).

2.4.1 THEOREM [A8].   *Suppose that T is a conservative, ergodic measure preserving transformation of the σ-finite, measure space $(X, \mathcal{B}, m)$, and let*

$$a(n) \uparrow \infty, \quad \frac{a(n)}{n} \downarrow 0 \text{ as } n \uparrow \infty,$$

then
(1)   If $\exists A \in \mathcal{B}$, $0 < m(A) < \infty$ such that

$$\int_A a(\varphi_A) dm < \infty,$$

then

$$\frac{S_n(f)}{a(n)} \to_{n \to \infty} \infty \text{ a.e. } \forall f \in L^1(m)_+.$$

(2)   Otherwise,

$$\liminf_{n \to \infty} \frac{S_n(f)}{a(n)} = 0 \text{ a.e. } \forall f \in L^1(m)_+.$$

PROOF.   We begin by proving (1). Suppose that $A \in \mathcal{B}$, $0 < m(A) < \infty$ and that

$$\int_A a(\varphi_A) dm < \infty,$$

then, by the proposition,

$$\frac{1}{n} a((\varphi_A)_n) \to 0 \text{ a.e. on } A, \text{ as } n \to \infty,$$

where

$$(\varphi_A)_n = \sum_{k=0}^{n-1} \varphi_A \circ T_A^k.$$

Since $S_{(\varphi_A)_n}(1_A) \equiv n$, we have, writing

$$(\varphi_A)_{k_n(x)}(x) \leq n < (\varphi_A)_{k_n(x)+1}(x),$$

that

$$\frac{S_n(1_A)}{a(n)} \geq \frac{S_{(\varphi_A)_{k_n}}}{a((\varphi_A)_{k_n+1})}$$

$$= \frac{k_n}{a((\varphi_A)_{k_n+1})}$$

$$\to \infty \text{ a.e. on } A, \text{ as } n \to \infty.$$

Clearly, the set on which this convergence occurs is T-invariant, and containing A, must be X mod m. Thus

$$\frac{S_n(1_A)}{a(n)} \to \infty \text{ a.e. on } X, \text{ as } n \to \infty,$$

and (1) follows from Hopf's ergodic theorem.

We now prove (2). To this end, we note first, that for $f \in L^1(m)_+$,

$$\frac{S_n(f) \circ T}{a(n)} \leq \left(1 + \frac{1}{n}\right) \frac{S_{n+1}(f)}{a(n+1)},$$

whence

$$\liminf_{n \to \infty} \frac{S_n(f)}{a(n)}$$

is T-subinvariant and constant a.e. on X by ergodicity.

Next, we suppose that the conclusion of (2) is not satisfied, and choose $A \in \mathcal{B}$, $0 < m(A) < \infty$. It follows from the above, and an Egorov type argument, that $\exists B \in \mathcal{B}_+ \cap A$, and $\epsilon > 0$ such that

$$S_n(1_A)(x) \geq \epsilon a(n), \quad \forall x \in B, n \geq 1.$$

whence

$$S_{\varphi_B(x)}(1_A)(x) \geq \epsilon a(\varphi_B(x)), \quad \forall x \in B.$$

This implies, using Hopf's ergodic theorem, that for a.e. $x \in B$,

$$\frac{1}{n} \sum_{k=0}^{n-1} a(\varphi_B(T_B^k x)) \leq \frac{1}{n\epsilon} \sum_{k=0}^{n-1} S_{\varphi_B(T_B^k x)}(1_A)(T_B^k x)$$

$$= \frac{1}{n\epsilon} S_{(\varphi_B)_n}(1_A)(x)$$

$$\to \frac{m(A)}{\epsilon m(B)} \text{ as } n \to \infty.$$

It follows that

$$\int_B a(\varphi_B) dm < \infty$$

contradicting the hypothesis of (2). $\square$

2.4.2 THEOREM.   *Suppose that T is a conservative, ergodic measure pre-serving transformation of the σ-finite, infinite measure space $(X, \mathcal{B}, m)$, and let $a_n > 0$, $(n \geq 1)$, then*

(1)   *either* $\displaystyle \liminf_{n \to \infty} \frac{S_n(f)}{a_n} = 0$ *a.e.* $\forall f \in L^1(m)_+$,

(2)   *or* $\exists n_k \uparrow \infty$ *such that* $\displaystyle \frac{S_{n_k}(f)}{a_{n_k}} \to_{n \to \infty} \infty$ *a.e.* $\forall f \in L^1(m)_+$.

PROOF. We restrict attention to the case $a_n \to \infty$ as $n \to \infty$, (else (2) holds as T is conservative, ergodic).

Suppose that (1) does not hold, then $a_n = o(n)$ as $n \to \infty$, and $\exists\, A \in \mathcal{B}$, $m(A) > 0$ such that

$$\liminf_{n \to \infty} \frac{S_n(f)}{a_n} > 0 \text{ a.e. on } A \,\, \forall f \in L^1(m)_+.$$

Set

$$\bar{a}_n = \max_{1 \leq k \leq n} a_k,$$

then $\bar{a}_n \uparrow$ as $n \uparrow$, $a_n \leq \bar{a}_n$, and, since $\bar{a}_n = a_{k(n)}$ where $1 \leq k(n) \leq n$, $k(n) \to \infty$ as $n \to \infty$, we have

$$\liminf_{n \to \infty} \frac{S_n(f)}{\bar{a}_n} = \liminf_{n \to \infty} \frac{S_n(f)}{a_{k(n)}}$$

$$\geq \liminf_{n \to \infty} \frac{S_{k(n)}(f)}{a_{k(n)}}$$

$$> 0 \text{ a.e. on } A \,\, \forall f \in L^1(m)_+.$$

Next, set $f_n = \bar{a}_n / n$, and let $1 = n_0 < n_1 < n_2 < \dots$ be defined by

$$\{n_k\}_{k \in \mathbb{N}} = \{j \geq 2 : f_i > f_j \,\, \forall 1 \leq i \leq j - 1\}.$$

For every $k \geq 0$,

$$f_{n_k} > f_{n_{k+1}}, \quad n_k f_{n_k} \leq n_{k+1} f_{n_{k+1}},$$

whence

$$0 < \frac{n_k}{n_{k+1}} \leq \frac{f_{n_{k+1}}}{f_{n_k}} < 1,$$

and there exists $\alpha_k \in (0, 1]$ such that

$$\left(\frac{n_k}{n_{k+1}} \right)^{\alpha_k} = \frac{f_{n_{k+1}}}{f_{n_k}}.$$

Define

$$f(x) = \frac{f_{n_k} n_k^{\alpha_k}}{x^{\alpha_k}}, \quad x \in [n_k, n_{k+1}], \,\, k \in \mathbb{N},$$

and

$$a(x) = x f(x).$$

Evidently

$$a(n_k) = \bar{a}_{n_k}, \,\, (k \in \mathbb{N}).$$

By the definition of the n_k, we have that for $k \in \mathbb{N}$, $n \in [n_k, n_{k+1})$,

$$f_n \geq f_{n_k}, \text{ hence } f_n \geq f(n),$$

whence
$$a(n) \leq \bar{a}_n, \ (n \in \mathbb{N})$$
and
$$\liminf_{n \to \infty} \frac{S_n(f)}{a(n)} > 0 \text{ a.e. on } A \ \forall f \in L^1(m)_+.$$
It is evident that
$$a(n) \uparrow, \ \frac{a(n)}{n} \downarrow \text{ as } n \uparrow \infty,$$
and so by theorem 2.4.1,
$$\frac{S_n(f)}{a(n)} \to \infty \text{ a.e. as } n \to \infty \ \forall f \in L^1(m)_+,$$
and (2) follows since $a_{n_k} \leq \bar{a}_{n_k} = a(n_k)$. $\square$

§2.5 Spectral Properties

Let T be an invertible, non-singular transformation of $(X, \mathcal{B}, m)$ and consider the invertible isometry $U_T : L^2(m) \to L^2(m)$ defined by

$$U_T f := \sqrt{\frac{dm \circ T}{dm}} f \circ T.$$

A *spectral property* of T is a property of U_T. These spectral properties are connected to certain spectral measures.

Spectral measures of Hilbert space isometries.

Let H br a separable Hilbert space, and let $U : H \to H$ be an invertible isometry. For $g \in H$, the function $n \mapsto \langle U^n g, g \rangle$ is *positive definite* in the sense that

$$\sum_{1 \leq i,j \leq k} z_i \langle U^{n_i - n_j} g, g \rangle \bar{z}_j = \left\| \sum_{1 \leq i \leq k} z_i U^{n_i} g \right\| \geq 0$$

$\forall \ k \geq 1$, $n_1, \ldots, n_k \in \mathbb{Z}$, $z_1, \ldots, z_k \in \mathbb{C}$.

By Herglotz's theorem (see [**Kat**]), $\exists$ a positive measure μ_g on $\mathbb{T}$ such that

$$\langle U^n g, g \rangle = \hat{\mu}_g(n) := \int_{\mathbb{T}} \chi_n d\mu$$

where $\chi_n(t) = e^{2\pi i n t}$. The measure μ_g is called the *spectral measure* of g under U.
The following theorem puts all the spectral measures together.

2.5.1 SCALAR SPECTRAL THEOREM [**Neu-v1**].
If $U : H \to H$ is an invertible isometry, then $\exists$ a positive measure $\sigma = \sigma_U$ on $\mathbb{T}$) such that $\mu_g << \sigma \ \forall \ g \in H$, and such that $\forall \ \mu << \sigma$, $\exists \ g \in H$ such that $\mu = \mu_g$. Moreover, there is a sesquilinear map $h : H \times H \to L^1(\sigma)$ such that

$$\langle U^n f, g \rangle = \int_{\mathbb{T}} \chi_n h(f, g) d\sigma \ \forall \ f, g \in H, \ n \in \mathbb{Z}.$$

Proofs of this can also be found in [**Cor-Sin-Fom**], [**Parr2**] and [**Rie-Sz.N**].

Clearly σ_U is defined uniquely up to equivalence of measures, and is known as the *spectral type* of U. The spectral type of a non-singular transformation T is the spectral type of U_T acting on $L^2(X_T)$.

The *restricted* spectral type of a probability preserving transformation T is the spectral type of U_T acting on $L^2(X_T)_0 := \{f \in L^2(X_T) : \int_{X_T} f \, dm_T = 0\}$. The spectral type μ, and the restricted spectral type ν of a probability preserving transformation are related by $\mu \sim \nu + \delta_0$.

Note that any atoms of σ_U are eigenvalues of U. Suppose that $\delta_z \ll \sigma_U$, then by the scalar spectral theorem $\exists\, f \in H$ with $\langle U^n f, f \rangle = \hat{\delta}_z = z^n$ $(n \in \mathbb{Z})$, whence $Uf = zf$.

It follows that if T is a non-singular transformation without an m-absolutely continuous, T-invariant probability, then σ_{U_T} is continuous. This is because otherwise $\exists\, |z| = 1$ and $f \in L^2(X_T)$ such that $U_T f = zf$, whence $\widehat{T}|f|^2 = |f|^2$ and $dp := |f|^2 dm_T$ defines a T-invariant probability.

Spectral measures of probability preserving transformations.

A probability preserving transformation T is called *weakly mixing* if

$$\sum_{k=0}^{n-1} |m(A \cap T^{-n}B) - m(A)m(B)| \to 0 \quad \forall\ A, B \in \mathcal{B};$$

and *mixing* if

$$m(A \cap T^{-n}B) \to m(A)m(B) \text{ as } n \to \infty \quad \forall\ A, B \in \mathcal{B}.$$

The following two propositions show that these properties are spectral properties.

2.5.2 PROPOSITION.   *Let T be a probability preserving transformation of a standard space, then T is weakly mixing if and only if σ_T, the restricted spectral type of T, is non-atomic.*

PROOF.

We use Wiener's theorem (see e.g. [**Kat**]) which says that for a complex measure μ on $\mathbb{T}$,

$$\frac{1}{n} \sum_{k=0}^{n-1} |\hat{\mu}(k)|^2 \ \to\ \sum_{t \in \mathbb{T}} |\mu(\{t\})|^2.$$

Suppose first that σ_T is non-atomic, then $\frac{1}{n} \sum_{k=0}^{n-1} |\hat{\sigma}_T(k)|^2 \to 0$, whence $\exists\, n_k \sim k$ such that $\hat{\sigma}_T(n_k) \to 0$, and $\frac{1}{n} \sum_{k=0}^{n-1} |\hat{\sigma}_T(k)| \to 0$.

We claim that

$$\frac{1}{n} \sum_{k=0}^{n-1} |\hat{\mu}(k)| \to 0 \ \forall\ \text{complex measures } \mu \ll \sigma_T.$$

This, being evident when $\frac{d\mu}{d\sigma_T}$ is a trigonometric polynomial, follows for general $\mu \ll \sigma_T$ by approximation. It now follows from the scalar spectral theorem that $\forall\, f, g \in L^2(m)_0$,

$$\frac{1}{n} \sum_{k=0}^{n-1} |\langle U_T^k f, g \rangle| \to 0.$$

Choosing $f = 1_A - m(A)$, $g = 1_B - m(B)$, we obtain the condition for weak mixing.

Conversely, suppose that T is weakly mixing, then

$$\frac{1}{n} \sum_{k=0}^{n-1} |\langle U_T^k f, g \rangle| \to 0$$

$\forall\ f, g \in L^2(m)_0$ simple functions, and hence (by approximation) $\forall\ f, g \in L^2(m)_0$. Choosing $f \in L^2(m)_0$ with $\langle U_T^k f, f \rangle = \hat{\sigma}_T(k)$, we have

$$\frac{1}{n} \sum_{k=0}^{n-1} |\sigma_T(k)| \to 0,$$

whence

$$\frac{1}{n} \sum_{k=0}^{n-1} |\sigma_T(k)|^2 \to 0$$

and σ_T is non-atomic. $\qquad\square$

2.5.3 PROPOSITION. *A probability preserving transformation T is mixing iff $\hat{\sigma}_T(n) \to 0$ as $n \to \infty$.*

PROOF.
Evidently T is mixing if and only if $U_T^n f \to 0\ \forall\ f \in L^2(m)_0$.
Suppose that T is mixing. Choosing $f \in L^2(m)_0$ with $\langle U_T^k f, f \rangle = \hat{\sigma}_T(k)$, we have

$$\hat{\sigma}_T(k) = \langle U_T^k f, f \rangle \to 0.$$

Conversely, suppose that $\hat{\sigma}_T(n) \to 0$ as $n \to \infty$. We claim that $\hat{\mu}(n) \to 0\ \forall$ complex measures $\mu \ll \sigma_T$. As before, this is evident when $\frac{d\mu}{d\sigma_T}$ is a trigonometric polynomial, follows for general $\mu \ll \sigma_T$ by approximation; and it follows from the scalar spectral theorem that

$$\langle U_T^n f, g \rangle \to \forall\ f, g \in L^2(m)_0,$$

whence T is mixing. $\qquad\square$

REMARK.
The question has arisen as to the generalisations of "mixing" and "weak mixing" for invertible, conservative, ergodic measure preserving transformations of infinite measure spaces. A generalisation of "weak mixing" is given in §2.7, whereas the discussion in [**Kre-Suc**] indicates that there is no reasonable generalisation of mixing.

Note that the spectral measure σ_T of a conservative, ergodic, measure preserving transformation T an infinite measure space $(X, \mathcal{B}, m)$ is always non-atomic (as there is no m-absolutely continuous T-invariant probability). It can be shown (see [**Haj-Kak1**]) that either $\hat{\sigma}_T(n) \to 0$ as $n \to \infty$, or $\limsup_{n \to \infty} m(A \cap T^{-n} B) > 0\ \forall\ A, B \in \mathcal{B}_+$. An example with the second property is given in [**Haj-Kak2**].

§2.6 Eigenvalues

Let T be a non-singular transformation. A complex number z is said to be an *eigenvalue* of T if there is a measurable function $g : X_T \to \mathbb{C}$ such that $g \circ T = zg$. The function g is called an *eigenfunction*. It is easily seen that if T is conservative, then all eigenvalues of T are of modulus 1, and eigenfunctions can be chosen in L^∞. If, in addition, T is ergodic, then eigenfunctions have constant absolute value.

We denote the multiplicative circle by $S^1 := \{z \in \mathbb{C} : |z| = 1\}$. It is homeomorphic with $\mathbb{T} = \mathbb{R}/\mathbb{Z}$ by $t \leftrightarrow e^{2\pi it}$.

Let
$$e(T) = \{t \in \mathbb{T} : e^{2\pi it} \text{ is an eigenvalue of } T\}.$$
Evidently, if T is conservative, ergodic, then $e(T)$ is a subgroup of $\mathbb{T}$.

If T is a probability preserving transformation, then eigenvalues for T are eigenvalues for U_T and by orthogonality of eigenfunctions, $e(T)$ is countable.

There are conservative, non-singular transformations T of standard spaces with $e(T)$ uncountable, however, in this situation, $e(T)$ is always a Borel subset of $\mathbb{T}$.

2.6.1 EIGENVALUE THEOREM. *Suppose that T is an invertible conservative, ergodic, non-singular transformation of a standard space, then $e(T)$ is a Borel set in $\mathbb{T}$, and*

$$\exists\, \psi : e(T) \times X_T \to S^1 \text{ such that } \psi(t, Tx) = e^{2\pi it}\psi(t, x)\ m_T - a.e.\ \forall\, t \in e(T).$$

PROOF.   Let G denote the collection of eigenfunctions of constant absolute value 1, and define a metric d on G by $L^2(X_T)$-distance. It follows that G is a complete, separable metric space, and a topological group under pointwise multiplication. The constant functions in G form a closed subgroup denoted by $\mathbb{K}$, and it follows that $G/\mathbb{K}$ is a complete, separable metric space with the metric $\rho(f\mathbb{K}, g\mathbb{K}) := \min_{|c|=1} d(f, cg)$. The map $P : G/\mathbb{K} \to e^{2\pi ie(T)}$ defined by $P(f\mathbb{K}) = \overline{f}f \circ T$ is continuous, 1-1, and onto; and $e(T) \cong e^{2\pi ie(T)}$ is a Borel set being the continuous 1-1 image of a complete, separable metric space.

To obtain the advertised ψ, let $\{h_n : n \in \mathbb{N}\}$ be a complete, orthonormal system for $L^2(X_T)$, and let

$$K_1 = \{g \in G : \langle g, h_1 \rangle \neq 0\}, \ K_n = \{g \in G : \langle g, h_n \rangle \neq 0, \langle g, h_k \rangle = 0, \ 1 \leq k \leq n-1\}.$$

Define $c : G \to \mathbb{K}$ by
$$c(f) = \frac{\overline{\langle f, h_n \rangle}}{|\langle f, h_n \rangle|} \text{ when } f \in K_n,$$
then c is measurable, and if $M : G \to G$ is defined by $M(f) = c(f)M(f)$, then $M(zf) = M(f)$. Defining $N(g\mathbb{K}) = M(g)$, we get

$$N : G/\mathbb{K} \to G \text{ measurable such that } N(g\mathbb{K}) = g.$$

To finish, set
$$\psi(t, x) = N(P^{-1}(e^{2\pi it}))(x).$$

□

A set $\Gamma \in \mathcal{B}(\mathbb{T})$ is called a *weak Dirichlet* set if

$$\liminf_{n\to\infty} \|\chi_n - 1\|_{L^2(\mu)} = 0 \ \forall \ \mu \in \mathcal{P}(\mathbb{T}).$$

2.6.2 THEOREM. *Let T be a conservative, ergodic, non-singular transformation of a standard space, then*
(i) ([**Schm2**]) $e(T)$ *is a weak Dirichlet set;*
   and
(ii) ([**A9**]) *if the Hausdorff dimension of $e(T)$ is larger than $\alpha \in [0,1)$, then*

$$\sum_{n=1}^{\infty} \frac{1}{n^{1-\alpha}} \widehat{T}^n h < \infty \quad \forall \ h \in L^1(m)_+.$$

PROOF.
Suppose that $\mu \in \mathcal{P}(e(T))$, and define a metric d_μ on G_0, the group of measurable functions $g : e(T) \to \mathbb{K}$ by $d_\mu(f,g) := \|f - g\|_{L^2(\mu)}$, then (G_0, d_μ) is a Polish group under pointwise multiplication.
Let

$$\psi : e(T) \times X_T \to \mathbb{C} \ |\psi| \equiv 1, \ni \psi(t, Tx) = e^{2\pi i t}\psi(t,x) \ m_T - \text{a.e.} \ \forall \ t \in e(T)$$

as in the eigenvalue theorem and define $\pi : X_T \to G$ by $\pi(x)(t) := \psi(t, x)$, then $\pi \circ T = \chi\pi$ where $\chi(t) := e^{2\pi i t}$.
   (i) Since T is conservative, $\forall \ \mu \in \mathcal{P}(e(T))$,

$$\liminf_{n\to\infty} \|\chi_n - 1\|_{L^2(\mu)} = \liminf_{n\to\infty} \|\pi \circ T^n - \pi\|_{L^2(\mu)} = 0 \text{ a.e. on } X.$$

   (ii) If H-dim. $(e(T)) > \alpha$ then by Frostman's theorem (see [**Kah-Sal**] or [**Ts**]), $\exists \ \mu \in \mathcal{P}(e(T))$ such that $\mu(J) \leq M|J|^\alpha \ \forall$ intervals $J \subset \mathbb{T}$, whence

$$\sum_{n=1}^{\infty} \frac{|\hat{\mu}(n)|^2}{n^{1-\alpha}} < \infty.$$

For $g \in G_0$ let $A(g) := \{x \in X_T : \ d_\mu(\pi(x), g) < \frac{1}{2}$. Since $d(\chi^n, 1) = 2(1 - \text{Re}\ \hat{\mu}(n))$, we have that

$$1_{A(g)}1_{A(g)} \circ T^n \leq 1_{(\frac{1}{2},1]}(|\mu(n)|),$$

whence

$$\sum_{n=1}^{\infty} \frac{1}{n^{1-\alpha}} m(A(g) \cap T^{-n}A(g)) \leq \sum_{n=1}^{\infty} \frac{1}{n^{1-\alpha}} 1_{(\frac{1}{2},1]}(|\mu(n)|)$$

$$\leq 4 \sum_{n=1}^{\infty} \frac{|\hat{\mu}(n)|^2}{n^{1-\alpha}} < \infty$$

whence

$$\sum_{n=1}^{\infty} \frac{1}{n^{1-\alpha}} \widehat{T}^n 1_{A(g)} < \infty \text{ a.e. on } A(g).$$

By Hurewicz's theorem,

$$\sum_{n=1}^{\infty} \frac{1}{n^{1-\alpha}} \widehat{T}^n h < \infty \text{ a.e. on } A(g) \ \forall \ h \in L^1(m)_+.$$

The theorem follows because $\exists\ g_1, g_2, \cdots \in G_0$ such that

$$\bigcup_{n=1}^{\infty} A(g_n) = X_T.$$

$\square$

REMARKS.

1) It follows from theorem 2.6.2 part (i) that $e(T)$ has zero Lebesgue measure. In what follows, we'll see that this is essentially the only metric limitation on the size of $e(T)$.

2) It was shown in [**Me**] that $e(T)$ also enjoys the property of *saturation*: if $P \in \mathcal{P}(e(T))$, $q \in \mathcal{P}(\mathbb{T})$ and $\hat{q}(n) \to 1$ as $\hat{p}(n) \to 1$, then $q(e(T)) = 1$.

Non-Polish, Borel, saturated subgroups of $\mathbb{T}$ were exhibited in [**A-Nad**], and non-saturated Polish subgroups of $\mathbb{T}$ were exhibited in [**Ho-Me-Par**]. As far as the author knows, there is as yet no intrinsic characterisation of eigenvalue groups.

Osikawa's examples.

These (appearing in [**Os**]) were the first examples of conservative, ergodic, non-singular transformations of standard spaces with uncountable eigenvalue groups.

Let $\Omega = \{0, 1\}^{\mathbb{N}}$ be the dyadic integers, and $\tau : \Omega \to \Omega$ be the adding machine preserving Haar measure m on Ω.

Fix $\gamma(n) \in \mathbb{N}$ $(n \geq 1)$ such that

$$\gamma(k+1) \geq \sum_{j=1}^{k} \gamma(j) + 1 \quad (k \geq 1)$$

and define $\varphi : \Omega \to \mathbb{N}$ by

$$\varphi(x) := \sum_{k=1}^{\infty} \gamma_k(\ (\tau x)_k - x_k\ ) = \gamma(\ell(x)) - \sum_{j=1}^{\ell(x)-1} \gamma(j)$$

where $\ell(x) := \inf \{n \geq 1 : x_n = 0\}$. Build the tower $(X, \mathcal{B}, \mu, T)$ over τ with height function φ:

$$X = \{(x, n) : 1 \leq n \leq \varphi(x)\}, \ \mu(A \times \{n\}) := m(A),$$

and

$$T(x, n) = \left\{ \begin{array}{ll} n+1 & n+1 \leq \varphi(x), \\ (\tau x, 1) & n+1 = \varphi(x). \end{array} \right.$$

Evidently, T is conservative and ergodic; and

$$s \in e(T) \ \Leftrightarrow \ \exists\ f : \Omega \to \mathbb{K} \text{ measurable, such that } f(\tau x) = e^{2\pi i s \varphi(x)} f(x).$$

We'll prove

2.6.3 THEOREM   [**A-Nad**].

$$e(T) = \{s \in \mathbb{T} : \sum_{k=1}^{\infty} \langle \gamma(k)s \rangle^2 < \infty\}$$

where $\langle s \rangle := \min_{n \in \mathbb{Z}} |s - n|.$

This is a Polish group continuously embedded in $\mathbb{T}$, and the Polish topology is given by the norm

$$\|s\|_{e(T)} = \langle s \rangle + \sqrt{\sum_{k=1}^{\infty} \langle \gamma(k)s \rangle^2}.$$

The main tool in the proof of this theorem is

2.6.4 OSIKAWA'S THEOREM [Os].

$s \in e(T)$ *if and only if* $\exists \, c_n \in \mathbb{R}$ *such that* $\sum_{k=1}^{n}(s\gamma_k x_k - c_k)$ *converges in* $\mathbb{T}$ *for a.e.* $x \in \Omega$.

PROOF.

Suppose that $\phi : \Omega \to \mathbb{T}$ and

$$\left\langle \sum_{k=1}^{n}(s\gamma_k x_k - c_k) - \phi(x) \right\rangle \;\to\; 0 \text{ for a.e. } x \in \Omega,$$

then

$$\phi(\tau x) - \phi(x) \overset{\mathbb{T}}{\leftarrow} s \sum_{k=1}^{n} \gamma_k\big(\ (\tau x)_k - x_k\ \big) \overset{\mathbb{T}}{\to} s\varphi(x)$$

whence

$$e^{2\pi i \phi(\tau x)} = e^{2\pi i s \varphi(x)} e^{2\pi i \phi(x)}$$

and $s \in e(T)$.

Conversely, if $s \in e(T)$, then $\exists \, \phi : X \to \mathbb{T}$ such that

$$\phi(\tau^n x) - \phi(x) = s \sum_{k=1}^{\infty} \gamma_k\big(\ (\tau^n x)_k - x_k\ \big) \;\;(n \in \mathbb{Z}),$$

whence, if $\sigma_n(x)_n := 1 - x_n$ and $\sigma_n(x)_k := x_k \; \forall \; k \neq n$, then

$$\phi(\sigma_n x) - \phi(x) = s \sum_{k=1}^{\infty} \gamma_k\big(\ (\sigma_n x)_k - x_k\ \big) = s \sum_{k=1}^{n} \gamma_k\big(\ (\sigma_n x)_k - x_k\ \big) \;\;(n \geq 1).$$

Set

$$\Phi_n(x) := s \sum_{k=1}^{n} \gamma_k x_k, \quad \Psi_n := \phi - \Phi_n,$$

then

$$\phi(\sigma_k x) - \phi(x) = \Phi_n(\sigma_k x) - \Phi_n(x) \quad \forall \; 1 \leq k \leq n$$

whence

$$\Psi_n(\sigma_k x) = \Psi_n(x) \quad \forall \; 1 \leq k \leq n,$$

Ψ_n is $\sigma(x_{n+1}, x_{n+2}, \dots)$-measurable, and it follows that

$$e^{2\pi i \Phi_n} E(e^{2\pi i \Psi_n}) = E(e^{2\pi i \phi} | x_1, \dots, x_n) \to e^{2\pi i \phi} \text{ a.e. as } n \to \infty$$

by the martingale convergence theorem (see [**Doo**]). Thus $\exists \, c_n$ such that

$$\left\langle \Phi_n - c_n - \phi \right\rangle \to 0 \text{ a.e. as } n \to \infty.$$

$\square$

PROOF OF THEOREM 2.6.3. We claim first that $\forall\, s \in \mathbb{T}\ \exists\, \epsilon_k = \pm 1$ such that

$$\Phi_n(x) := s\sum_{k=1}^{n}\gamma_k x_k = \sum_{k=1}^{n}\gamma_k x_k \epsilon_k \langle \gamma(k)s\rangle \qquad \text{mod } 1.$$

Indeed, choosing $\nu_1, \nu_2, \cdots \in \mathbb{Z}$ and $\epsilon_1, \epsilon_2, \cdots \in \{-1, 1\}$ such that
$\gamma(k)s = \epsilon_k\langle\gamma(k)s\rangle + \nu_k$ we have,

$$\Phi_n(x) := s\sum_{k=1}^{n}\gamma_k x_k = \sum_{k=1}^{n}x_k(\epsilon_k\langle\gamma(k)s\rangle + \nu_k) = \sum_{k=1}^{n}\gamma_k x_k \epsilon_k\langle\gamma(k)s\rangle \qquad \text{mod } 1.$$

Suppose that $\sum_{k=1}^{\infty}\langle\gamma(k)s\rangle^2 < \infty$, then since $\operatorname{Var}(\gamma_k x_k \epsilon_k\langle\gamma(k)s\rangle) = \frac{\langle\gamma(k)s\rangle^2}{4}$,
by Kolmogorov's three series theorem (see [**Kol**]) $\exists\, c_k \in \mathbb{R}$, $f : \Omega \to \mathbb{R}$ measurable
such that

$$\sum_{k=1}^{n}(\gamma_k x_k \epsilon_k\langle\gamma(k)s\rangle - c_k) \to f(x)$$

a.e., whence

$$\left\langle \sum_{k=1}^{n}(s\gamma_k x_k - c_k) - f(x)\right\rangle \to 0 \text{ for a.e. } x \in \Omega$$

and $s \in e(T)$ by Osikawa's theorem.

Conversely, if $s \in e(T)$ then by Osikawa's theorem $\exists\, c_k \in \mathbb{R}$ and $f : \Omega \to \mathbb{R}$
measurable such that

$$\left\langle \sum_{k=1}^{n}(\gamma_k x_k \epsilon_k\langle\gamma(k)s\rangle - c_k) - f(x)\right\rangle \to 0$$

for m-a.e. $x \in \Omega$. It follows that

$$\left\langle \sum_{k=1}^{n}\gamma_k(x_k - y_k)\epsilon_k\langle\gamma(k)s\rangle - (f(x) - f(y))\right\rangle \to 0$$

for $m \times m$-a.e. $(x, y) \in \Omega \times \Omega$.

Now $|\gamma_k(x_k - y_k)\epsilon_k\langle\gamma(k)s\rangle| \le \langle\gamma(k)s\rangle \le \frac{1}{2}$, and the convergence implies

$$|(x_k - y_k)\epsilon_k\langle\gamma(k)s\rangle| = \langle(x_k - y_k)\epsilon_k\langle\gamma(k)s\rangle \to 0$$

whence

$$\sum_{k=1}^{n}(x_k - y_k)\epsilon_k\langle\gamma(k)s\rangle$$

converges in $\mathbb{R}$. Again by the Kolmogorov three series theorem,

$$\sum_{k=1}^{\infty}\langle\gamma(k)s\rangle^2 = \sum_{k=1}^{\infty}\operatorname{Var}(\gamma_k(x_k - y_k)\epsilon_k\langle\gamma(k)s\rangle) < \infty.$$

$\square$

2.6.5 COROLLARY [**Fu-W**].

*Suppose that $\mu \in \mathcal{P}(\mathbb{T})$ and suppose that $\exists$ a weak Dirichlet set $\Gamma \subset \mathbb{T}$ such that
$\mu(\Gamma) = 1$, then $\exists$ a conservative, ergodic measure preserving transformation T of a
standard σ-finite measure space such that $\mu(e(T)) = 1$.*

PROOF.   Since Γ is a weak Dirichlet set, $\exists\, n_j \to \infty$ such that $\chi_{n_j} \xrightarrow{\mu} 1$. Choose a subsequence $\gamma(k) = n_{j_k}$ such that

$$\gamma(k+1) \geq \sum_{j=1}^{k} \gamma(j) + 1 \quad (k \geq 1)$$

and such that

$$\sum_{k=1}^{\infty} |1 - \chi_{\gamma(k)}|^2 < \infty \quad \mu - \text{ a.e..}$$

It follows from theorem 2.6.3 that $\mu(e(T)) = 1$ for T the tower, over τ with height function

$$\varphi(x) = \sum_{k=1}^{\infty} \gamma_k (\, (\tau x)_k - x_k \,).$$

$\square$

REMARK.

Suppose $\gamma(k) = 2^{\beta(k)}$ where $\beta(k) < \beta(k+1) \in \mathbb{N}$, then (see [**A9**]) the Hausdorff dimension of

$$\{s \in \mathbb{T} : \sum_{k=1}^{\infty} |1 - \chi_{\gamma(k)}(s)|^2 < \infty\}$$

is given by

$$1 - \limsup_{n \to \infty} \frac{|K \cap [1, n]|}{n}$$

where $K := \{\beta(k) : k \geq 1\}$.

Thus, using Osikawa's constructions, it is possible to construct conservative, ergodic measure preserving transformation T whose eigenvalue group has prescribed Hausdorff dimension. Indeed, in [**A9**], it is shown that theorem 2.6.2, part (ii) is sharp:

for each $\alpha \in (0, 1)$ there is a conservative, ergodic measure preserving T whose eigenvalue group has Hausdorff dimension α, and such that

$$\sum_{n=1}^{\infty} \frac{1}{n^{1-\alpha}} h \circ T^n = \infty \quad \text{a.e.} \quad \forall\, h \in L^1(m)_+.$$

§2.7 Ergodicity of Cartesian products

Let S be a probability preserving transformation, and let T be a conservative, non-singular transformation. Consider the Cartesian product transformation $S \times T$ on the measure space $(X_S \times X_T, \mathcal{B}_S \otimes \mathcal{B}_T, m_S \times m_T)$.

By proposition 1.2.5, $S \times T$ is conservative. The following is a standard criterion for ergodicity (due to M. Keane).

2.7.1 ERGODIC MULTIPLIER THEOREM.

Let S be an invertible ergodic probability preserving transformation of a standard space, and let T be a conservative, ergodic, non-singular, invertible transformation of a standard space, then

$$S \times T \text{ is ergodic } \Leftrightarrow \sigma_0(e(T)) = 0$$

where σ_0 is the restricted spectral type of S.

PROOF. Suppose first that $S \times T$ is not ergodic, and let $F : X_S \times X_T \to \mathbb{R}$ be a bounded, non-constant invariant function. We'll show that $\sigma_0(e(T)) > 0$. Define $\phi : X_T \to L^1(\sigma_0)$ by

$$\phi(y)(x) = F(x, y).$$

It follows that

$$\phi \circ T = U_S^{-1} \phi,$$

whence, by the scalar spectral theorem, for $g \in L^2(X_S)_0$,

$$h(\phi(T^n y), g) = h(U_S^{-n} \phi(y), g) = \chi_n h(\phi(y), g).$$

By Fubini's theorem, if, for $s \in \mathbb{T}$, $g \in L^2(X_S)_0$, we set $f_s(y) = h(\phi(y), g)(s)$, then for σ_0-a.e. $s \in \mathbb{T}$

$$f_s \circ T = e^{2\pi i s} f_s \; m_T - \text{a.e.}$$

whence $|f_s| = c_s$ is constant m_T-a.e. and to show that $\sigma_0(e(T)) > 0$ it is sufficient to show that

$$\exists \; g \in L^2(X_S)_0 \ni \; \sigma_0(\{s \in \mathbb{T} : c_s > 0\}) > 0.$$

If this where not the case, then

$$h(\phi(y), g) = 0 \; \forall \; g \in L^2(X_S)_0$$

with the consequence that

$$F(x, y) = \int_{X_S} F(u, y) dm_S(u) := E(y) \;\; m_S \times m_T - \text{a.e..}$$

This contradicts the ergodicity of T, as $E \circ T = E$.

Now suppose that $g \in L^2(X_S)_0$ is such that $\mu_g(e(T)) = \|g\|^2$, and let $V : H_g \to L^2(\mu_g)$ be the Hilbert space isometry defined by

$$V(U_S^n g) = \chi_{-n}.$$

Let $\psi : e(T) \times X_T \to \mathbb{C}$ be as in the eigenvalue theorem. Define

$$F(x, y) = V^{-1}(\psi(\cdot, y))(x),$$

then

$$
\begin{aligned}
F(S^n x, T^n y) &= U_S^n V^{-1}(\psi(\cdot, T^n y))(x) \\
&= U_S^n V^{-1}(\chi_n \psi(\cdot, y))(x) \\
&= V^{-1}(\psi(\cdot, y))(x) \\
&= F(x, y).
\end{aligned}
$$

$\square$

REMARK. An *ergodic multiplier property* of a probability preserving transformation S is a property of form

$$S \times T \text{ is ergodic } \forall \; T \in \; P$$

where P is some collection of non-singular transformations. The ergodic multiplier theorem shows that ergodic multiplier properties are spectral properties, being determined by the spectral measures of the probability preserving transformations involved. See [A12].

DEFINITION: WEAK MIXING. A non-singular transformation T is called *weakly mixing* if

$$f \in L^\infty(X_T), \ z \in \mathbb{C}, \ f \circ T = zf \implies f \text{ is constant.}$$

2.7.2 PROPOSITION [A-Lin-W]. *A non-singular transformation T is weakly mixing $\Leftrightarrow$ $T \times S$ is ergodic $\forall$ ergodic probability preserving transformations S.*

PROOF.
Evidently, T is weakly mixing if and only if T is ergodic and $e(T) = \{0\}$. $\Rightarrow$ now follows from the ergodic multiplier theorem. Conversely, if T is ergodic, but not weakly mixing $\exists \ 0 \neq \alpha \in e(T)$. $\exists$ an ergodic probability preserving transformations S with $\alpha \in e(S)$ (rotation of $\mathbb{T}$ by α if $\alpha \notin \mathbb{Q}$ or a permutation of a finite set if $\alpha \in \mathbb{Q}$) whence $T \times S$ is not ergodic. $\square$

REMARK.
A non-singular transformation T was called weakly mixing in [A-Lin-W] if

$$\frac{1}{n} \sum_{k=0}^{n-1} | \int_X \widehat{T}^k u \, f dm | \to 0 \text{ as } n \to \infty \ \forall \ u \in L^1(m)_0, \ f \in L^\infty(m);$$

a condition shown in [A-Lin-W] to be equivalent to that of our definition.

DEFINITION: MILD MIXING. A non-singular transformation T is called *mildly mixing* if $f \in L^\infty$, $n_k \to \infty$, $f \circ T^{n_k} \to f$ weak $*$ in $L^\infty(m) \implies f$ is constant.

2.7.3 PROPOSITION.
A probability preserving transformation T is mildly mixing if and only if T is ergodic, and $\sigma_T(\Gamma) = 0$ for every weak Dirichlet set $\Gamma \in \mathcal{B}(\mathbb{T})$.

PROOF.
Let $f \in (L^\infty)_0$ be non-constant, $n_k \to \infty$, $f \circ T^{n_k} \to f$ weak $*$ in L^∞, then $\langle f \circ T^{n_k}, f \rangle \to \|f\|_2^2$ whence $\hat{\mu}_f(n_k) \to \|f\|_2^2$ and $\exists \ m_\ell = n_{k_\ell}$ such that

$$\chi_{m_\ell} \to 1 \quad \mu_f - \text{a.e.,}$$

whence $\Gamma := \{t \in \mathbb{T} : \ \chi_{m_\ell}(t) \to 1\} \in \mathcal{B}(\mathbb{T})$ is a weak Dirichlet set and $\mu_f(\Gamma) > 0$ whence $\sigma_T(\Gamma) > 0$.
Conversely, let $\Gamma \in \mathcal{B}(\mathbb{T})$ be a weak Dirichlet set with $\sigma_T(\Gamma) > 0$. We'll show that T is not mildly mixing.
Choosing $f \in L^2(m)_0$ such that $\mu_f(\Gamma^c) = 1$ we have that $\exists \ n_k \to \infty$ such that $\chi_{n_k} \xrightarrow{\mu_f} 1$, whence $f \circ T^{n_k} \to f$ in $L^2(m)$. If $f \in L^\infty(m)$, this contradicts mild mixing.
If not, let $J \subset \mathbb{C}$ be open, and such that $0 < m(A) < 1$ where $A := f^{-1}J$. Evidently, $m(A \triangle T^{-n_k} A) \to 0$ and if $F := 1_A - m(A)$, then $F \in L^\infty(m)_0$ is non-constant, and $F \circ T^{n_k} \to F$ weak $*$ in $L^\infty(m)$. Thus T is not mildly mixing. $\square$

Proposition 2.7.2 shows that among probability preserving transformations, mixing $\Rightarrow$ mild mixing $\Rightarrow$ weak mixing.
Examples are known which show that weak mixing $\not\Rightarrow$ mild mixing $\not\Rightarrow$ mixing.

2.7.4 PROPOSITION.
An exact, non-singular transformation is mildly mixing.

PROOF.   If $(X, \mathcal{B}, m, T)$ is an exact, non-singular transformation; and $f \in L^\infty$, $n_k \to \infty$, $f \circ T^{n_k} \to f$ weak $*$ in $L^\infty(m)$, then f is $\bigcap_{n \geq 1} R^{-n} \mathcal{B}$-measurable and hence constant. $\square$

2.7.5 THEOREM   [Fu-W].

Let S be an ergodic probability preserving transformation, then $S \times T$ is ergodic for every conservative, ergodic, non-singular transformation T if, and only if S is mildly mixing.

PROOF.   We use the spectral characterisation of mild mixing in proposition 2.7.3.

If $S \times T$ is not ergodic, then, by the ergodic multiplier theorem, there exists $f \in L^2(X_S)_0$ such that
$$\mu_f(e(T)) = \mu(\mathbb{T}) = 1.$$
By theorem 2.6.2, $e(T)$ is a weak Dirichlet set and $\sigma_S(e(T)) > 0$.

Conversely, suppose that $\sigma_S(\Gamma) > 0$ for some weak Dirichlet set $\Gamma \in \mathcal{B}(\mathbb{T})$, then $\exists \, f \in L^2(X_S)_0$ such that $\mu_f(\Gamma) = 1$. By corollary 2.6.5, $\exists$ a conservative, ergodic, measure preserving transformation T such that $\mu_f(e(T)) = 1$ and by the ergodic multiplier theorem, $S \times T$ is not ergodic. $\square$

2.7.6 THEOREM   [A-Lin-W].

Suppose that R is a mildly mixing, non-singular transformation, then $S \times R$ is ergodic for any conservative, ergodic, invertible non-singular transformation S.

PROOF.

Let $f \in L^\infty(X_S \times X_R)$ satisfy $f(Sx, Ry) = f(x, y)$, $\|f\|_\infty \leq 1$.

Let $B = \{g \in L^\infty(X_R) : \|g\|_\infty \leq 1$, then B is a compact metric space under the weak $*$ topology. Let d be a metric on B generating the weak $*$ topology.

Define $\pi : X_S \to B$ by $\pi(x)(y) := f(x, y)$. Evidently $\pi(S^{-n}x) = \pi(x) \circ R^n$. By conservativity of S,
$$\liminf_{n \to \infty} d(\pi \circ S^{-n}, \pi) = 0 \quad \text{a.e.},$$
whence for a.e. $x \in X_R$, $\exists \, n_k(x) \to \infty$ such that $f_x \circ R^{n_k(x)} \to f_x$ weak $*$ in $L^\infty(m_R)$ where $f_x(y) := f(x, y)$. By mild mixing of R, f_x is constant for a.e. $x \in X_S$. Thus $\exists \, g \in L^\infty(m_S)$ such that $f(x, y) = g(x) = g(Sx)$. By ergodicity of S, g is constant. $\square$

REMARK.

We'll see in chapter 5 that if T is a conservative, ergodic non-singular transformation without a m_T-absolutely continuous, T-invariant probability, then $\exists$ an invertible, conservative ergodic measure preserving transformation S such that $S \times T$ is totally dissipative. Thus (as in [A-Lin-W]) if T is invertible and mildly mixing, $\exists$ a m_T-absolutely continuous, T-invariant probability. Otherwise if S is as above, $S \times T$ would be ergodic, totally dissipative and invertible which is not possible.

For information on invertible conservative ergodic measure preserving transformations of σ-finite, infinite measure spaces related to mildly mixing transformations, see corollary 3.1.8.

CHAPTER 3

Transformations with infinite invariant measures

The point of this chapter is to investigate special properties of for infinite measure preserving transformations paying special attention to their structure (§3.1). The theorems are special in the sense that they are not valid under conservativity and ergodicity alone: they require additional assumptions (specified below).

Let T be a conservative ergodic measure preserving transformation of the σ-finite measure space $(X, \mathcal{B}, m)$, then, for $f \in L^1(m)$, $f \geq 0, \int_X f dm > 0,$

$$S_n(f) := \sum_{k=1}^{n-1} f \circ T^k \uparrow \infty \text{ a.e as } n \uparrow \infty.$$

We are interested in the "rate" at which this divergence takes place.

Hopf's ergodic theorem says that the growth rate of $S_n(f)(x)$ to ∞ does not depend on $f \in L^1(m)_+$, in the sense that there are measurable functions $a_n(x)$ such that

$$\frac{S_n(f)(x)}{a_n(x)} \longrightarrow_{n \to \infty} \int_X f dm \text{ for a.e.} x \in X.$$

This leaves open the possibility of whether the growth rate depends on the point $x \in X$.

By Birkhoff's theorem,

$$\frac{S_n(f)(x)}{n} \to \frac{1}{m(X)} \int_X f dm \text{ a.e. as } n \to \infty \ \forall \ f \in L^1(m),$$

identifying precisely the growth rate when $m(X) < \infty$,
but only establishing $S_n(f) = o(n)$ a.e. $(f \in L^1(m)_+)$ when $m(X) = \infty$.

Indeed in case $m(X) = \infty$, by theorem 2.4.2, for no sequence of constants a_n and $f \in L^1(m)_+$ is

$$S_n(f)(x) \asymp a_n \text{ a.e. as } n \to \infty.$$

However, as will be seen below in §3.3, for certain infinite measure preserving transformations, there exist constants a_n such that

$$\frac{S_n(f)(x)}{a_n} \longrightarrow \int_X f dm \ \forall \ f \in L^1(m)$$

in weaker senses. This sequence of constants will be "invariant", and their existence presupposes a law of large numbers (§3.2).

The modes of convergence considered include almost sure Cesàro convergence along subsequences, various kinds of local weak convergence and strong distributional convergence.

Diametrically opposed to these transformations are the Maharam transformations (§3.4) which do not have laws of large numbers.

§3.1 Isomorphism, Factors, and Similarity

We shall use the the quadruple $(X, \mathcal{B}, m, T)$ to denote a measure preserving transformation T of the standard measure space $(X, \mathcal{B}, m)$. We shall sometimes abbreviate this to $T = (X_T, \mathcal{B}_T, m_T, T)$.

In this section, we define the notions of factor, extension and isomorphism for measure preserving transformations. The reader should note that these notions are related to, but different from analogous notions with similar names defined for non-singular transformations in §1.0.

DEFINITION: c-FACTOR MAP.

For $c \in \mathbb{R}_+ \cup \{\infty\}$, a c-*factor map* from a measure preserving transformation S onto a measure preserving transformation T is a map $\pi : X'_S \to X'_T$ (where $X'_S \in \mathcal{B}_S$ and $X'_T \in \mathcal{B}_T$ are sets of full measure such that $S : X'_S \to X'_S$, and $T : X'_T \to X'_T$) satisfying

$$\pi^{-1}\mathcal{B}_T \subset \mathcal{B}_S, \quad m_S \circ \pi^{-1}(A) = cm_T(A) \; \forall \, A \in \mathcal{B}_T, \text{ and } \pi \circ S = T \circ \pi.$$

We shall denote this situation by $\pi : S \xrightarrow{c} T$.

It is necessary to consider c-factor maps with $c \neq 1$ as our measure spaces are not normalised (being infinite). The constant c can be thought of as a relative normalisation of the transformations concerned. We'll return to this idea in the next section.

DEFINITION: c-ISOMORPHISM.

If $\pi : S \xrightarrow{c} T$, and π is invertible, then $\pi^{-1} : T \xrightarrow{1/c} S$. We shall denote this situation by

$$\pi : S \xleftrightarrow{c} T,$$

and call π a c-*isomorphism* from S onto T.

DEFINITION: FACTOR, EXTENSION. If, for some $c \in \mathbb{R}_+$ there exists $\pi : S \xrightarrow{c} T$, we shall call T a *factor* of S, and S an *extension* of T, denoting this by $S \to T$.

Clearly if $\pi : S \xrightarrow{c} T$, then $\pi^{-1}\mathcal{B}_T$ is a σ-finite, sub-σ-algebra of $\mathcal{B}_S$ which is T-*invariant* in the sense that $S^{-1}(\pi^{-1}\mathcal{B}_T) \subset \pi^{-1}\mathcal{B}_T$. In case T is invertible, $\pi^{-1}\mathcal{B}_T$ is *strictly T-invariant* : $S^{-1}(\pi^{-1}\mathcal{B}_T) = \pi^{-1}\mathcal{B}_T$.

3.1.1 FACTOR PROPOSITION.   *Suppose that T is a measure preserving transformation of the σ-finite, standard measure space $(X, \mathcal{B}, m)$.*

For every T-invariant, sub-σ-algebra of $\mathcal{F} \subset \mathcal{B}_T$, there is a factor U, and a factor map $\pi : T \xrightarrow{1} U$ with $\mathcal{F} = \pi^{-1}\mathcal{B}_U$.

The factor is invertible iff the sub-σ-algebra is strictly T-invariant.

PROOF. Let $\mathcal{F}_0 = \{F_n : n \in \mathbb{N}\}$ be a T-invariant, countable ring of sets of finite measure generating $\mathcal{F}$. Define $\pi : X \to \{0,1\}^{\mathbb{N}}$ by $(\pi x)_n = 1_{F_n}(x)$. This map is clearly measurable as

$$\pi^{-1}[a_1, \ldots, a_n] = \bigcap_{k=1}^{n} F_{a_k} \in \mathcal{B}_T,$$

indeed $\pi^{-1} : \mathcal{B}(\{0,1\}^{\mathbb{N}}) \cap \pi(X_T) \to \mathcal{F}$ is a Boolean isomorphism.

Define a σ-finite measure $\mu : \mathcal{B}(\{0,1\}^{\mathbb{N}}) \cap \pi(X_T) \to [0, \infty]$ by $\mu = m_T \circ \pi^{-1}$.

Clearly, $\pi(x) = \pi(y)$ if and only if $1_F(x) = 1_F(y) \; \forall \, F \in \mathcal{F}$. By the T-invariance of $\mathcal{F}$, if $\pi(x) = \pi(y)$ then $\pi(Tx) = \pi(Ty)$ and we can define a transformation $U : \pi(X_T) \to \pi(X_T)$ by $U\pi(x) := \pi(Tx)$. The transformation U is measurable since for $A \in \mathcal{B}(\{0,1\}^{\mathbb{N}})$,

$$U^{-1}(A \cap \pi(X_T)) = U^{-1}\pi(\pi^{-1}A) = \pi(T^{-1}\pi^{-1}A) \in \mathcal{B}(\{0,1\}^{\mathbb{N}}) \cap \pi(X_T).$$

The measure μ is U-invariant because m_T is T-invariant. It remains to represent U as a measure preserving transformation of a standard measure space.

By the universal measurability theorem, $\pi(X_T) \in \overline{\mathcal{B}(\{0,1\}^{\mathbb{N}})}_{\mu}$, and

$$\exists \, X_U \in \mathcal{B}(\{0,1\}^{\mathbb{N}}), \; X_U \subset \pi(X_T) \text{ such that } \mu(\{0,1\}^{\mathbb{N}} \setminus X_U) = 0.$$

The second part of the proposition has the same proof as the non-singular factor proposition in §1.0. $\qquad \square$

REMARK.

Conservative, ergodic, measure preserving transformations of infinite measure spaces which are *prime* in the sense of having no (non-trivial σ-finite) factors are constructed in [A-Nad] (indeed the example of [Haj-Kak2] is one).

A conservative, ergodic, measure preserving transformations of infinite measure spaces without non-trivial invertible (non-singular) factors is constructed in [Rudo-Sil]. See the remark in §1.0.

DEFINITION: SIMILARITY, STRONG DISJOINTNESS.

Two measure preserving transformations are called *strongly disjoint* if they have no common extension, and *similar* otherwise.

No two probability preserving transformations are strongly disjoint. Indeed, if S, T are probability preserving transformations, the Cartesian product transformation $R = S \times T$ (see §1.2)

$$X_R = X_S \times X_T, \; \mathcal{B}_R = \mathcal{B}_S \otimes \mathcal{B}_T, \; m_R = m_S \times m_T, \text{ and } R(x,y) = (Sx, Ty)$$

is a probability preserving transformation, and

$$S \leftarrow R \to T.$$

Recall from [Fu] that the probability transformations S, T are called *disjoint* if any common extension R has the Cartesian product as factor (i.e. $R \to S \times T$).

For arbitrary measure preserving transformations S, T,

$$S \times T \to S \iff m_T(X_T) < \infty.$$

In particular, when S, and T are infinite measure preserving transformations, the Cartesian product is not an extension of either transformation. We shall see in the sequel that strong disjointness is not uncommon among infinite measure preserving transformations.

3.1.2 PROPOSITION.   *Suppose that S and T are similar measure preserving transformations, then S is conservative if, and only if T is conservative.*

PROOF.

It is sufficient to establish the result in case $S \to T$. Suppose that $\pi : S \overset{c}{\to} T$ (where $c \in \mathbb{R}_+$), and let $f \in L^1(m_T)$, $f > 0$, then $f \circ \pi \in L^1(m_S)$), $f \circ \pi > 0$, and by proposition 1.1.6

$$\mathfrak{C}(T) = \left[\sum_{n=0}^{\infty} f \circ T^n = \infty \right] \quad \mod m_T,$$

and

$$\mathfrak{C}(S) = \left[\sum_{n=0}^{\infty} f \circ \pi \circ S^n = \infty \right] = \pi^{-1}\mathfrak{C}(T) \quad \mod m_S.$$

$\square$

DEFINITION: FIBRE EXPECTATION. Next, suppose that $\pi : R \overset{c}{\longrightarrow} T$. For $f \in L^1(X_R)$, define the *fibre expectation* $E(f|\pi) \in L^1(X_T)$ using the Radon-Nikodym theorem by

$$\int_{X_T} E(f|\pi)g \, dm_T = \int_{X_R} fg \circ \pi \, dm_R.$$

It follows that $E(\cdot|\pi)$ is positive, linear, and satisfies

$$E(f \circ R|\pi) = E(f|\pi) \circ T, \text{ and } \int_{X_T} E(f|\pi) \, dm_T = \int_{X_R} f \, dm_R.$$

Moreover, the domain of definition of $E(\cdot|\pi)$ extends to include all non-negative measurable functions, and $E(1|\pi) \equiv c$.

Note that for $f \in L^1(X_R)$

$$E(f|\pi) \circ \pi = cE(f|\pi^{-1}\mathcal{B}_T)$$

where $E(f|\pi^{-1}\mathcal{B}_T) \in L^1(m_R)$ is the conditional expectation with respect to $\pi^{-1}\mathcal{B}_T$.

3.1.3 PROPOSITION. *Suppose that S, and T are measure preserving transformations of standard spaces, then S and T are similar if, and only if there is a positive linear operator*

$$P : L^1(X_S) \to L^1(X_T)$$

such that

(i)
$$P(f \circ S) = (Pf) \circ T,$$

(ii)
$$P1 = c \ m_T\text{-a.e. where } c \in \mathbb{R}_+,$$

and

(iii) $\qquad \int_{X_T} Pf dm_T = c' \int_{X_S} f dm_S \quad \forall\, f \in L^1(X_S)$ *where* $c' \in \mathbb{R}_+$.

PROOF.

Suppose first that P is an operator as above. We define a measure preserving transformation R by

$$X_R = X_S \times X_T, \quad \mathcal{B}_R = \mathcal{B}_S \otimes \mathcal{B}_T, \quad R(x,y) = (Sx, Ty),$$

and $m_R : \mathcal{B}_S \otimes \mathcal{B}_T \to [0, \infty]$ is the measure satisfying

$$m_R(A \times B) = \int_B P 1_A dm_T \quad \forall\, A \in \mathcal{B}_S, \ B \in \mathcal{B}_T.$$

It follows that $R \xrightarrow{c'} S$ and $R \xrightarrow{c} T$.

Now suppose that $\pi : R \xrightarrow{c_T} T$ and $\phi : R \xrightarrow{c_S} S$. The required $P : L^1(X_S) \to L^1(X_T)$ is defined by

$$Pf = E(f \circ \phi | \pi).$$

$\qquad\qquad\qquad\qquad\qquad\qquad\qquad\qquad\qquad\qquad\qquad\qquad\qquad\qquad$ □

COROLLARY.   *Similarity is an equivalence relation.*

3.1.4 LEMMA. *Suppose*
that R and T are measure preserving transformations of standard spaces,
that T is conservative and ergodic,
and that R is an extension of T,
    then almost all ergodic components of R are extensions of T.

PROOF.

Suppose that $\pi : R \xrightarrow{c} T$, and let the ergodic decomposition of R be $\{\mu_\omega : \omega \in \Omega\}$, the probability on Ω being p. For $\omega \in \Omega$, let R_ω be the conservative, ergodic, measure preserving transformation $(X_R, \mathcal{B}_R, R, \mu_\omega)$. We'll show that $R_\omega \to T$ for a.e. $\omega \in \Omega$.

For $A \in \mathcal{B}(\Omega)$, set $\mu_A(\cdot) = \int_A \mu_\omega(\cdot) dp(\omega)$, then $\mu_A \circ R^{-1} = \mu_A$ whence

$$\mu_A \circ \pi^{-1} \circ T^{-1} = \mu_A \circ \pi^{-1}.$$

Evidently, $\mu_A \circ \pi^{-1} \le m_R \circ \pi^{-1} = c m_T$, so by ergodicity of T, $\exists\, \nu(A) \ge 0$ such that $\mu_A \circ \pi^{-1} = \nu(A) m_T$. It follows from this that ν is a probability on Ω and $\nu \sim p$, whence $\exists\, c : \Omega \to \mathbb{R}_+$ such that $\mu_\omega \circ \pi^{-1} = c(\omega) m_T$ a.e. In other words: $\pi : R_\omega \xrightarrow{c(\omega)} T$.
$\qquad\qquad\qquad\qquad\qquad\qquad\qquad\qquad\qquad\qquad\qquad\qquad\qquad\qquad$ □

COROLLARY. *Any two similar conservative, ergodic, measure preserving transformations of standard spaces, have a conservative, ergodic, common extension.*

DEFINITION: NATURAL EXTENSION.

Let T be a measure preserving transformation of a σ-finite, standard measure space. A *natural extension* of T is an invertible extension T' of T which is minimal in the sense that

$$\pi^{-1}\mathcal{B}_T \subset \mathcal{B}_{T'}, \ \pi \circ T' = T \circ \pi, \ m_{T'} \circ \pi^{-1} = m_T,$$

and

$$\bigvee_{n=1}^{\infty} T'^n \pi^{-1}\mathcal{B}_T = \mathcal{B}_{T'} \quad \mod \ m_{T'}$$

where $\pi : T' \overset{1}{\to} T$ is the extension map.

3.1.5 THEOREM EXISTENCE OF NATURAL EXTENSIONS ([**Ro2**]). *Any measure preserving transformation of a standard, σ-finite measure space has a natural extension (also on a standard space).*

PROOF.

Suppose that T is a measure preserving transformation of the standard, σ-finite measure space $(X, \mathcal{B}, m)$.

There is a sequence of mutually refining partitions $\alpha_n \subset \mathcal{B}$ such that

$$\bigcap_{n=1}^{\infty} A_n$$

consists of at most one point whenever $A_n \in \alpha_n$, $(n \in \mathbb{N})$.

Define $T' : X^{\mathbb{N}} \to X^{\mathbb{N}}$ by

$$T'(x_1, x_2, ...) = (Tx_1, x_1, ...).$$

To see that $T'^{-1}\mathcal{B}^{\mathbb{N}} \subseteq \mathcal{B}^{\mathbb{N}}$, we first note that $\mathcal{B}^{\mathbb{N}}$ is generated by sets of the form

$$[A_1, ..., A_n] = \{(x_1, x_2, ..) \in X' : x_k \in A_k \ \forall 1 \le k \le n\},$$

where $A_1, ..., A_n \in \mathcal{B}$. It is easy to check that

$$T'^{-1}[A_1, ..., A_n] = [T^{-1}A_1 \cap A_2, A_3, ..., A_n],$$

whence $T'^{-1}\mathcal{B}^{\mathbb{N}} \subseteq \mathcal{B}^{\mathbb{N}}$ as desired.

We define the measure $m' : \mathcal{B}' \to [0, \infty]$, by

$$m'([A_1, ..., A_n]) = m(\bigcap_{k=1}^{n} T^{-(n-k)}A_k).$$

Such a measure exists by Kolmogorov's existence theorem. Since

$$T'^{-1}[A_1, ..., A_n] = [T^{-1}A_1 \cap A_2, A_3, ..., A_n],$$

it follows that $m'(T'^{-1}A) = m'(A)$ for all sets

$$A = [A_1, ..., A_n], \ \ A_1, ..., A_n \in \mathcal{B},$$

whence T' is a measure preserving transformation of $(X^{\mathbb{N}}, \mathcal{B}^{\mathbb{N}}, m')$.

If $\pi : X' \to X$ is defined by

$$\pi(x_1, ...) = x_1,$$

then π is an extension map.

It remains to show that T' is invertible.

To see this, we define

$$X' = \{(x_1, x_2, ..) \in X^{\mathbb{N}} : Tx_{n+1} = x_n \ \forall n \in \mathbb{N}\},$$

and note that, $T' : X' \to X'$ is $1 - 1$, with

$$T'^{-1}(x_1, x_2, ...) = (x_2, ...), \quad \forall (x_1, ..) \in X'.$$

We show that $m'(X'^c) = 0$, whence T' is an invertible measure preserving transformation of $(X', \mathcal{B}^{\mathbb{N}} \cap X', m')$. To prove $m'(X'^c) = 0$, set

$$Y_n = \bigcup_{A \in \alpha_n} [A, T^{-1}A, ..., T^{-n}A],$$

then

$$\bigcap_{n=1}^{\infty} Y_n = X'.$$

Since

$$m'([A_1, T^{-1}A_2, ..., T^{-(n-1)}A_n]) = m(\bigcap_{k=1}^{n} A_k) = 0$$

in case $A_1, ..., A_n \in \alpha_n$ are not all the same, we obtain

$$m'(Y_n^c) = 0 \ \ \forall n \in \mathbb{N},$$

whence

$$m'(X'^c) = 0.$$

Thus, if $\mathcal{B}' = \mathcal{B}^{\mathbb{N}} \cap X'$ then T' is an invertible measure preserving transformation of $(X', \mathcal{B}', m')$ and $\pi : X' \to X$ is an extension map. Clearly

$$\pi^{-1}\mathcal{B} = \{[A] : A \in \mathcal{B}\}$$

and, since

$$T'[A_1, ..., A_n] = [X, A_1, ..., A_n]$$

it follows that

$$\bigvee_{n=0}^{\infty} T'^n \pi^{-1}\mathcal{B} = \mathcal{B}'.$$

$\square$

3.1.6 THEOREM [**Ro2**] (UNIQUENESS OF NATURAL EXTENSION). *Suppose that T_1 and T_2 are both natural extensions of the measure preserving transformation T, then T_1 and T_2 are isomorphic.*

PROOF. Suppose that $(X_i, \mathcal{B}_i, m_i, T_i)$ $(i = 1, 2)$ are both invertible measure preserving transformations σ-finite, standard measure spaces, equipped with extension maps $\pi_i : X_i \to X$ $(i = 1, 2)$ satisfying

$$\pi_i^{-1}\mathcal{B} \subset \mathcal{B}_i, \ \pi_i \circ T_i = T \circ \pi_i, \ m_i \circ \pi^{-1} = m,$$

and

$$\bigvee_{n=1}^{\infty} T_i^n \pi_i^{-1}\mathcal{B} = \mathcal{B}_i \quad \text{mod } m_i \ (i = 1, 2).$$

For $n \geq 0$, $i = 1, 2$, set $\mathcal{A}_n^{(i)} := T_i^n \pi_i^{-1} \mathcal{B}$, and define $\Phi_n : \mathcal{A}_n^{(1)} \to \mathcal{A}_n^{(2)}$ by

$$\Phi_n(T_1^n \pi_1^{-1} A) := T_2^n \pi_2^{-1} A.$$

Clearly: $\Phi_n : \mathcal{A}_n^{(1)} \to \mathcal{A}_n^{(2)}$ is a bijection,

$$m_2 \circ \Phi_n = m_1, \ \Phi_{n+1}\big|_{\mathcal{A}_n^{(1)}} = \Phi_n, \ \Phi_{n+1} \circ T_1 = T_2 \circ \Phi_n,$$

and Φ_n preserves countable Boolean operations.

It follows that $\exists$ a bijection $\Phi : \mathcal{B}_1 \to \mathcal{B}_2$ preserving countable Boolean operations and such that

$$m_2 \circ \Phi = m_1, \ \& \ \Phi \circ T_1 = T_2 \circ \Phi,$$

whence $\exists$ a measure space isomorphism $\phi : X_2 \to X_1$ such that

$$\Phi = \phi^{-1}.$$

$\square$

3.1.7 THEOREM [**Parr1**].

The natural extension of a conservative, ergodic measure preserving transformation is conservative, and ergodic.

PROOF.

The natural extension T' of T is clearly conservative. Choose $p \in L^1(X_T)$, $p > 0$, $\int_X p \, dm = 1$, and let $p' = p \circ \pi^{-1}$, then, by Hopf's theorem for T',

$$\frac{\sum_{k=0}^{n-1} f \circ T'^k(x)}{\sum_{k=0}^{n-1} p' \circ T'^k(x)} \to E_{m'_{p'}}(f | \mathcal{I}(T')) \text{ for a.e. } x \in X' \text{ as } n \to \infty.$$

By Hopf's theorem for T, this limit is constant for

$$f \in \bigcup_{n=0}^{\infty} L^1(T'^n \pi^{-1} \mathcal{B}),$$

whence

$$E_{m'_{p'}}(f | \mathcal{I}) = \int_{X'} f \, dm' \text{ a.e. } \forall f \in \overline{\bigcup_{n=0}^{\infty} L^1(T'^n \pi^{-1} \mathcal{B})} = L^1(\mathcal{B}')$$

and $\mathcal{I}(T') = \{\emptyset, X'\} \bmod m'$. $\square$

3.1.8 COROLLARY [**A-Lin-W**].

Suppose that T is a conservative, ergodic measure preserving transformation, and that T is the natural extension of a mildly mixing factor, then for any conservative, ergodic non-singular transformation S:

$$S \times T \text{ conservative} \implies S \times T \text{ ergodic}.$$

PROOF.

Suppose that T is the natural extension of the conservative mildly mixing measure preserving transformation R, and let S be a conservative, ergodic measure preserving transformation such that $S \times T$ is conservative. Evidently, $S \times R$ (as a factor of $S \times T$) is conservative.

By theorem 3.1.7, we may assume that S is invertible. By theorem 2.7.5, $S \times R$ is ergodic, and again by Parry's theorem, $S \times T$ (as the natural extension of $S \times R$) is ergodic.   □

REMARK. In the non-singular category, there may be more than one "natural extension", and there may be totally dissipative "natural extensions" of a conservative, ergodic non-singular transformation. Indeed it is not known ([**Sil**]) whether every conservative, ergodic non-singular transformation has a conservative, invertible extension.

§3.2 Intrinsic normalising constants and laws of large numbers

Although infinite measure spaces are not canonically normalised, it may be that a measure preserving transformation T is intrinsically normalised, for example, in the sense that $T \overset{c}{\nleftrightarrow} T$ for each $c \neq 1$. To this end, we consider (for T a measure preserving transformation)

$$\Delta_0(T) := \{c \in (0, \infty) : \ T \overset{c}{\leftrightarrow} T\}$$

first introduced in [**Haj-It-Kak**]. Clearly, this is a multiplicative subgroup of $\mathbb{R}_+$, and if S and T are isomorphic, then $\Delta_0(S) = \Delta_0(T)$.

3.2.1 PROPOSITION. *If T is a conservative, ergodic measure preserving transformation of $(X, \mathcal{B}, m)$, then $\Delta_0(T)$ is a Borel subset of $\mathbb{R}$.*

PROOF.

Consider the group $\mathfrak{A}_1(X, \mathcal{B}, m)$ of invertible, measure multiplying transformations of $(X, \mathcal{B}, m)$. Recall from §1.5 that $\mathfrak{A}_1(X, \mathcal{B}, m)$ is a Polish group. The *centraliser* of T is defined by $C(T) := \{Q \in \mathfrak{A}_1(X, \mathcal{B}, m) : \ QT = TQ\}$, a closed subgroup of $\mathfrak{A}_1$ and hence Polish. We have that $\Delta_0(T) = D(C(T))$ where $D : \mathfrak{A}_1(X, \mathcal{B}, m) \to \mathbb{R}_+$ is the dilation function- a continuous homomorphism. Thus, $\Delta_0(T) = \tilde{D}(C(T)/\mathrm{Ker}\, D)$ where $\tilde{D}(Q\mathrm{Ker}\, D) := D(Q)$, a continuous injection from the Polish group $C(T)/\mathrm{Ker}\, D$. By Souslin's theorem, $\Delta_0(T)$ as the continuous injective image of a Polish space, is Borel.   □

EXAMPLE.

1) Let T be a totally dissipative measure preserving transformation of the standard σ-finite, non-atomic measure space $(X, \mathcal{B}, m)$. By proposition 1.1.2, there is a standard σ-finite, non-atomic measure space $(W, \mathcal{B}(W), \mu)$ such that (up to isomorphism)

$$X = W \times \mathbb{Z}, \ \mathcal{B} = \sigma(\{A \times \{n\} : A \in \mathcal{B}(W), \ n \in \mathbb{Z}\}),$$

$$T(x, n) = (x, n + 1), \ \& \ m(A \times \{n\}) = \mu(A).$$

It follows that if $Q : T \overset{c}{\hookleftarrow} T$, then $Q(x, n) = (qx, n)$ where $q : W \to W$ is an invertible nonsingular map with $\mu \circ q^{-1} = c\mu$, whence

$$\Delta_0(T) = \begin{cases} \{1\} & \mu(W) < \infty, \\ \mathbb{R}_+ & \mu(W) = \infty. \end{cases}$$

The first conservative, ergodic example with $\Delta_0(T) = \{1\}$ was given in [**Haj-Kak2**] (see the next example below).

Unfortunately, it is not the case that $\Delta_0(S) = \Delta_0(T)$ for similar conservative, ergodic measure preserving transformations (see [**A11**]).

Accordingly, we consider the *intrinsic normalising constants* of a conservative, ergodic, measure preserving transformation T, namely the collection

$$\Delta_\infty(T) := \{c \in (0, \infty] : \exists \text{ a c.e.m.p.t. } R \text{ such that } R \overset{1}{\to} T, \ \& \ R \overset{c}{\to} T\}.$$

Note that in particular if $T \overset{c}{\to} T$, then $c \in \Delta_\infty(T)$ so that $\Delta_0(T) \subset \Delta_\infty(T)$.

It is shown in [**A11**] that $\Delta_\infty(T) \cap \mathbb{R}_+$ is an analytic, subset of $\mathbb{R}_+$.

EXAMPLE.

The conservative, ergodic measure preserving transformation $(X, \mathcal{B}, m, T)$ in [**Haj-Kak2**] enjoys the property that there is a weakly wandering set $W \in \mathcal{B}$, $m(W) = 1$ and a sequence $n_k \uparrow \infty$ such that $\{T^{-n_k} W : k \geq 1\}$ are disjoint and

$$X = \bigcup_{k=1}^{\infty} T^{-n_k} W \quad \text{mod } m.$$

It can be checked that if $W' \in \mathcal{B}$ also has this property with the same $n_k \uparrow \infty$, then $m(W') = 1$ and it can deduced from this that $\Delta_\infty(T) = \{1\}$.

3.2.2 PROPOSITION [**A11**]. *If S and T are similar conservative ergodic measure preserving transformations, then*

$$\Delta_\infty(S) = \Delta_\infty(T).$$

PROOF. It suffices to prove the proposition in case $S \to T$. Therefore suppose $\pi : S \overset{a}{\to} T$ where $a \in \mathbb{R}_+$.

We show first that $\Delta_\infty(S) \subset \Delta_\infty(T)$. Indeed if $c \in \Delta_\infty(S)$ then $\exists$ a conservative ergodic measure preserving transformation U such that $U \overset{1}{\to} S$ and $U \overset{c}{\to} S$. Composition of factor maps shows that $U \overset{a}{\to} T$ and $U \overset{ac}{\to} T$ showing $c \in \Delta_\infty(T)$.

We now turn to the less trivial $\Delta_\infty(T) \subset \Delta_\infty(S)$. Let $c \in \Delta_\infty(T)$, then $\exists$ a conservative ergodic measure preserving transformation U equipped with factor maps $\alpha : U \overset{1}{\to} T$ and $\beta : U \overset{c}{\to} T$.

Define a $S \times S$-invariant measure on $X_S \times X_S$ by

$$\overline{\mu}(A \times B) := \int_{X_U} E(1_A | \pi) \circ \alpha E(1_B | \pi) \circ \beta \, dm_U$$

and let $\overline{V} = (X_S \times X_S, \mathcal{B}_S \otimes \mathcal{B}_S, \overline{\mu}, S \times S)$.

Define maps $p_i : X_S \times X_S \to X_S$ by $p_i(x_1, x_2) = x_i$ $(i = 1, 2)$, then

$p_i : \overline{V} \overset{ac^{i-1}}{\to} S$ $(i = 1, 2)$ (where in case $c = \infty$, c^{i-1} is defined by $\infty^0 := 1$ and $\infty^1 := \infty$).

Let the ergodic decomposition of $\overline{V}$ be $\{\overline{V}_\omega : \omega \in \Omega\}$. By lemma 3.1.4, for a.e. $\omega \in \Omega$, we have $p_i : W \overset{t_i(\omega)}{\to} S$ $(i = 1, 2)$ for some $t_1(\omega) \in \mathbb{R}_+$, $t_2(\omega) \in (0, \infty]$.

It is sufficient to show that $t_2 = ct_1$ a.e..

To this end, define a $T \times T$-invariant measure on $X_T \times X_T$ by

$$\mu(A \times B) := m_U(\alpha^{-1} A \cap \beta^{-1} B).$$

Clearly $V := (X_T \times X_T, \mathcal{B}_T \otimes \mathcal{B}_T, \mu, T \times T)$ is conservative and ergodic being a factor of U (in fact $(\alpha, \beta) : U \overset{1}{\to} V$).

If $\pi_i : X_T \times X_T \to X_T$ $(i = 1, 2)$, is defined by $\pi_i(x_1, x_2) = x_i$ $(i = 1, 2)$, then $\mu \circ \pi_i^{-1} = c^{i-1} m_T$ $(i = 1, 2)$.

Define $\psi : X_S \times X_S \to X_T \times X_T$ by $\psi(x_1, x_2) = (\pi(x_1), \pi(x_2))$, then $\psi : \overline{V} \overset{a^2}{\to} V$.

Again by lemma 3.1.4, for a.e. $\omega \in \Omega$, we have $\psi : \overline{V}_\omega \overset{s(\omega)}{\to} V$ for some $s(\omega) \in \mathbb{R}_+$.

To calculate t_i $(i = 1, 2)$, we compose factor maps to obtain for a.e. $\omega \in \Omega$,
$$\pi_i \circ \psi : \overline{V}_\omega \overset{s(\omega)c^{i-1}}{\to} T, \quad \pi \circ p_i : \overline{V}_\omega \overset{at_i(\omega)}{\to} T \quad (i = 1, 2) \text{ and then note that}$$

$$\pi_i \circ \psi = \pi \circ p_i \quad (i = 1, 2),$$

whence for a.e. $\omega \in \Omega$

$$t_i(\omega) = \frac{s(\omega)c^{i-1}}{a}, \text{ and } c \in \Delta_\infty(S).$$

$\square$

3.2.3 COROLLARY [A11].
If T is a conservative, ergodic, measure preserving transformation;
then $\Delta(T) := \Delta_\infty(T) \cap \mathbb{R}_+$ is a multiplicative subgroup of $\mathbb{R}_+$.

PROOF. Evidently, $c \in \Delta(T) \implies c^{-1} \in \Delta(T)$. We show $c, d \in \Delta(T) \implies cd \in \Delta(T)$.

Suppose that there is a conservative, ergodic, measure preserving transformation U such that $U \overset{1}{\to} T$ and $U \overset{c}{\to} T$. By proposition 3.2.2, $d \in \Delta(U)$, so $\exists$ a conservative, ergodic, measure preserving transformation V such that $V \overset{1}{\to} U$ and $V \overset{d}{\to} U$. Composition of the relevant factor maps shows that $V \overset{1}{\to} T$ and $V \overset{cd}{\to} T$.
$\square$

3.2.4 RELATIVE NORMALISATION LEMMA   [A11].
Let T be a conservative, ergodic, measure preserving transformation, and suppose $\Delta_\infty(T) = \{1\}$.

For each conservative, ergodic S which is similar to T, $\exists\ c(S) \in \mathbb{R}_+$ such that if U is conservative, ergodic; and

$$U \overset{a}{\to} T, \ U \overset{b}{\to} S \ (a, b \in \mathbb{R}_+),$$

then $\frac{b}{a} = c(S)$.

PROOF.

Suppose that U_i $(i = 1, 2)$ are conservative, ergodic, measure preserving transformations satisfying

$$U_i \overset{a_i}{\rightsquigarrow} S, \quad U_i \overset{b_i}{\rightsquigarrow} T, \quad (a_i, b_i \in \mathbb{R}_+, \quad i = 1, 2).$$

By proposition 3.1.4, $\exists$ a conservative, ergodic common extension V of U_1 and U_2. Assume that $V \overset{c_i}{\rightsquigarrow} U_i$ where $c_i \in \mathbb{R}_+$ $(i = 1, 2)$.

Composing the factor maps involved, we see that:

$V \overset{c_i a_i}{\rightsquigarrow} T$ $(i = 1, 2)$, whence $c_1 a_1 = c_2 a_2$; and $V \overset{c_i b_i}{\rightsquigarrow} S$ $(i = 1, 2)$, whence $c_1 b_1 = c_2 b_2$; with the conclusion that $\frac{a_1}{b_1} = \frac{a_2}{b_2}$. □

DEFINITION: LAW OF LARGE NUMBERS. A *law of large numbers* for a conservative, ergodic, measure preserving transformation T of $(X, \mathcal{B}, m)$ is a function $L : \{0, 1\}^{\mathbb{N}} \to [0, \infty]$ such that $\forall A \in \mathcal{B}$, for a.e. $x \in X$,

$$L(1_A(x), 1_A(Tx), \dots) = m(A).$$

REMARKS.

1) If T has a law of large numbers, then $T \overset{c}{\not\rightsquigarrow} T$ $\forall c \neq 1$, and in particular, $\Delta_0(T) = \{1\}$.

To see this, let $L : \{0, 1\}^{\mathbb{N}} \to [0, \infty]$ be a law of large numbers for T, and let $Q : X_T \to X_T$ be nonsingular, commuting with T, then $\forall A \in \mathcal{B}$, for a.e. $x \in X$,

$$m_T(A) = L(1_A(Qx), 1_A(TQx), \dots) = L(1_{Q^{-1}A}(x), 1_{Q^{-1}A}(Tx), \dots) = m(Q^{-1}A).$$

2) If L is a law of large numbers for U, and $U \overset{c}{\rightsquigarrow} T$, then $\frac{1}{c}L$ is a law of large numbers for T.

Indeed, supposing that $\pi : U \overset{c}{\rightsquigarrow} T$, we have for $A \in \mathcal{B}_T$,

$$L(1_A \circ \pi, 1_A \circ T \circ \pi, \dots) = L(1_{\pi^{-1}A}, 1_{\pi^{-1}A} \circ U, \dots) = cm_T(A) \quad m_U - \text{a.e.}$$

whence

$$\frac{1}{c}L(1_A, 1_A \circ T, \dots) = m_T(A) \quad m_T - \text{a.e.}$$

3.2.5 THEOREM. *If* $\Delta_\infty(T) = \{1\}$, *then* T *has a law of large numbers.*

PROOF.

Let $\Omega = \{0, 1\}^{\mathbb{N}}$ equipped with its product discrete topology, let S be the shift map and let $\mathfrak{M}$ denote the collection of σ-finite, S-invariant measures μ on Ω such that $\mu([1]) = 1$, and let $\mathfrak{M}_e$ denote those $\mu \in \mathfrak{M}$ such that $S_\mu := (\Omega, \mathcal{B}(\Omega), \mu, S)$ is a conservative, ergodic measure preserving transformation.

For $\mu \in \mathfrak{M}_e$, let

$$\Omega_\mu := \{\omega \in \Omega : \frac{\sum_{k=0}^{n-1} 1_{[\epsilon]}(S^k \omega)}{\sum_{k=1}^{n} \omega_k} \to \mu([\epsilon]) \quad \forall N \geq 1, \ \epsilon \in \{0, 1\}^N\}.$$

Clearly if $\mu \neq \mu'$ then $\Omega_\mu \cap \Omega_{\mu'} = \emptyset$, and by Hopf's ergodic theorem $\mu(\Omega_\mu^c) = 0$.

By lemma 3.2.4, $\exists\, c(S) \in \mathbb{R}_+$ such that if U is a conservative, ergodic, measure preserving transformation such that $U \overset{a}{\to} T$, and $U \overset{b}{\to} S$ $(a, b \in \mathbb{R}_+)$, then $\frac{b}{a} = c(S)$.

Define $L : \Omega \to [0, \infty]$ by

$$
L(\omega) = \begin{cases}
0, & \omega_n = 0 \ \forall\, n \geq 1, \\
c(S_\mu), & \omega \in \Omega_\mu \text{ where } S_\mu \text{ and } T \text{ are similar}, \\
\infty, & \text{else}.
\end{cases}
$$

We claim that L is a law of large numbers for T. For $A \in \mathcal{B}$, define $\pi_A : X_T \to \Omega$ by $\pi_A(x)_n = 1_A(T^{n-1}x)$, $\ (n \geq 1)$.

We must show that $L \circ \pi_A = m_T(A)$ a.e. $\forall\, A \in \mathcal{B}_T$.

If $0 < m_T(A) < \infty$, we have that $\pi_A : T \overset{m_T(A)}{\to} S_{\mu^A}$ where

$$
\mu^A := \frac{1}{m_T(A)} m_T \circ \pi_A^{-1} \in \mathfrak{M}_e.
$$

Consequently $c(S_{\mu^A}) = m_T(A)$, and for a.e. $x \in X$, $\pi_A(x) \in \Omega_{\mu^A}$ whence $L \circ \pi_A = m_T(A)$ a.e..

If $m_T(A) = 0$, then $\pi_A(x)_n = 0$ a.e. $\forall\, n \geq 1$ whence $L \circ \pi_A = 0 = m_T(A)$ a.e..

Lastly, suppose that $m_T(A) = \infty$.

Since T is conservative and ergodic, $m_T([(\pi_A)_n = 0 \ \forall\, n \geq 1]) = 0$ whence $L \circ \pi_A > 0$ a.e..

Let

$$
F := \{ x \in X : \ \exists\, \mu \in \mathfrak{M}_e \text{ such that } \pi_A x \in \Omega_\mu, \text{ and } S_\mu \text{ and } T \text{ are similar} \}.
$$

We have that $L \circ \pi_A = \infty$ a.e. on F^c and we must show that $m_T(F) = 0$.

To this end, note that $F \subset \pi_A^{-1} \Omega_1$ where $\Omega_1 :=$

$$
\{ \omega \in \Omega : \sum_{k=1}^{\infty} \omega_k = \infty, \ \exists\, \lim_{n \to \infty} \frac{\sum_{k=0}^{n-1} 1_{[\underline{\epsilon}]}(S^k \omega)}{\sum_{k=1}^{n} \omega_k} := \mu_\omega([\underline{\epsilon}]) \ \forall\, N \geq 1, \ \underline{\epsilon} \in \{0, 1\}^N \}.
$$

For $\omega \in \Omega$, μ_ω is a S-invariant measure, and $\mu_{S\omega} = \mu_\omega$; and the map $\omega \mapsto \mu_\omega$ is measurable $\Omega_1 \to \mathfrak{M}$.

From this we see that the map $x \mapsto \mu_{\pi_A(x)}$ is measurable $\pi_A^{-1}\Omega_1 \to \mathfrak{M}$ and T-invariant.

If $m_T(F) > 0$, then by ergodicity of T, $\exists\, \mu \in \mathfrak{M}$ such that $\mu_{\pi_A(x)} = \mu$ for a.e. $x \in F$. It follows that $\mu \in \mathfrak{M}_e$ and that S_μ and T are similar. But also $\pi_A : T \overset{\infty}{\to} S_\mu$, whence $\infty \in \Delta_\infty(T)$ contradicting the assumption that $\Delta_\infty(T) = \{1\}$. $\qquad \square$

3.2.6 COROLLARY.

$\Delta_\infty(T) = \{1\}$ *iff every conservative, ergodic measure preserving transformation similar to T has a law of large numbers.*

PROOF. If $\Delta_\infty(T) = \{1\}$, and S is conservative, ergodic and similar to T, then $\Delta_\infty(S) = \{1\}$ and by theorem 3.2.5, S has a law of large numbers.

Conversely, suppose that every conservative, ergodic measure preserving transformation similar to T has a law of large numbers, and let $c \in \Delta_\infty(T)$, U be conservative, ergodic, and $U \xrightarrow{1} T$, $U \xrightarrow{c} T$. Let L be a law of large numbers for U. By remark 2, both L and $c^{-1}L$ are laws of large numbers for T and $c = 1$. $\square$

REMARK.

It was shown in [**A11**] that $\exists\, L : \{0,1\} \to [0,\infty)$ such that for every conservative, ergodic measure preserving transformation T,

$$L(1_A, 1_A \circ T, \dots) = m_T(A) \quad \mod \Delta(T)\ m_T - \text{ a.e. } \forall\, A \in \mathcal{B}_T,\ m_T(A) < \infty.$$

Thus if $\Delta(T) = \{1\}$, then

$$L(1_A, 1_A \circ T, \dots) = m_T(A)\ m_T - \text{ a.e. } \forall\, A \in \mathcal{B}_T,\ m_T(A) < \infty.$$

This does not entail existence of a law of large numbers for T. A suitable example is given in chapter 8.

§3.3 Rational ergodicity

A conservative, ergodic, measure preserving transformation T of $(X, \mathcal{B}, m)$ is called *rationally ergodic* if there is a set $A \in \mathcal{B}$, $0 < m(A) < \infty$ satisfying a *Renyi inequality*: $\exists\, M > 0$ such that

$$\int_A (S_n(1_A))^2\, dm \le M \left(\int_A S_n(1_A) dm \right)^2 \quad \forall\, n \ge 1.$$

In this section, we prove special ergodic theorems for rationally ergodic transformations.

3.3.1 THEOREM. *Suppose that T is rationally ergodic, then there is a sequence of constants $a_n \uparrow \infty$, unique up to asymptotic equality, such that whenever $A \in \mathcal{B}$ satisfies a Renyi inequality,*

$$(3.3.1) \qquad \frac{1}{a_n} \sum_{k=0}^{n-1} m(B \cap T^{-k}C) \to m(B)m(C) \text{ as } n \to \infty\ \forall\, B, C \in \mathcal{B} \cap A;$$

$$(3.3.2) \qquad \varlimsup_{n \to \infty} \frac{1}{a_n} \sum_{k=0}^{n-1} m(B \cap T^{-k}C) \ge m(B)m(C)\ \forall\, B, C \in \mathcal{B},$$

and

$$\forall\, m_\ell \uparrow \infty\ \exists\, n_k = m_{\ell_k} \uparrow \infty$$

such that

$$(3.3.3) \qquad \frac{1}{N} \sum_{k=1}^{N} \frac{1}{a_{n_k}} \sum_{j=0}^{n_k-1} f \circ T^j \to \int_X f dm \text{ a.e. as } n \to \infty\ \forall\, f \in L^1(m).$$

DEFINITION: RETURN SEQUENCE, ASYMPTOTIC TYPE. The sequence a_n defined by (3.3.1) is called a *return sequence* of T and sometimes denoted $a_n(T)$. Its asymptotic proportionality class $\mathcal{A}(T) := \{(a'_n)_{n \in \mathbb{N}} : \exists \lim_{n \to \infty} \frac{a'_n}{a_n(T)} \in \mathbb{R}_+\}$ is called the *asymptotic type* of T.

REMARKS.

1) The convergence (3.3.1) (established in [**A1**]) is a "ratio limit theorem" in the sense of [**Fo-Lin**] (whence the term "rational ergodicity").

2) A conservative, ergodic, measure preserving transformation satisfying (3.3.3) with respect to some sequence of constants $\{a_n\}$ is called *weakly homogeneous*. The sequence of constants $\{a_n\}$ appearing in (3.3.3) is unique up to asymptotic equality. As before, it is called the *return sequence* of T. The second part of the theorem (see [**A3**]) shows that there is no ambiguity in this choice of name.

3) If T is weakly homogeneous, then T has a law of large numbers. Indeed in case n_k is as in (3.3.3), we may take

$$L(\epsilon_1, \epsilon_2, \dots) := \varlimsup_{N \to \infty} \frac{1}{N} \sum_{k=1}^{N} \frac{1}{a_{n_k}} \sum_{j=1}^{n_k} \epsilon_j.$$

Note that this law of large numbers is *monotone*:

$$\epsilon_n \le \epsilon'_n \; \forall \, n \ge 1 \implies L(\underline{\epsilon}) \le L(\underline{\epsilon}').$$

Theorem 3.2.5 does not promise a monotone law of large numbers.

PROOF OF THEOREM 3.3.1.

We assume without loss of generality that T is invertible (passing if necessary to T's natural extension which is also rationally ergodic).

Suppose that $A \in \mathcal{B}$, $0 < m(A) < \infty$ satisfies a Renyi inequality, then, writing $S_n := S_n(1_A)$ and setting

$$a_n = a_n(A) = \frac{1}{m(A)^2} \sum_{k=0}^{n-1} m(A \cap T^{-k}A),$$

we have

$$\int_A \left(\frac{S_n}{a_n}\right) dm = m(A)^2$$

and $\exists\, M > 0$ such that

$$\int_A \left(\frac{S_n}{a_n}\right)^2 dm \le M \forall\, n \in \mathbb{N}.$$

Now let $m_\ell \uparrow \infty$, then there is a subsequence ν_j and $\Psi \in L^2(A)$ such that

$$\int_A \Psi dm = m(A)^2 \text{ and } \frac{S_{\nu_j}}{a_{\nu_j}} \to \Psi \text{ weakly in } L^2(A).$$

There is a further subsequence n_k such that

$$\left| \int_A \left(\frac{S_{n_k}}{a_{n_k}} - \Psi\right) \left(\frac{S_{n_j}}{a_{n_j}} - \Psi\right) dm \right| \le \frac{1}{2^k} \; \forall\, 1 \le j < k.$$

By the exercise after lemma 1.4.5,

$$\frac{1}{N} \sum_{k=1}^{N} \frac{S_{n_k}}{a_{n_k}} \to \Psi \text{ a.e. on } A \text{ as } N \to \infty.$$

The set on which this convergence takes place is T-invariant, as is the limit Ψ, therefore $\Psi \equiv m(A)$, and

$$\frac{1}{N} \sum_{k=1}^{N} \frac{S_{n_k}}{a_{n_k}} \to m(A) \text{ a.e. as } N \to \infty.$$

This is (3.3.3) for $f = 1_A$.

By Hopf's theorem, if (3.3.3) holds for some $f \in L^1(m)$, then it holds for every $f \in L^1(m)$.

We now turn to (3.3.1). The above shows that

$$\frac{S_n^T(1_A)}{a_n} \to m(A) \text{ weakly in } L^2(A),$$

and (substituting T^{-1} for T)

$$\frac{S_n^{T^{-1}}(1_A)}{a_n} \to m(A) \text{ weakly in } L^2(A),$$

whence

$$\int_A \left(\frac{S_n^T(1_B)}{a_n} \right) dm \to m(A)m(B) \ \forall \ B \in \mathcal{B} \cap A.$$

Fix $B \in \mathcal{B} \cap A$, then

$$\sup_n \int_A \left(\frac{S_n^T(1_B)}{a_n} \right)^2 dm \leq \sup_n \int_A \left(\frac{S_n^T(1_A)}{a_n} \right)^2 dm < \infty$$

and an analogous argument establishes

$$\frac{S_n^T(1_B)}{a_n} \to m(B)) \text{ weakly in } L^2(A)$$

which implies (3.3.1).

To demonstrate (3.3.2), note that for $B \in \mathcal{B}$, $B = \bigcup_{n=0}^{\infty} B_n$ where $B_0 = A \cap B$ and $B_n = B \cap T^{-n}A \setminus \bigcup_{j=0}^{n-1} T^{-j}A$. If $B, C \in \mathcal{B}$, then $\forall \ N \geq 1$

$$\frac{1}{a_n} \sum_{k=0}^{n-1} m(B \cap T^{-k}C) \geq \sum_{i,j=0}^{N} \sum_{k=0}^{n-1} m(B_i \cap T^{-k}C_j)$$

$$\to \sum_{i,j=0}^{N} m(B_i)m(C_j) \quad \text{as } n \to \infty$$

which implies (3.3.2) because

$$\sum_{i,j=0}^{N} m(B_i)m(C_j) \to m(B)m(C) \quad \text{as } N \to \infty.$$

$\square$

3.3.2 PROPOSITION   [**A3**]. *Suppose that S and T are similar measure preserving transformations, both conservative and ergodic.*
If S is weakly homogeneous, then so is T and

$$\exists \lim_{n \to \infty} \frac{a_n(S)}{a_n(T)} \in \mathbb{R}_+.$$

PROOF. We show that if $S \to T$, then S is weakly homogeneous iff T is weakly homogeneous.

Suppose that $\pi : S \xrightarrow{c} T$. By the ratio ergodic theorem, S is weakly homogeneous iff T is weakly homogeneous, and in this case

$$\frac{a_n(S)}{a_n(T)} \to \frac{1}{c} \text{ as } n \to \infty.$$

$\square$

3.3.3 COROLLARY. *If T is weakly homogeneous, then $\Delta_\infty(T) = \{1\}$.*

DEFINITION: FEEBLE CONVERGENCE.
Let $(X, \mathcal{B}, m)$ be a measure space, and suppose that $f_n, f : X \to \mathbb{R}_+$. We'll say that f_n *tends to f feebly* (written $f_n \rightsquigarrow f$) if

$$\forall \, m_\ell \uparrow \infty \; \exists \, n_k = m_{\ell_k} \uparrow \infty$$

such that $\forall \, p_j = n_{k_j} \uparrow \infty$, we have

$$\frac{1}{N} \sum_{j=1}^{N} f_{p_j} \to f \text{ a.e. as } N \to \infty.$$

REMARK. The proof of (3.3.3) in theorem 3.3.1 actually shows that

$$\frac{S_n(f)(x)}{a_n} \rightsquigarrow \int_X f dm \;\; \forall \, f \in L^1(m).$$

It follows from the exercise after lemma 1.4.5 that if Ω is a probability space, $f_n, f \in L^2(\Omega)$ and $f_n \to f$ weakly in $L^2(\Omega)$, then $f_n \rightsquigarrow f$.

Weak convergence in $L^1(\Omega)$ also implies feeble convergence, and by Komlos' theorem [**Kom**], every bounded sequence in $L^1(\Omega)$ has a feebly convergent subsequence.

Note that if $f_n \geq 0$, and $f_n \rightsquigarrow 0$, then $f_n \to 0$ in measure (see the proof to theorem 1.4.4).

3.3.4 PROPOSITION [**A7**].
Suppose that T is a conservative, ergodic measure preserving transformation which does not have a law of large numbers, then $\forall \, A \in \mathcal{B}$, $0 < m(A) < \infty$,

$$\frac{1}{a_n(A)} S_n(f) \to 0 \text{ in measure } \;\; \forall \, f \in L^1(m)$$

where $a_n(A) = \frac{1}{m(A)^2} \sum_{k=0}^{n-1} m(A \cap T^{-k}A)$.

PROOF. Fix $A \in \mathcal{B}$, $0 < m(A) < \infty$, then

$$\left\| \frac{S_n(1_A)}{a_n(A)} \right\|_{L^1(A)} = m(A)^2 \ \forall \ n \geq 1$$

and by Komlos' theorem,

$$\forall \ m_\ell \uparrow \infty \ \exists \ n_k = m_{\ell_k} \uparrow \infty$$

and $\psi \in L^1(A)$ such that $\forall \ p_j = n_{k_j} \uparrow \infty$, we have

$$\frac{1}{N} \sum_{j=1}^N \frac{S_{p_j}(1_A)}{a_{p_j}(A)} \to \psi \text{ a.e. on } A \text{ as } N \to \infty.$$

As in the proof of theorem 3.3.1, the set on which this convergence takes place is T-invariant as is the limit; whence for some constant $c \geq 0$,

$$\frac{1}{N} \sum_{j=1}^N \frac{S_{p_j}(1_A)}{a_{p_j}(A)} \to c \text{ a.e. as } N \to \infty.$$

By Hopf's theorem,

$$\frac{1}{N} \sum_{j=1}^N \frac{S_{p_j}(f)}{a_{p_j}(A)} \to \frac{c}{m(A)} \int_X f dm \text{ a.e. as } N \to \infty \ \forall \ f \in L^1(m),$$

and the absence of a law of large numbers for T forces $c = 0$.

Thus $\frac{1}{a_n(A)} S_n(1_A) \rightsquigarrow 0$, hence $\frac{1}{a_n(A)} S_n(1_A) \to 0$ in measure and the result follows from Hopf's theorem. $\qquad \square$

§3.4 Maharam transformations

A Maharam transformation is a special kind of skew product transformation (see chapter 7 for more on skew products).

Let R be a conservative, ergodic, invertible, non-singular transformation of the probability space
$(X, \mathcal{B}, m)$, and set

$$R' := \frac{dm \circ R}{dm}.$$

The *Maharam* $\mathbb{R}$-extension of R is the transformation $\widetilde{R} : X \times \mathbb{R} \to X \times \mathbb{R}$ defined by

$$\widetilde{R}(x, y) := (Rx, y - \log R'(x)).$$

The Maharam $\mathbb{R}$-extension has a natural invariant measure $\mu \sim m \times \lambda$ on $(X \times \mathbb{R}, \mathcal{B} \otimes \mathcal{B}(\mathbb{R}))$ given by

$$d\mu(x, y) = e^y dm(x) dy,$$

the dilation of R on the first coordinate being cancelled by the translation on the second.

Maharam $\mathbb{R}$-extensions are connected with the existence of absolutely continuous, invariant measures. It is routine to check that $\log R'$ is a coboundary in the sense that $\log R' = h - h \circ R$ for some $h : X \to \mathbb{R}$ is measurable iff $\exists$ a R-invariant, m-absolutely continuous measure (with m-derivative e^h).

In some cases $R' = a^\phi$ where $a > 1$ and $\phi : X \to \mathbb{Z}$ is measurable. Here, we can define the *Maharam* $\mathbb{Z}$-extension of R which is the transformation $R_\phi : X \times \mathbb{Z} \to X \times \mathbb{Z}$ defined by

$$R_\phi(x, y) := (Rx, y - \phi(x)).$$

The Maharam $\mathbb{Z}$-extension preserves the measure ν on $(X \times \mathbb{Z}, \mathcal{B} \otimes \mathcal{B}(\mathbb{Z}))$ given by

$$\nu(A \times \{n\}) = c^n m(A).$$

and is conservative if R is.

3.4.1 THEOREM [**Mah**]. *The Maharam extension of a conservative nonsingular transformation is conservative.*

PROOF. We shall only consider the Maharam $\mathbb{R}$-extension $\widetilde{R}$ of a conservative nonsingular transformation R. The proof for $\mathbb{Z}$-extensions is analogous.

Writing $\frac{dm \circ R^n}{dm} = R^{n'}$ and setting $F(x) = \sup_{n \geq 0} R^{n'}(x)$ we see that $F \circ R \leq \frac{F}{R'}$, equivalently $\widehat{(R^{-1})} F \leq F$.

Since conservativity implies dual incompressibility (see exercise 1.3.1)

$$F \circ R = \frac{F}{R'}.$$

For $a \in \mathbb{R}$,

$$\mu(X \times (-\infty, a)) = e^a < \infty.$$

Set

$$\Omega_a := \bigcup_{n=0}^{\infty} R^{-n}(X \times (-\infty, a)),$$

then

$$\Omega_a = \{(x, y) \in X \times \mathbb{R} : y < a + \log F(x)\},$$

whence

$$\begin{aligned}
\widetilde{R}^{-1} \Omega_a &= \{(x, y) \in X \times \mathbb{R} : (Rx, y - \log R'(x)) \in \Omega_a\} \\
&= \{(x, y) \in X \times \mathbb{R} : y < a + \log F(Rx) + \log R'(x)\} \\
&= \Omega_a \quad \because \ \log F(Rx) + \log R'(x) = \log F(x).
\end{aligned}$$

By Maharam's recurrence theorem $(1.1.7)$, $R|_{\Omega_a}$ is conservative and the theorem is established because

$$\mathfrak{C}(R) \supseteq \bigcup_{a \in \mathbb{Z}} \Omega_a = X \times \mathbb{R}.$$

$\square$

No Maharam transformation T can have $\Delta_0(T) = \{1\}$, so the existence of ergodic Maharam transformations is of interest to us, an ergodic Maharam transformation having no law of large numbers.

The first ergodic Maharam $\mathbb{R}$-extension was found by [**Kri**]. We present here the ergodic Maharam $\mathbb{Z}$-extension of [**Haj-It-Kak**]. Remarks on the construction of an ergodic Maharam $\mathbb{R}$-extension can be found at the end of §8.4.

Let Ω be the group of dyadic integers, let $\tau x = x + \underline{1}$, and (as in example 1.2.8) for $p \in (0, 1)$, define a probability μ_p on Ω by

$$\mu_p([\epsilon_1, ..., \epsilon_n]) = \prod_{k=1}^{n} p(\epsilon_k).$$

Recall that τ is ergodic with respect to μ_p, and that

$$\frac{d\mu_p \circ \tau}{d\mu_p} = \left(\frac{1-p}{p}\right)^{\phi}$$

where

$$\phi(x) = \min\{n \in \mathbb{N} : x_n = 0\} - 2.$$

For $0 < p < 1$, the Maharam $\mathbb{Z}$-extension of $(\Omega, \mathcal{B}(\Omega), \mu_p, \tau)$ is

$$T_p := (X, \mathcal{B}, m_p, T)$$

where

$$X = \Omega \times \mathbb{Z}, \ \mathcal{B} := \mathcal{B}(\Omega) \otimes 2^{\mathbb{Z}},$$

$$m_p(A \times \{n\}) = \mu_p(A)\left(\frac{1-p}{p}\right)^n, \text{ and } T(x,n) = (\tau x, n - \phi(x)).$$

3.4.2 THEOREM [**Haj-It-Kak**]. T_p *is a conservative, ergodic measure preserving transformation.*

PROOF.

The method of proof is to show that a bounded, measurable T-invariant function $f(x, n)$ does not depend on n, and hence, by ergodicity of τ, is constant.

Suppose $f \in L^\infty$ and $f \circ T = f$, then

$$f(x, a) = f(T^{2^n}(x, a)) = f(\tau^{2^n} x, a - \phi(S^n x)),$$

or, equivalently,

$$f(\tau^{2^n} x, a) = f(x, a + \phi(S^n x)).$$

By lemma 1.2.10, there is a subsequence $n_k \to \infty$ such that

$$f(\tau^{2^{n_k}} x, a) \to f(x, a) \text{ a.e. as } n \to \infty \ \forall a \in \mathbb{Z}.$$

Using the Borel-Cantelli lemma (as in the proof of Arnold's theorem) we see that

$$\phi \circ S^{n_k} = -1 \text{ i.o. a.e.}$$

whence,

$$f(x, a) \leftarrow f(\tau^{2^{n_k}} x, a) \text{ a.e. as } k \to \infty$$
$$= f(x, a + \phi(S^{n_k} x))$$
$$= f(x, a - 1) \text{ i.o. a.e. },$$

and so

$$f(x, a) = f(x, 0) = f(\tau x, 0) \text{ a.e.}$$

whence f is constant, and T ergodic. $\qquad\qquad\qquad\qquad\qquad\qquad\square$

REMARK. Note that

$$T^{-1}(x,n) = (\tau^{-1}x, n + \phi(\tau^{-1}x)) = (\tau^{-1}x, n + \psi(x))$$

where

$$\psi(x) = \min\{n \in \mathbb{N} : x_n = 1\} - 2.$$

If $\pi : \Omega \to \Omega$ is the involution defined by $\pi(x)_n := 1 - x_n$, then

$$\pi \circ \tau \circ \pi = \tau^{-1}, \ \ \phi \circ \pi = \psi \text{ and } \mu_p \circ \pi = \mu_{1-p}.$$

Define $\overline{\pi} : X \to X$ is by $\overline{\pi}(x,n) = (\pi x, -n)$, then

$$\begin{aligned}
\overline{\pi} \circ T^{-1} \circ \overline{\pi}(x,n) &= \overline{\pi} \circ T^{-1}(\pi x, -n) \\
&= \overline{\pi}(\tau^{-1}\pi x, -n + \psi(\pi x)) \\
&= (\pi\tau^{-1}\pi x, n - \psi(\pi x).
\end{aligned}$$

Since $\pi x = -\underline{1} - x$, we have that $\pi\tau^{-1}\pi x = \tau x$, whence, since $\psi(\pi x) = \phi(x)$, we obtain that

$$\overline{\pi} \circ T^{-1} \circ \overline{\pi} = T.$$

In view of $m_p \circ \overline{\pi} = m_{1-p}$, this implies that $\overline{\pi} : T_p^{-1} \overset{1}{\leftrightarrow} T_{1-p}$ and in particular $T_{\frac{1}{2}} \leftrightarrow T_{\frac{1}{2}}^{-1}$.

3.4.3 PROPOSITION [A11]. *For $p \neq \frac{1}{2}$, T_p is strongly disjoint from T_p^{-1}.*

PROOF.
By the above remark, it is sufficient to show that for $p \neq \frac{1}{2}$, T_p is strongly disjoint from T_{1-p}.

If this is not the case, then T_p and T_{1-p} have a common conservative ergodic extension R. Let the factor maps be given by $\alpha : R \to T_p$ and $\beta : R \to T_{1-p}$.
Choose a probability $P : \mathcal{B}_R \to [0,1]$, $P \sim m_R$ and define a measure μ on $\Omega \times \Omega$ by

$$\mu(A \times B) := P\left(\alpha^{-1}(A \times \mathbb{Z}) \cap \beta^{-1}(B \times \mathbb{Z})\right).$$

It follows that $\tau \times \tau$ is a conservative, ergodic nonsingular transformation of $(\Omega \times \Omega, \mathcal{B}(\Omega \times \Omega), \mu)$ and that if $\gamma_i : \Omega \times \Omega \to \Omega$ $(i = 1,2)$ is defined by $\gamma_i(x_1, x_2) := x_i$, then

$$\mu \circ \gamma_1^{-1} \sim \mu_p, \ \& \ \mu \circ \gamma_2^{-1} \sim \mu_{1-p}.$$

Next, define $f : \Omega \times \Omega \to \Omega$ by $f(x,y) := x - y$. Evidently $f \circ (\tau \times \tau) = f$ whence $\exists \, \eta \in \Omega$ such that $f = \eta$ μ-a.e. on $\Omega \times \Omega$. Thus, $\mu(A \times B) = \nu(A \cap \eta B)$ where $\nu \sim \mu_p$, and we obtain

$$\mu_p \circ \eta \sim \mu_{1-p}.$$

We complete the proof by showing that this latter is impossible. We only treat the case where $p > \frac{1}{2}$, (the case $p < \frac{1}{2}$ being analogous and left to the reader).
Setting $\kappa := \frac{d\mu_p \circ \eta}{d\mu_{1-p}}$, noting that

$$\eta\tau^{2^n} = \tau^{2^n}\eta,$$

and using the chain rule for Radon-Nikodym derivatives,

$$\kappa \frac{d\mu_p \circ \tau^{2^n}}{d\mu_p} \circ \eta = \frac{d\mu_p \circ \tau^{2^n} \eta}{d\mu_{1-p}} = \frac{d\mu_p \circ \eta \tau^{2^n}}{d\mu_{1-p}} = \kappa \circ \tau^{2^n} \frac{d\mu_{1-p} \circ \tau^{2^n}}{d\mu_{1-p}}.$$

Substituting in the formulae for $\frac{d\mu_p \circ \tau^{2^n}}{d\mu_p}$ $(n \geq 1, 0 < p < 1)$ we obtain

$$\left(\frac{1-p}{p}\right)^{\phi \circ \sigma^n \circ \eta + \phi \circ \sigma^n} = \frac{\kappa \circ \tau^{2^n}}{\kappa},$$

whence

$$\left(\frac{1-p}{p}\right)^{\phi \circ \sigma^n} \geq (\frac{1-p}{p})^2 \frac{\kappa \circ \tau^{2^n}}{\kappa}.$$

By lemma 1.2.10, $\exists\, n_k \to \infty$ such that $n_{k+1} \geq n_k + 6$ and

$$\frac{\kappa \circ \tau^{2^{n_k}}}{\kappa} \to 1 \text{ a.e.}$$

The events

$$A_k = [\phi \circ S^n = 3] = \{x \in \Omega : x_{n_k+j} = 0 \ j = 1, 2, 3, 4, \ \& \ x_{n_k+5} = 0\}$$

are independent, and $\mu_p(A_k) = (1-p)^4 p$. By the Borel-Cantelli lemma,

$$\phi \circ S^{n_k} = 3 \text{ i.o. a.e.}$$

whence

$$\left(\frac{1-p}{p}\right)^3 \geq (\frac{1-p}{p})^2$$

contradicting $p > \frac{1}{2}$. $\square$

3.4.4 PROPOSITION [A11].

$$\Delta_0(T_p) = \Delta_\infty(T_p) = \left\{ \left(\frac{1-p}{p}\right)^n : n \in \mathbb{Z} \right\}.$$

3.4.5 LEMMA   [Ham]. *Suppose that $p \neq \frac{1}{2}$ and that $\sigma : \Omega \to \Omega$ is μ_p-non-singular, and commutes with τ, then*

$$\sigma = \tau^n \text{ for some } n \in \mathbb{Z}.$$

PROOF.
Since σ commutes with τ, we have $\sigma \circ \tau^{2^n} = \tau^{2^n} \circ \sigma$ and hence

$$\sigma' \circ \tau^{2^n} \cdot \tau^{2^n \prime} = \tau^{2^n \prime} \circ \sigma \cdot \sigma'$$

where for $g : \Omega \to \Omega$ a nonsingular transformation, $g' := \frac{d\mu_p \circ g}{d\mu_p}$.

Next, define $f : \Omega \to \Omega$ by setting $f(x) := \sigma(x) - x$, then $f(\tau x) = x$ and by ergodicity of τ, $\exists\, \eta \in \Omega$ such that $\sigma(x) = x + \eta$ a.e..

In order to show that $\sigma = \tau^n$ for some $n \in \mathbb{Z}$, we show that $\exists\, \lim_{n \to \infty} \eta_n$.

If this is not the case, $\exists\ n_k \uparrow \infty$ such that $n_{k+1} \geq n_k + 2$ and $\eta_{n_k} = 0$ and $\eta_{n_{k+1}} = 1\ \forall\ k$. Set $A_k := \{x \in \Omega:\ x_{n_k} = 0,\ x_{n_k+1} = 1,\ x_{n_k+2} = 0\}$.

Fix $x \in A_k$, we have $\tau^{2^{n_k}\prime}(x) = \left(\frac{1-p}{p}\right)^{\phi(S^{n_k}x)} = 1$, $(\sigma x)_{n_k+1} = (\eta+x)_{n_k+1} = 0$, and hence $\tau^{2^{n_k}\prime}(\sigma x) = \left(\frac{1-p}{p}\right)^{\phi(S^{n_k}\sigma x)} = \frac{p}{1-p}$, whence

$$\frac{\sigma'(\tau^{2^{n_k}}x)}{\sigma(x)} = \frac{p}{1-p}.$$

By lemma 1.2.10, there is a subsequence $m_\ell = n_{k_\ell} \to \infty$ such that

$$\sigma' \circ \tau^{2^{m_\ell}} \to \sigma' \text{ a.e. }.$$

The sets $\{A_k\ :\ k \geq 1\}$ are independent, and $\mu_p(A_k) = p(1-p)$. By the Borel-Cantelli lemma, for a.e. $x \in \Omega$,

$$x \in A_{k_\ell} \text{ i.o.}$$

with the conclusion that for a.e. $x \in \Omega$,

$$1 \leftarrow \frac{\sigma'(\tau^{2^{n_k}}x)}{\sigma(x)} \overset{\text{i.o.}}{=} \frac{p}{1-p}$$

contradicting $p \neq \frac{1}{2}$.   □

PROOF OF PROPOSITION 3.4.4. Define $Q : X \to X$ by $Q(x,a) = (x,a+1)$. Clearly $Q : T_p \overset{\frac{p}{1-p}}{\to} T_p$, whence

$$\left\{\left(\frac{1-p}{p}\right)^n\ :\ n \in \mathbb{Z}\right\} \subset \Delta_0(T_p) \subset \Delta_\infty(T_p).$$

Conversely, suppose that $c \in \Delta_\infty(T_p)$, and let U be a conservative, ergodic measure preserving transformation equipped with factor maps $\alpha : U \overset{1}{\to} T_p$ and $\beta : U \overset{c}{\to} T_p$. Define a measure μ on $X \times X$ by

$$\mu(A \times B) := m_U(\alpha^{-1}A \cap \beta^{-1}B).$$

If $V = (X \times X, \mathcal{B} \otimes \mathcal{B}, \mu, T)$ then $\pi_i : V \overset{c^{i-1}}{\to} T_p$ where $\pi_i(x_1, x_2) := x_i\ (i = 1, 2)$. Writing $x \in X = \Omega \times \mathbb{Z}$ as $x = (\omega(x), a(x))$, we see by the lemma that $\exists\ \nu \in \mathbb{Z}$ such that $\omega(x_2) = \omega(x_1) + \nu\underline{1} = \tau^\nu \omega(x_1)$ a.e..

Define $f : X \times X \to \mathbb{Z}$ by

$$f(x_1, x_2) := a(x_2) - a(x_1) + \phi_\nu(\omega(x_1))$$

to obtain that

$$\begin{aligned}
f \circ (T \times T)(x_1, x_2) &= f(Tx_1, Tx_2) \\
&= a(Tx_2) - a(Tx_1) + \phi_\nu(\omega(Tx_1)) \\
&= a(x_2) - \phi(\omega(x_2)) - a(x_1) + \phi(\omega(x_1)) + \phi_\nu(\omega(Tx_1)) \\
&= a(x_2) - a(x_1) - \phi(\tau^\nu \omega(x_1)) + \phi(\omega(x_1)) + \phi_\nu(\tau\omega(x_1)) \\
&= a(x_2) - a(x_1) + \phi_\nu(\omega(x_1)) \\
&= f(x_1, x_2),
\end{aligned}$$

whence $\exists\, v \in \mathbb{Z}$ such that $f = v$ a.e..

The conclusion is that for μ-a.e. $(x_1, x_2) \in X \times X$, $x_2 = Q^v T^\nu x_2$, whence

$$\mu(A \times B) = m_p(A \cap R^{-1}B) \text{ where } R = Q^v T^\nu x_2$$

and $c = \left(\frac{1-p}{p}\right)^v$.

For further results on the Maharam $\mathbb{Z}$-extensions T_p, see [**A-W3**].

In [**A11**], examples of conservative, ergodic, measure preserving transformations T with $\Delta(T) = \Delta_0(T)$ of arbitrary Hausdorff dimension were given, also strongly disjoint from their inverses. The constructions generalise those of this section, and to each T constructed, was constructed a similar T' with $\Delta_0(T') = \{1\}$.

An conservative, ergodic, measure preserving transformation T with $\Delta_\infty(T) = \{1, \infty\}$ is given in §8.6.

$\square$

§3.5 Category theorems

Let $(X, \mathcal{B}, m)$ be a standard σ-finite, infinite measure space and let $\mathcal{F}$ denote the ring of measurable subsets with finite measure.

Recall from §1.0 that the group $\mathfrak{A}_0(X, \mathcal{B}, m)$ of invertible, measure preserving transformations of $(X, \mathcal{B}, m)$ is a Polish group when equipped with the weak topology inherited from $\mathfrak{B}(L^2(m))$ (being a closed subgroup of $\mathfrak{A}(X, \mathcal{B}, m)$). A convenient neighbourhood base for this topology is

$$\{N(T; \epsilon; A_1, \ldots, A_N) : T \in \mathfrak{A}_0, \ \epsilon > 0, \ A_1, \ldots, A_N \in \mathcal{F}\}$$

where

$$N(T; \epsilon; A_1, \ldots, A_N) := $$
$$\{S \in \mathfrak{A}_0 : m(SA_i \Delta T A_i), \ m(S^{-1}A_i \Delta T^{-1}A_i) < \epsilon \ \forall \ 1 \le i \le N\}.$$

We show here that "in general" a measure preserving transformation is conservative, ergodic, and has a law of large numbers.

3.5.1 THEOREM [**A4**].
There is a residual set in $\mathfrak{H} \subset \mathfrak{A}_0(X, \mathcal{B}, m)$ such that $\forall\, T \in \mathfrak{H}$, $\exists\, n_k, \ d_k \to \infty$ so that

$$\frac{1}{d_k} S_{n_k}^T(f) \to \int_X f \, dm \ a.e. \ \forall \ f \in L^1(m).$$

3.5.2 CONJUGACY LEMMA   [**Sach**].
Suppose that $T \in \mathfrak{A}_0$ is ergodic, then

$$\{\pi^{-1} \circ T \circ \pi : \ \pi \in \mathfrak{A}_0\} \text{ is dense in } \mathfrak{A}_0.$$

PROOF.

Call $R \in \mathfrak{A}_0$ *n-cyclic* if $R^n =$ Id, and $\exists$ a partition $\{A_1, \ldots, A_n\} \subset \mathcal{F}$ disjoint and a cyclic permutation $p : \{1, \ldots, n\} \to \{1, \ldots, n\}$ such that $R^n =$ Id and $RA_i = R_{p(i)}$; and call $R \in \mathfrak{A}_0$ *cyclic* if it is n-cyclic for some $n \geq 1$.

By Rokhlin's tower theorem (theorem 1.5.9), if $T \in \mathfrak{A}_0$ is ergodic, then $\forall \ \epsilon > 0$, $\exists \ E \in \mathcal{F}$ such that $\{T^k E : \ 0 \leq k \leq n-1\}$ are disjoint, and

$$m\left(X \setminus \bigcup_{k=0}^{n-1} T^k E\right) = \epsilon.$$

Let $\{F_1, \ldots, F_n\} \subset \mathcal{B}$ be a partition such that $F_k = T^{k-1} E \cup E_k$ where $T^{k-1} E \cap E_k = \emptyset$ and $m(E_k) = \frac{\epsilon}{n}$ $(1 \leq k \leq n)$; and let $R \in \mathfrak{A}_0$ be defined by

$$Rx = \begin{cases} Tx & x \in \bigcup_{k=0}^{n-2} T^k E, \\ T^{-(n-1)}x & x \in T^{n-1}x, \\ R_k x & x \in E_k \ (1 \leq k \leq n) \end{cases}$$

where $R_k : E_k \to E_{k+1}$ $(1 \leq k \leq n-1)$ are invertible, measure preserving maps, and $R_n = R_{n-1}^{-1} \circ \cdots \circ R_1^{-1} : E_n \to E_1$, then R is n-cyclic, and

$$\{x \in X : \ Tx \neq Rx\} \subset T^n E \cup \bigcup_{k=1}^{n} E_k.$$

From this, it follows that if $T \in \mathfrak{A}_0$ is ergodic, $R \in \mathfrak{A}_0$ is n-cyclic, $A_1, \ldots, A_N \in \mathcal{F}$ and $\epsilon > 0$ then $\exists \ \pi \in \mathfrak{A}_0$ such that

$$m(T' A_i \Delta R A_i), \ m(T'^{-1} A_i \Delta R^{-1} A_i) < \epsilon \ \forall \ 1 \leq i \leq N$$

where $T' := \pi^{-1} \circ T \circ \pi$.

It now suffices to show that the cyclic transformations are dense in $\mathfrak{A}_0$.

We claim that if $T \in \mathfrak{A}_0$ and $\epsilon > 0$, $A_1, \ldots, A_N \in \mathcal{F}$ then $\exists \ R$ cyclic such that

$$m(T A_i \Delta R A_i), \ m(T^{-1} A_i \Delta R^{-1} A_i) < \epsilon \ \forall \ 1 \leq i \leq N.$$

Given $\epsilon > 0$, $\exists$ a partition $\alpha \subset \mathcal{F}$ with $m(a) = c \ \forall \ a \in \alpha$ and subsets

$$\mathfrak{b}_1, \ldots, \mathfrak{b}_N, \mathfrak{c}_1, \ldots, \mathfrak{c}_N, \mathfrak{d}_1, \ldots, \mathfrak{d}_N \subset \alpha$$

such that $m(B_i) = m(C_i) = m(D_i)$ and

$$m(B_i \Delta A_i), \ m(C_i \Delta T^{-1} A_i), \ m(D_i \Delta T A_i) < \epsilon \ \ (1 \leq i \leq N)$$

where

$$B_i := \bigcup_{a \in \mathfrak{b}_i} a, \ C_i := \bigcup_{a \in \mathfrak{c}_i} a, \ D_i := \bigcup_{a \in \mathfrak{d}_i} a, \ \ (1 \leq i \leq N).$$

There is a bijection

$$\rho : \bigcup_{i=1}^{N} (\mathfrak{b}_i \cup \mathfrak{c}_i) \to \bigcup_{i=1}^{N} (\mathfrak{b}_i \cup \mathfrak{d}_i)$$

such that $\rho(\mathfrak{c}_i) = \mathfrak{a}_i$, and $\rho(\mathfrak{a}_i) = \mathfrak{d}_i$, $(1 \leq i \leq N)$.

Since the domain of definition of ρ is a finite subset of the countable α, we can find a bijection $\tilde{\rho} : \alpha \to \alpha$ extending ρ in the sense that

$$\tilde{\rho}(a) = \rho(a) \ \forall \ a \in \bigcup_{i=1}^{N}(\mathfrak{b}_i \cup \mathfrak{c}_i);$$

and such that $\exists \, n \geq 1$, a partition $\{\mathfrak{e}_1, \ldots, \mathfrak{e}_n\}$ of α with $\tilde{\rho}^n = \mathrm{Id}$ and $\tilde{\rho}\mathfrak{e}_k = \mathfrak{e}_{k+1}$ for $1 \leq k \leq n-1$, and $\tilde{\rho}\mathfrak{e}_n = \mathfrak{e}_1$.

To conclude, we define $R \in \mathfrak{A}_0$ n-cyclic according to $\tilde{\rho}$. Let $E_k := \bigcup_{a \in \mathfrak{e}_k} a$ $(1 \leq k \leq n)$, then $\{E_1, \ldots, E_n\}$ is a partition of X. For $a \in \bigcup_{k=1}^{n-1} \mathfrak{e}_k$, choose an invertible, measure preserving map $r_a : a \to \tilde{\rho}(a)$, and define invertible, measure preserving maps $R_k : E_k \to E_{k+1}$ $(1 \leq k \leq n-1)$ by $R_k x := r_a x$ for $x \in a \in \mathfrak{e}_k$. Finally, define $R \in \mathfrak{A}_0$ by

$$Rx = \left\{ \begin{array}{ll} R_k x & x \in E_k, \ 1 \leq k \leq n-1, \\ R_{n-1} \circ \cdots \circ R_1^{-1} x & x \in E_n. \end{array} \right.$$

Evidently, R is n-cyclic, but also $R^{-1}B_i = C_i$ and $RB_i = D_i$ $(1 \leq i \leq N)$ whence

$$m(TA_i \Delta RA_i), \ m(T^{-1}A_i \Delta R^{-1}A_i) < 3\epsilon \ \forall \ 1 \leq i \leq N.$$

$\square$

3.5.3 EXISTENCE LEMMA.
$\exists \, T \in \mathfrak{A}_0$ such that

$$\frac{\log n}{n} S_n^T(f) \overset{\text{in measure}}{\longrightarrow} \int_X f \, dm \ \text{as } n \to \infty \ \forall \ f \in L^1(m).$$

Here $g_n \overset{\text{in measure}}{\longrightarrow} g$ means

$$m(A \cap [|g_n - g| > \epsilon]) \to 0 \text{ as } n \to \infty \ \forall \ \epsilon > 0, \ A \in \mathcal{F}.$$

PROOF.
We show first that $\exists$ an ergodic probability preserving transformation S of $(\Omega, \mathcal{A}, p)$ and $\varphi : \Omega \to \mathbb{N}$ so that

$$\frac{1}{n \log n} \sum_{k=0}^{n-1} \varphi \circ S^k \to 1 \text{ in measure.}$$

The desired $T \in \mathfrak{A}_0$ will be the tower over S with height function φ.

Let S be an ergodic probability preserving transformation of the standard probability space $(\Omega, \mathcal{A}, p)$ and $\varphi : \Omega \to \mathbb{N}$ so that $\{\varphi \circ S^n : n \geq 0\}$ are independent, and $p([\varphi \geq x]) \sim \frac{1}{x}$ as $x \to \infty$.

As in [**Fe2**] set

$$\varphi_n = \sum_{k=0}^{n-1} \varphi \circ S^k, \quad \varphi_n' = \sum_{k=0}^{n-1} \varphi \circ S^k 1_{[\varphi \circ S^k \leq n \log n]},$$

then

$$E(\varphi_n') \sim n \log n, \ \& \ E(\varphi_n'^2) \sim (n \log n)^2$$

whence $\frac{\varphi'_n}{n \log n} \to 1$ in $L^2(p)$ and so in measure. On the other hand,

$$p([\varphi_n \neq \varphi'_n]) \leq \sum_{k=0}^{n-1} p([\varphi \circ S^k > n \log n]) = np([\varphi > n \log n]) \sim \frac{1}{\log n} \to 0$$

and $\frac{\varphi_n}{n \log n} \to 1$ in measure.

We now build T, the tower transformation over S with height function φ. By exercise 1.5.1, T is a conservative, ergodic measure preserving transformation (of a standard σ-finite measure space).

Let $\Psi_n := S_n^T(1_\Omega)$, then for a.e. $x \in \Omega$

$$\Psi_{\varphi_n(x)} = n$$

and

$$\Psi_n \geq N \Leftrightarrow \varphi_N(x) \leq N.$$

Write $b(n) = n \log n$ and $a(n) := b^{-1}(n) \sim \frac{n}{\log n}$. For $\kappa > 0$ write $N_\kappa(n) := \kappa a(n)$, then $b(N_\kappa(n)) \sim \kappa n$ and by step 1,

$$p([\varphi_{N_\kappa(n)} \leq n]) \to \begin{cases} 1 & \kappa < 1, \\ 0 & \kappa > 1. \end{cases}$$

It follows that

$$p([\Psi_n \geq \kappa a(n)]) = p([\varphi_{N_\kappa(n)} \leq n]) \to \begin{cases} 1 & \kappa < 1, \\ 0 & \kappa > 1 \end{cases}$$

and

$$\frac{\Psi_n}{a(n)} \to 1 \text{ in measure on } \Omega.$$

Step 2 is now established by ergodicity of T and Hopf's theorem. □

PROOF OF THEOREM 3.5.1. Set $a(n) = \frac{n}{\log n}$, and let

$$\mathfrak{H} := \{T \in \mathfrak{A}_0 : \exists n_k \to \infty \text{ such that } \frac{S_{n_k}^T(f)}{a(n_k)} \to \int_X f dm \text{ a.e. } \forall f \in L^1(m)\}.$$

Clearly,

$$T \in \mathfrak{H} \Rightarrow \pi^{-1} \circ T \circ \pi \in \mathfrak{H} \ \forall \ \pi \in \mathfrak{A}_0,$$

and any $T \in \mathfrak{H}$ is evidently ergodic. By the existence lemma $\mathfrak{H} \neq \emptyset$, hence by the conjugacy lemma $\mathfrak{H}$ is dense in $\mathfrak{A}_0$.

We prove the theorem by showing that $\mathfrak{H}$ is a G_δ set in $\mathfrak{A}_0$.

Let $P \sim m$ be a probability. Sets of form

$$\{T \in \mathfrak{A}_0 : P([S_n^T(1_A) \in [a,b]]) \in [c,d])\}$$

are closed in $\mathfrak{A}_0$ for $a, b, c, d \in \mathbb{R}_+$.

Fix $\{A_n : n \in \mathbb{N}\} \subset \mathcal{F} := \{A \in \mathcal{B} : m(A) < \infty\}$ which generates $\mathcal{B}$ in the sense that $\sigma(\{A_n : n \in \mathbb{N}\}) = \mathcal{B}$, and let

$$\mathfrak{H}' := \bigcap_{k=1}^{\infty} \bigcup_{n=k}^{\infty} \bigcap_{\nu=1}^{k} \left\{ T \in \mathfrak{A}_0 : P\left(\left[\left|\frac{S_n(1_{A_\nu})}{a(n)} - m(A_\nu)\right| > \frac{1}{k}\right]\right) < \frac{1}{2^k} \right\},$$

then $\mathfrak{H}'$ is a G_δ. We claim $\mathfrak{H}' = \mathfrak{H}$.

Evidently,

$$\mathfrak{H}' := \{T \in \mathfrak{A}_0 : \exists\, n_k \to \infty \text{ such that } \frac{S_{n_k}^T(1_{A_\nu})}{a(n_k)} \to m(A_\nu) \text{ a.e. } \forall\, \nu \geq 1\}$$

whence $\mathfrak{H}' \supset \mathfrak{H}$.

Now suppose that $T \in \mathfrak{H}'$, then

$$\sum_{n=1}^{\infty} 1_{A_\nu} \circ T^n = \infty \text{ a.e. } \forall\, \nu \geq 1$$

and so T is conservative.

By Hopf's theorem, $\forall\, f \in L^1(m)$, $\nu \geq 1$ and a.e. $x \in X$,

$$\frac{S_n(f)(x)}{S_n(1_{A_\nu})(x)} \to h_\nu(f)$$

where $h_\nu(f) \circ T = h_\nu(f)$ and $\int_{A_\nu} h_\nu(f) dm = \int_X f dm$.

If $\frac{S_{n_k}^T(1_{A_\nu})}{a(n_k)} \to m(A_\nu)$ a.e. $\forall\, \nu \geq 1$, then

$$\frac{S_{n_k}^T(f)}{a(n_k)} \to h(f) \text{ a.e. } \forall\, f \in L^1(m)$$

where $h(f) \circ T = h(f)$ and $\int_{A_\nu} h(f) dm = m(A_\nu) \int_X f dm$.

Since $\{A_n : n \in \mathbb{N}\}$ generates $\mathcal{B}$, we have $h(f) = \int_X f dm$ and $T \in \mathfrak{H}$. $\square$

§3.6 Asymptotic Distributional Behaviour

Let Y be a Polish space and let $p_n, p \in \mathcal{P}(Y)$ be probabilities on Y. Recall that the probabilities p_n *converge weakly* to p (denoted $p_n \xrightarrow{w} p$) if

$$\int_Y f dp_n \to \int_Y f dp \quad \forall\, f : Y \to \mathbb{R} \text{ bounded and continuous.}$$

Now let $(\Omega, \mathcal{A}, P)$ be a probability space, and suppose that $U : \Omega \to Y$ is a random variable (= measurable function). The *distribution of* U is the probability measure on Y defined by $P \circ U^{-1}(A) = P(\{\omega \in \Omega : U(\omega) \in A\})$.

The random variables $U_n : \Omega \to Y$ *converge in distribution* to a random variable U with distribution p_U (but possibly defined on a different probability space) if $P \circ U_n^{-1} \xrightarrow{w} p_U$.

We'll be interested in a stronger form of distributional convergence.

DEFINITION: STRONG DISTRIBUTIONAL CONVERGENCE. We'll say that the random variables $U_n : \Omega \to Y$ converge *strongly in distribution* to U if

$$q \circ U_n^{-1} \xrightarrow{w} p_U \quad \forall\, q \in \mathcal{P}(\Omega, \mathcal{A}),\ q << P$$

and denote this by

$$U_n \xrightarrow{\mathcal{L}} U.$$

We have an analogous definition in case $(X, \mathcal{B}, m)$ is a measure space: the measurable functions $U_n : X \to Y$ converge strongly to the random variable U ($U_n \xrightarrow{\mathcal{L}} U$) if

$$q \circ U_n^{-1} \xrightarrow{w} p_U \quad \forall\, q \in \mathcal{P}(\Omega, \mathcal{A}),\ q << m.$$

REMARK.

Note that if U is constant then $U_n \xrightarrow{\mathfrak{L}} U$ if and only if

$$q([|U_n - U| \geq \epsilon]) \to 0 \text{ as } n \to \infty \; \forall \, \epsilon > 0, \; q \in \mathcal{P}(\Omega, \mathcal{A}), \; q << m,$$

equivalently, U_n converges to U in measure on subsets of finite measure.

This strong distributional convergence arises naturally in ergodic theory.

3.6.1 PROPOSITION (COMPACTNESS).

Let T be a conservative, ergodic non-singular transformation of the σ-finite, standard measure space $(X, \mathcal{B}, m)$, and let $f : X \to \mathbb{R}$ be bounded and measurable.

Suppose that $n_k \to \infty$, and $d_k > 0$, then $\exists \; m_\ell := n_{k_\ell} \to \infty$, and a random variable Y on $[-\infty, \infty]$ such that

$$\frac{1}{d_{k_\ell}} \sum_{j=0}^{m_\ell - 1} f \circ T^j \xrightarrow{\mathfrak{L}} Y.$$

PROOF. Write $S_n = \sum_{j=0}^{n-1} f \circ T^j$. It is sufficient to obtain a random variable Y on $[-\infty, \infty]$, and $m_\ell := n_{k_\ell} \to \infty$ such that for each continuous function $g : [\infty, \infty] \to \mathbb{R}$

$$g\left(\frac{S_{m_\ell}}{d_{k_\ell}}\right) \to E(g(Y)) \text{ weak } * \text{ in } L^\infty(m) \text{ as } \ell \to \infty.$$

For fixed $g \in C([-\infty, \infty])$, we have

$$\left\| g\left(\frac{S_{n_k}}{d_k}\right) \right\|_{L^\infty(m)} \leq \|g\|_{C([-\infty, \infty])} \; \forall \, k \geq 1.$$

There is a subsequence $k' \to \infty$ and a function $\lambda(g) \in L^\infty(m)$ so that

$$g\left(\frac{S_{n_{k'}}}{d_{k'}}\right) \to \lambda(g) \text{ weak } * \text{ in } L^\infty(m) \text{ as } k' \to \infty.$$

We claim that the limit function $\lambda(g)$ is T-invariant. This is because of the uniform continuity of g:

$$\omega_g(\epsilon) := \sup_{y, h \in \mathbb{R}, \; |h| < \epsilon} |g(y + h) - g(y)| \to 0 \text{ as } \epsilon \to 0;$$

whence

$$\left\| g\left(\frac{S_{n_k}}{d_k}\right) \circ T - g\left(\frac{S_{n_k}}{d_k}\right) \right\|_\infty = \left\| g\left(\frac{S_{n_k}}{d_k} + \frac{f \circ T^{n_k} - f}{d_k}\right) - g\left(\frac{S_{n_k}}{d_k}\right) \right\|_\infty$$

$$\leq \omega_g\left(\frac{2\|f\|_\infty}{d_k}\right)$$

$$\to 0$$

as $k \to \infty$. By ergodicity of T, $\lambda(g)$ is constant.

Let $\mathcal{G} \subset C([-\infty, \infty])$ be a countable, dense set. A diagonalisation argument yields $m_\ell := n_{k_\ell} \to \infty$ and constants $\{\lambda(g) \in \mathbb{R} : g \in \mathcal{G}\}$ such that

$$g\left(\frac{S_{m_\ell}}{d_{k_\ell}}\right) \to \lambda(g) \text{ weak } * \text{ in } L^\infty(m) \text{ as } \ell \to \infty \; \forall \, g \in \mathcal{G}.$$

We claim that $\forall\ h \in C([-\infty,\infty])$, $\exists\ \lambda(h) \in \mathbb{R}$ such that

$$h\left(\frac{S_{m_\ell}}{d_{k_\ell}}\right) \to \lambda(h) \text{ weak } * \text{ in } L^\infty(m) \text{ as } \ell \to \infty\ \ \forall\ h \in C([-\infty,\infty]).$$

To prove this, let $h \in C([-\infty,\infty])$. There is a sequence $g_j \in \mathcal{G}$, $\|g_j - h\|_\infty \to 0$. For any probability $p << m$,

$$\left|\lambda(g_i) - \lambda(g_j)\right| \leftarrow \left|\int_X \left(g_i\left(\frac{S_{m_\ell}}{d_{k_\ell}}\right) - g_j\left(\frac{S_{m_\ell}}{d_{k_\ell}}\right)\right)dp\right| \leq \|g_i - g_j\|_\infty$$

whence $\exists \lim_{j\to\infty} \lambda(g_j) := \lambda(h) \in \mathbb{R}$, and

$$|\lambda(g_j) - \lambda(h)| \leq \|g_j - h\|_\infty\ \forall\ j \geq 1.$$

For any probability $p << m$ and $j \geq 1$,

$$\left|\int_X h\left(\frac{S_{m_\ell}}{d_{k_\ell}}\right)dp - \lambda(h)\right| \leq$$

$$\left|\int_X h\left(\frac{S_{m_\ell}}{d_{k_\ell}}\right)dp - \int_X g_j\left(\frac{S_{m_\ell}}{d_{k_\ell}}\right)dp\right| + \left|\int_X g_j\left(\frac{S_{m_\ell}}{d_{k_\ell}}\right)dp - \lambda(g_j)\right| + |\lambda(g_i) - \lambda(h)|$$

$$\leq 2\|h - g_j\|_\infty + \left|\int_X g_j\left(\frac{S_{m_\ell}}{d_{k_\ell}}\right)dp - \lambda(g_j)\right|$$

$$\to 2\|h - g_j\|_\infty$$

as $\ell \to \infty$. Letting $j \to \infty$ we see that

$$h\left(\frac{S_{m_\ell}}{d_{k_\ell}}\right) \to \lambda(h) \text{ weak } * \text{ in } L^\infty(m) \text{ as } \ell \to \infty.$$

Clearly $\lambda : C([-\infty,\infty]) \to \mathbb{R}$ is linear and positive, whence $\exists$ a random variable Y on $[-\infty,\infty]$ such that

$$\lambda(g) = E(g(Y)).$$

$\square$

3.6.2 COROLLARY   [**A7**]. *If T is a conservative, ergodic measure preserving transformation of $(X,\mathcal{B},m)$, and $n_k \to \infty$, and $d_k > 0$, then $\exists\ m_\ell := n_{k_\ell} \to \infty$, and a random variable Y on $[0,\infty]$ such that*

(*) $$\frac{S_{m_\ell}^T(f)}{d_{k_\ell}} \xrightarrow{\mathfrak{L}} \mu(f)Y\ \forall\ f \in L^1(m),\ f \geq 0$$

where $\mu(f) := \int_X f\,dm$.

PROOF. Choose $A \in \mathcal{B}$, $m(A) = 1$. By proposition 3.6.1, and positivity $\exists\ m_\ell := n_{k_\ell} \to \infty$, and a random variable Y on $[0,\infty]$ such that

$$\frac{S_{m_\ell}^T(f)}{d_{k_\ell}} \xrightarrow{\mathfrak{L}} Y.$$

The result follows from Hopf's theorem. $\square$

In the situation of (*), we'll write

$$\frac{S_{m_\ell}^T}{d_{k_\ell}} \xrightarrow{\mathfrak{d}} Y,$$

and call Y a *distributional limit of T along $\{m_\ell\}$.*

Clearly $Y = 0$ and $Y = \infty$ are distributional limits of any conservative, ergodic, measure preserving transformation. We'll consider a random variable Y on $[0, \infty]$ *trivial* if it is supported on $\{0, \infty\}$.

A conservative, ergodic, measure preserving transformation is constructed in [**A-W1**] which has as distributional limit every random variable on $[0, \infty]$, and it follows from theorem 2 in [**A7**] that the collection of such transformations is residual in $\mathfrak{A}_0$.

3.6.3 PROPOSITION [**A7**].

If a conservative, ergodic measure preserving transformation has a nontrivial distributional limit, then it has a law of large numbers.

PROOF. Suppose that T is a conservative, ergodic measure preserving transformation and that

$$\frac{S^T_{n_k}}{d_k} \xrightarrow{\mathfrak{d}} Y,$$

where Prob. $(0 < Y < \infty) > 0$.

Let $g \in C([0, \infty])$ be positive and strictly increasing, and let $H : \mathbb{R}_+ \to \mathbb{R}_+$ be defined by $H(x) = E(g(xY))$, then H is also continuous, positive and strictly increasing. We have

$$g\left(\frac{S^T_{n_k}(f)}{d_k}\right) \to H(\mu(f)) \text{ weak} * \text{ in } L^\infty(m) \text{ as } k \to \infty \ \forall \ f \in L^1(m) \ f \geq 0,$$

and there is a subsequence $m_\ell = n_{k_\ell}$ such that

$$\frac{1}{N} \sum_{\ell=1}^{N} g\left(\frac{S^T_{m_\ell}(f)}{d_{k_\ell}}\right) \to H(\mu(f)) \text{ a.e.}$$

whence

$$L(\epsilon_1, \epsilon_2, \dots) := \limsup_{n \to \infty} H^{-1} \circ g\left(\frac{1}{d_{k_\ell}} \sum_{j=1}^{m_\ell} \epsilon_j\right)$$

is a law of large numbers for T. □

By proposition 3.6.3, ergodic Maharam transformations only have trivial distributional limits. It is shown in [**A7**] that a conservative, ergodic measure preserving transformation whose only limits are trivial indeed has all trivial random variables as distributional limits.

On the other hand some transformations have only one distributional limit (up to positive constant multiplication).

The Darling-Kac Theorem.

To complete this section, we prove the Darling-Kac distributional limit theorem. This theorem is heavily dependent on the theory of regular variation (as are many other distributional limit theorems). We begin with a review of regular variation.

DEFINITION: REGULARLY VARYING FUNCTION.

A measurable function $f : \mathbb{R}_+ \to \mathbb{R}_+$ is *regularly varying* at ∞ if

$$\forall \ y > 0, \ \exists \ \lim_{x \to \infty} \frac{f(xy)}{f(x)} \in \mathbb{R}_+.$$

In this case $\exists\, \alpha \in \mathbb{R}$ such that

$$\lim_{x \to \infty} \frac{f(xy)}{f(x)} = y^\alpha \quad (y > 0).$$

The number α is called the *index* of regular variation (of f) and f is said to be regularly varying at ∞ *with index* α.

Similarly, a measurable function $f : \mathbb{R}_+ \to \mathbb{R}_+$ is *regularly varying at 0 with index α* if

$$\frac{f(xy)}{f(x)} \to y^\alpha \text{ as } x \to 0 \,\forall\, y > 0.$$

Evidently $x \mapsto f(x)$ is regularly varying at 0 if and only if $x \mapsto f(\frac{1}{x})$ is regularly varying at ∞.

DEFINITION: SLOWLY VARYING FUNCTION.

A measurable function $f : \mathbb{R}_+ \to \mathbb{R}_+$ is *slowly varying* at 0 or ∞ if it is regularly varying there with index 0.

A function $f : \mathbb{R}_+ \to \mathbb{R}_+$ is regularly varying at ∞ with index α if and only if

$$f(x) = x^\alpha L(x)$$

where $L : \mathbb{R}_+ \to \mathbb{R}_+$ is slowly varying at ∞.

REPRESENTATION OF SLOWLY VARYING FUNCTIONS   [**Fe3**] [**Kar**].

Suppose that $L : \mathbb{R}_+ \to \mathbb{R}_+$ is slowly varying at ∞, then $\exists\, M > 0$ and measurable functions $C,\ \epsilon : [M, \infty) \to \mathbb{R}$ such that $C > 0$, $C(x) \to c \in \mathbb{R}_+$, $\epsilon(x) \to 0$ as $x \to \infty$ and

$$L(x) \;=\; C(x) e^{\int_M^x \frac{\epsilon(t)}{t} dt} \quad \forall\, x \geq M.$$

KARAMATA'S TAUBERIAN THEOREM   [**Fe3**] [**Kar**].

Suppose that $u : \mathbb{R}_+ \to \mathbb{R}_+$ is measurable and locally integrable. Let

$$a(x) := \int_0^x u(t)dt, \quad \overline{u}(p) := \int_0^\infty e^{-pt} u(t)dt,$$

then
1) the following are equivalent:
   a) $x \mapsto a(x)$ is regularly varying at ∞,
   b) $p \mapsto \overline{u}(p)$ is regularly varying at 0,
   c) $\exists\, \lim_{p \to 0} \frac{\overline{u}(p)}{a(\frac{1}{p})} \in \mathbb{R}_+$.
In this case, $\overline{u}(p) \sim \Gamma(1+\alpha) a(\frac{1}{p})$ as $p \to 0$ where $\alpha \geq 0$ is the (mutual) index of regular variation.
2) Now suppose that $\alpha > 0$.
   d) If $u(t)$ is regularly varying at ∞ with index $\alpha - 1$, then $a(t) \sim \frac{t u(t)}{\alpha}$ as $t \to \infty$.
   e) Conversely, if a is regularly varying at ∞ with index α, and u is monotone, then $u(t) \sim \frac{\alpha a(t)}{t}$ as $t \to \infty$.

INVERSE THEOREM   [**Fe3**] [**Kar**].

Suppose that $a : \mathbb{R}_+ \to \mathbb{R}_+$ is continuous, strictly increasing and regularly varying with index $\alpha > 0$, then $a^{-1} : \mathbb{R}_+ \to \mathbb{R}_+$ is regularly varying with index $\frac{1}{\alpha}$.

DEFINITION: MITTAG-LEFFLER DISTRIBUTION. Let $\alpha \in [0,1]$. The random variable Y_α on $\mathbb{R}_+$ has the *normalised Mittag-Leffler distribution of order α* if

$$E(e^{zY_\alpha}) = \sum_{p=0}^{\infty} \frac{\Gamma(1+\alpha)^p z^p}{\Gamma(1+p\alpha)}.$$

Note that $E(Y_\alpha) = 1$. Evidently $Y_1 = 1$ and the densities of Y_0 and $Y_{\frac{1}{2}}$ are given by

$$f_{Y_0}(y) = e^{-y}, \quad f_{Y_{\frac{1}{2}}}(y) = \frac{2}{\pi} e^{-\frac{y^2}{\pi}}.$$

Let T be a conservative, ergodic, measure preserving transformation of $(X, \mathcal{B}, m)$. Let $A \in \mathcal{B}$, $0 < m(A) < \infty$ and set

$$a_n(A) := \sum_{k=0}^{n-1} \frac{m(A \cap T^{-k}A)}{m(A)^2}, \quad u^A(\lambda) := \sum_{n=0}^{\infty} e^{-\lambda n} \frac{m(A \cap T^{-n}A)}{m(A)^2}.$$

The set A is called a *moment* set for T if

$$\sum_{n=0}^{\infty} e^{-\lambda n} \int_A S_n(1_A)^p dm \sim p! m(A)^{p+1} \frac{u(\lambda)^p}{\lambda} \text{ as } \lambda \to 0 \ \forall \, p \in \mathbb{N}.$$

3.6.4 DARLING KAC THEOREM [**Da-Kac**]. *Suppose that A is a moment set for T, and that $u^A(\lambda)$ is regularly varying with index $\alpha \in [0,1]$ as $\lambda \to 0$, then*

$$\frac{S_n^T}{a_n(A)} \xrightarrow{\mathfrak{d}} Y_\alpha$$

where Y_α has the normalised Mittag-Leffler distribution of order α.

PROOF. It follows from Karamata's Tauberian theorem (see [**Kar**], [**Fe3**]) that

$$\int_A S_n(1_A) dm \sim a(n) \text{ as } n \to \infty,$$

and that

$$\sum_{n=0}^{N} \int_A S_n(1_A)^p dm \sim \frac{p! m(A)^{p+1}}{\Gamma(2+p\alpha)} N u(\frac{1}{N})^p \text{ as } N \to \infty.$$

Since $\int_A S_n(1_A)^p dm$ increases in $n \geq 1$, we have that

$$\int_A S_n(1_A)^p dm \sim \frac{p! m(A)^{p+1}}{\Gamma(2+p\alpha)} (1+p\alpha) u(\frac{1}{n})^p \text{ as } n \to \infty,$$

$$= \frac{p! \Gamma(1+\alpha)^p m(A)^{p+1}}{\Gamma(1+p\alpha)} a(n);$$

equivalently

$$(\ddagger) \qquad \int_A \left(\frac{S_n(1_A)}{a(n)} \right)^p dm \to m(A)^{p+1} E(Y_\alpha^p) \text{ as } n \to \infty \ \forall \, p \geq 1.$$

Suppose that

$$\frac{S_{n_k}^T}{a(n_k)} \xrightarrow{\mathfrak{d}} Z.$$

We claim that $E(Z^p) = E(Y_\alpha^p) \ \forall \, p \geq 1$.

To prove our claim, we note that $\forall\, x > 0$, and $p \geq 1$,

$$\int_A \left(\frac{S_{n_k}^T}{a(n_k)} \right)^p \wedge x\, dm \to m(A)E((m(A)Z)^p \wedge x) \uparrow m(A)^{p+1}E(Z^p)$$

which shows that $E(Z^p) \leq E(Y_\alpha^p)\ \forall\, p \geq 1$.

To get the reverse inequality, note that by (‡) $\left\{ \left(\frac{S_n(1_A)}{a(n)} \right)^p : n \geq 1 \right\}$ is uniformly integrable on A for each $p \geq 1$. Thus given $\epsilon > 0$ and $p \geq 1$, $\exists\, x = x(p,\epsilon) > 0$ such that

$$E(Z^p \wedge x) > E(Z^p) - \epsilon,\ \&\ \int_A \left(\frac{S_n(1_A)}{a(n)} \right)^p \wedge x\, dm > \int_A \left(\frac{S_n(1_A)}{a(n)} \right)^p dm - \epsilon\ \forall\, n \geq 1$$

and we have

$$m(A)^{p+1}E(Y_\alpha^p) \leftarrow \int_A \left(\frac{S_{n_k}^T}{a(n_k)} \right)^p dm$$

$$< \int_A \left(\frac{S_{n_k}^T}{a(n_k)} \right)^p \wedge x\, dm + \epsilon$$

$$\to m(A)E((m(A)Z)^p \wedge x) + \epsilon$$

$$\leq m(A)^{p+1}E(Z^p) + \epsilon.$$

Our claim is established, and it follows that

$$E(e^{zZ}) = E(e^{zY_\alpha})$$

whence $Z = Y_\alpha$. □

REMARK. There is a converse to this result. It is shown in [**Da-Kac**] that if A is a moment set for the conservative, ergodic, measure preserving transformation T and for some random variable Y on $(0, \infty)$ and constants $a(n) \to \infty$,

$$\frac{S_n^T}{a(n)} \overset{\mathfrak{d}}{\to} Y,$$

then $u^A(\lambda)$ is regularly varying as $\lambda \to 0$.

§3.7 Pointwise dual ergodicity

A conservative, ergodic, measure preserving transformation is called *pointwise dual ergodic* if there are constants a_n such that

$$(3.7.1) \qquad \frac{1}{a_n} \sum_{k=0}^{n-1} \widehat{T}^k f \to \int_X f\, dm_T \text{ a.e. as } n \to \infty\ \forall\, f \in L^1(X_T).$$

REMARK. By exercise 1.3.4, Boole's transformation $(x \mapsto x - \frac{1}{x})$ is pointwise dual ergodic with return sequence $\frac{\sqrt{2n}}{\pi}$.

3.7.1 PROPOSITION [**A5**].
Suppose that T is pointwise dual ergodic, then T is rationally ergodic and the sequence a_n in (3.7.1) ia a return sequence for T.

PROOF.   Let $A \in \mathcal{B}$, $m(A) = 2$, then by Egorov's theorem, $\exists \, B \in \mathcal{B} \cap A$, and $M \in \mathbb{R}_+$ such that $m(B) = 1$, and

$$\frac{1}{a_n} \sum_{k=0}^{n-1} \widehat{T}^k 1_A \leq M \text{ on } B \,\forall\, n \in \mathbb{N}.$$

It follows that

$$\frac{1}{a_n} \sum_{k=0}^{n-1} \widehat{T}^k 1_A \leq \frac{1}{a_n} \sum_{k=0}^{n-1} \widehat{T}^k 1_B \leq M \text{ on } B \,\forall\, n \in \mathbb{N}.$$

We'll show that Renyi's inequality holds for $f = 1_B$. Set

$$S_n = \sum_{k=0}^{n-1} 1_B \circ T^k,$$

then

$$\int_B S_n \, dm = \int_B \left(\sum_{k=0}^{n-1} \widehat{T}^k 1_B \right) dm \leq M a_n$$

whence by bounded convergence,

$$\int_B S_n \, dm \sim a_n \text{ as } n \to \infty.$$

Also,

$$\int_B S_n^2 \, dm = \sum_{k,\ell=0}^{n-1} m(B \cap T^{-k} B \cap T^{-\ell} B)$$

$$= \sum_{k=0}^{n-1} m(B \cap T^{-k} B) + 2 \sum_{k=0}^{n-2} \sum_{\ell=k+1}^{n-1} m(B \cap T^{-k} B \cap T^{-\ell} B)$$

$$= \sum_{k=0}^{n-1} m(B \cap T^{-k} B) + 2 \sum_{k=0}^{n-2} \sum_{\ell=0}^{n-k-1} m(B \cap T^{-k} B \cap T^{-(k+\ell)} B)$$

$$= \int_B S_n \, dm + 2 \int_B \left(\sum_{k=0}^{n-2} \widehat{T}^k 1_B S_{n-k} \right) dm$$

$$\leq M a_n + M a_n^2 \leq M' a_n^2.$$

This implies that

$$\int_B \left(\frac{S_n}{a_n} \right) dm \to 1 \text{ as } n \to \infty$$

and

$$\int_B \left(\frac{S_n}{a_n} \right)^2 dm \leq M'' \forall \, n \in \mathbb{N}.$$

$\square$

3.7.2 THEOREM   [**A7**]. *Suppose that T is pointwise dual ergodic, and that $A \in \mathcal{B}_+$ satisfies*

$$\left\| \frac{1}{a_n} \sum_{k=0}^{n-1} \widehat{T}^k 1_A \right\|_{L^\infty(A)} \le M \;\; \forall\, n \ge 1,$$

then A is a moment set for T.

PROOF.

Define $a(p,n) : X \to \mathbb{Z}$ $(n, p \in \mathbb{N})$ by $a(0,n) \equiv 1$, and

$$a(p+1, n)(x) = \sum_{k=1}^{n} 1_A(T^k x) a(p, n-k)(T^k x).$$

Some elementary combinatorics show that

$$S_n(1_A)^p = \sum_{q=1}^{p} \gamma_p(q)\, a(q, n)$$

where $\gamma_1(q) = \delta_{1,q}$, and $\gamma_{p+1}(q) = q(\gamma_p(q) + \gamma_p(q-1))$. We see that $\gamma_p(q)$ is the number of ways of putting p balls into q cells without leaving any cell empty, and that in particular $\gamma_p(p) = p!$.

We have that

$$\sum_{n=1}^{\infty} e^{-\lambda n} \int_A a(p,n)\, dm = \int_A \sum_{n=1}^{\infty} e^{-\lambda n} \sum_{k=1}^{n} 1_A \circ T^k a(p-1, n-k) \circ T^k\, dm$$

$$= \sum_{n=1}^{\infty} e^{-\lambda n} \sum_{k=1}^{n} \int_A \widehat{T}_A^k a(p-1, n-k)\, dm$$

$$= \int_A \left(\sum_{k=1}^{\infty} e^{-\lambda k} \widehat{T}_A^k \right) \left(\sum_{n=0}^{\infty} e^{-\lambda n} a(p-1, n) \right) dm.$$

Under the assumptions, we may assume that $a_n = a_n(A)$, and we have

$$\sum_{n=0}^{\infty} e^{-\lambda n} \widehat{T}^n 1_A \sim m(A) u^A(\lambda) \;\; \text{a.e. as } \lambda \to 0$$

and

$$\sum_{n=0}^{\infty} e^{-\lambda n} \widehat{T}^n 1_A \le M u^A(\lambda) \text{ on } A \;\forall\; \lambda > 0.$$

It now follows that

$$\sum_{n=1}^{\infty} e^{-\lambda n} \int_A a(p,n)\, dm \le M u^A(\lambda) \sum_{n=0}^{\infty} e^{-\lambda n} \int_A a(p-1, n)\, dm,$$

whence

$$\sum_{n=1}^{\infty} e^{-\lambda n} \int_A a(p,n)\, dm = O\left(\frac{u(\lambda)^p}{\lambda} \right) \text{ as } \lambda \to 0.$$

From all this, we conclude that for $p \geq 1$,

$$\sum_{n=0}^{\infty} e^{-\lambda n} | \int_A S_n(1_A)^p dm - p! \int_A a(p,n) dm | \leq \sum_{q=1}^{p} \gamma_p(q) \sum_{n=0}^{\infty} e^{-\lambda n} \int_A a(q,n) dm$$

$$= O\left(\frac{u(\lambda)^{p-1}}{\lambda}\right) \text{ as } \lambda \to 0,$$

whence in order to show that A is a moment set for T, it suffices to establish
$u_p(\lambda) := \sum_{n=0}^{\infty} e^{-\lambda n} \int_A a(p,n) dm \sim m(A)^{p+1} \frac{u(\lambda)^p}{\lambda}$ as $\lambda \to 0$ $\forall p \in \mathbb{N}$.

We do this by means of two inequalities:

(1) $$\liminf_{\lambda \to 0} \frac{\lambda u_p(\lambda)}{u(\lambda)^p} \geq m(A)^{p+1} \quad \forall p \in \mathbb{N}$$

and

(2) $$\limsup_{\lambda \to 0} \frac{u_{p+1}(\lambda)}{u(\lambda) u_p(\lambda)} \leq m(A) \quad \forall p \in \mathbb{N}.$$

To prove (1), choose $\epsilon > 0$. By Egorov's theorem, $\exists$ sets $A_j \in \mathcal{B}$ $(j \geq 0)$ and $\lambda_j \downarrow 0$ such that $A = A_0 \supset A_1 \supset \ldots$, $m(A_j) > (1-\epsilon)m(A)$ $\forall j \geq 1$ and $\sum_{n=0}^{\infty} e^{-\lambda n} \widehat{T}^n 1_{A_j} > (1-\epsilon)m(A)u^A(\lambda)$ on A_{j+1} $\forall \lambda < \lambda_j$, $j \geq 1$.

We have for $p \geq 2$,

$$u_p(\lambda) \sim \int_A \left(\sum_{k=1}^{\infty} e^{-\lambda k} \widehat{T}_A^k\right)\left(\sum_{n=0}^{\infty} e^{-\lambda n} a(p-1,n)\right) dm$$

$$\geq \int_{A_1} \left(\sum_{k=1}^{\infty} e^{-\lambda k} \widehat{T}_A^k\right)\left(\sum_{n=0}^{\infty} e^{-\lambda n} a(p-1,n)\right) dm$$

$$> (1-\epsilon)m(A)u(\lambda) \int_{A_1} \left(\sum_{n=0}^{\infty} e^{-\lambda n} a(p-1,n)\right) dm \quad \forall \lambda < \lambda_1.$$

Continuing, we have for any $j \geq 1$ and $q \geq 1$,

$$\sum_{n=1}^{\infty} e^{-\lambda n} \int_{A_j} a(q,n) dm$$

$$= \int_{A_j} \sum_{n=1}^{\infty} e^{-\lambda n} \sum_{k=1}^{n} 1_A \circ T^k a(q-1,n-k) \circ T^k dm$$

$$= \sum_{n=1}^{\infty} e^{-\lambda n} \sum_{k=1}^{n} \int_A \widehat{T}^k 1_{A_j} a(q-1,n-k) dm$$

$$= \int_A \left(\sum_{k=1}^{\infty} e^{-\lambda k} \widehat{T}^k 1_{A_j}\right)\left(\sum_{n=0}^{\infty} e^{-\lambda n} a(q-1,n)\right) dm$$

$$\geq \int_{A_{j+1}} \left(\sum_{k=1}^{\infty} e^{-\lambda k} \widehat{T}^k 1_{A_j}\right)\left(\sum_{n=0}^{\infty} e^{-\lambda n} a(q-1,n)\right) dm$$

$$> (1-\epsilon)m(A)u(\lambda) \int_{A_{j+1}} \left(\sum_{n=0}^{\infty} e^{-\lambda n} a(q-1,n)\right) dm \quad \forall \lambda < \lambda_j.$$

Putting it all together, we obtain that for $\lambda < \lambda p + 1$,

$$u_p(\lambda) > ((1-\epsilon)m(A)u(\lambda))^{p-1} \int_{A_p} \left(\sum_{n=0}^{\infty} e^{-\lambda n} a(1,n)\right) dm > ((1-\epsilon)m(A))^p \frac{u(\lambda)^p}{\lambda}$$

establishing (1).

To prove (2), again choose $\epsilon > 0$ and obtain (using Egorov's theorem) a set $A' \in \mathcal{B}$ $A' \subset A$ with $m(A \setminus A') < \epsilon m(A)$ and $\lambda_0 > 0$ such that

$$\sum_{n=0}^{\infty} e^{-\lambda n} \widehat{T}^n 1_A < (1+\epsilon)m(A)u^A(\lambda) \text{ on } A' \ \forall \ \lambda < \lambda_0.$$

We have

$$u_{p+1}(\lambda)$$
$$\sim \int_{A'} \left(\sum_{k=1}^{\infty} e^{-\lambda k} \widehat{T}_A^k\right)\left(\sum_{n=0}^{\infty} e^{-\lambda n} a(p,n)\right) dm +$$
$$\int_{A \setminus A'} \left(\sum_{k=1}^{\infty} e^{-\lambda k} \widehat{T}_A^k\right)\left(\sum_{n=0}^{\infty} e^{-\lambda n} a(p,n)\right) dm$$
$$\leq (1+\epsilon)m(A)u^A(\lambda) \int_{A'} \left(\sum_{n=0}^{\infty} e^{-\lambda n} a(p,n)\right) dm +$$
$$Mu^A(\lambda) \int_{A \setminus A'} \left(\sum_{n=0}^{\infty} e^{-\lambda n} a(p,n)\right) dm$$
$$\leq (1+\epsilon)m(A)u^A(\lambda)u_p(\lambda) +$$
$$Mu^A(\lambda) \int_{A \setminus A'} \left(\sum_{n=0}^{\infty} e^{-\lambda n} a(p,n)\right) dm$$

and we must show that

$$\frac{1}{u_p(\lambda)} \int_B \left(\sum_{n=0}^{\infty} e^{-\lambda n} a(p,n)\right) dm$$

(3) $\qquad \to 0$ as $m(B) \to 0$, $B \subset A$ uniformly in $\lambda > 0 \ \forall \ p \geq 1$

Note that

$$\left(\sum_{n=0}^{\infty} e^{-\lambda n} a(p,n)\right)^2 \leq 2 \sum_{0 \leq \ell \leq k} e^{-\lambda k} e^{-\lambda \ell} a(p,k)a(p,\ell)$$
$$\leq 2 \sum_{0 \leq \ell \leq k} e^{-\lambda k} e^{-\lambda \ell} a(p,k)^2$$
$$\leq \frac{2e}{\lambda} \sum_{k \geq 0} e^{-\lambda k} a(p,k)^2.$$

Now,

$$a(p,k)^2 \leq S_k(1_A)^{2p} = \sum_{q=1}^{2p} \gamma_{2p}(q)a(2q,k) \leq M_p a(2p,k),$$

where $M_p := \sum_{q=1}^{2p} \gamma_{2p}(q)$, whence

$$\left(\sum_{n=0}^{\infty} e^{-\lambda n} a(p,n)\right)^2 \leq \frac{2eM_p}{\lambda} \sum_{k \geq 0} e^{-\lambda k} a(2p,k),$$

and

$$\int_A \left(\sum_{n=0}^{\infty} e^{-\lambda n} a(p,n)\right)^2 \leq \frac{2eM_p}{\lambda} \int_A \sum_{k \geq 0} e^{-\lambda k} a(2p,k) dm$$

$$= \frac{2eM_p}{\lambda} u_{2p}(\lambda) \leq \frac{2eM_p}{\lambda} M^{2p} \frac{u(\lambda)^{2p}}{\lambda}$$

$$\leq M_p' u_p(\lambda)^2$$

Thus by the Cauchy-Schwartz inequality,

$$\frac{1}{u_p(\lambda)} \int_B \left(\sum_{n=0}^{\infty} e^{-\lambda n} a(p,n)\right) dm \leq \sqrt{m(B)} \left\| \frac{1}{u_p(\lambda)} \sum_{n=0}^{\infty} e^{-\lambda n} a(p,n) \right\|_{L^2(A)}$$

$$\leq M_p'^{\frac{1}{2}} \sqrt{m(B)}.$$

$\square$

3.7.3 COROLLARY [A7].

Suppose that T is pointwise dual ergodic, and that $a_n(T)$ is regularly varying with index $\alpha \in [0,1]$ as $n \to \infty$, then

$$\frac{S_n^T}{a_n(T)} \xrightarrow{\mathfrak{d}} Y_\alpha$$

where Y_α has the normalised Mittag-Leffler distribution of order α.

PROOF.

This follows from theorem 3.7.2 and the Darling-Kac theorem. $\square$

REMARK. Corollary 3.7.3 completes the proof of the distributional convergence advertised for Boole's transformation in the preface, pointwise dual ergodicity having been established, and the return sequence found, in exercise 1.3.4.

DEFINITION: DARLING-KAC SETS. A set $A \in \mathcal{B}$, $0 < m(A) < \infty$ is called a *Darling- Kac set* (for the conservative, ergodic, measure preserving transformation T of $(X, \mathcal{B}, m)$)if there are constants $a_n > 0$ such that

$$\frac{1}{a_n} \sum_{k=0}^{n-1} \widehat{T}^k 1_A \to m(A) \text{ almost uniformly on } A.$$

One way a set $A \in \mathcal{B}$, $0 < m(A) < \infty$ can be a Darling- Kac set for T is when $\widehat{T}^n 1_A$ is constant on $A \, \forall \, n \geq 1$. This is the case when T is a one-sided Markov shift and A is the the event of being in a certain state at a certain time (see [**Da-Kac**] and §4.5).

The situation $\widehat{T}^n 1_A$ constant on $A \, \forall \, n \geq 1$ is characterised by the (statistical) independence of φ_A and $T_A^{-1}(\mathcal{B} \cap A)$. Sufficiency of this condition can be shown as in the next lemma which gives a more general condition for a set to be a Darling-Kac set. For necessity see §5.2.

DEFINITION: CONTINUED FRACTION MIXING. Let T be a probability preserving transformation of $(X, \mathcal{B}, m)$, and let $\alpha \subset \mathcal{B}$ be a countable generating partition.

The process $(X, \mathcal{B}, m, T, \alpha)$ is called *continued fraction mixing* if $\exists \, \epsilon_n \downarrow 0$ such that

$$m(a \cap T^{-(k+n)}B) \leq (1 + \epsilon_n)m(a)m(B) \; \forall \; k \geq 1, \; a \in \alpha_0^{k-1}, \; B \in \mathcal{B}, \; n \geq 1;$$

and $\exists \, n_0 \geq 1$ such that

$$m(a \cap T^{-(k+n)}B) \geq (1 - \epsilon_n)m(a)m(B) \; \forall \; k \geq 1, \; a \in \alpha_0^{k-1}, \; B \in \mathcal{B}, \; n \geq n_0.$$

3.7.4 LEMMA [A10].

Let T be a conservative, ergodic measure preserving transformation of $(X, \mathcal{B}, m)$, and let $A \in \mathcal{B}$, $0 < m(A) < \infty$.

Suppose $\alpha \subset \mathcal{B} \cap A$ is a countable partition generating $\mathcal{B}$ under T_A, and is such that $\varphi = \varphi_A$ is α-measurable, and $(A, \mathcal{B} \cap A, m|_A, T_A, \alpha)$ is continued fraction mixing,

then A is a Darling-Kac set for T.

PROOF. Assume that $A \in \mathcal{B}$ and $\alpha \subset \mathcal{B} \cap A$ are as assumed. There is no loss of generality in assuming that $m(A) = 1$.

The continued fraction mixing of $(A, \mathcal{B} \cap A, m|_A, T_A, \alpha)$ implies that

$$\widehat{T}_A^{k+n} 1_a \leq (1 + \epsilon_n)m(a) \; \forall \; k \geq 1, \; a \in \alpha_0^{k-1}, \; B \in \mathcal{B}, \; n \geq 1;$$

and

$$\widehat{T}_A^{k+n} 1_a \geq (1 - \epsilon_n)m(a) \; \forall \; k \geq 1, \; a \in \alpha_0^{k-1}, \; B \in \mathcal{B}, \; n \geq n_0.$$

To calculate $\widehat{T}^n 1_A$ in terms of powers of $\widehat{T}_A$, note that for $B \in \mathcal{B} \cap A$, $n \geq 1$,

$$A \cap T^{-n}B = \bigcup_{k=1}^{n} [\varphi_k = n] \cap T_A^{-k}B$$

whence

$$\widehat{T}^n 1_A = \sum_{k=1}^{n} \widehat{T}_A^k 1_{[\varphi_k = n]}$$

and

$$\sum_{k=1}^{n} \widehat{T}^k 1_A = \sum_{k=1}^{n} \widehat{T}_A^k 1_{[\varphi_k \leq n]}$$

The Darling-Kac property is established by means of two inequalities, in which $a(n) := \sum_{k=1}^{n} m(A \cap T^{-k}A)$:

(1)
$$\sum_{k=1}^{n} \widehat{T}^k 1_A \leq p + (1 + \epsilon_{p+1})a(n) \; \forall \; n, p \geq 1;$$

(2)
$$\sum_{k=1}^{n} \widehat{T}^k 1_A \geq \left(1 - \epsilon_{p+1} - (1 + \epsilon_1)^2 m([\varphi_A \geq q]) \right) a(n) - (1 + \epsilon_1)^2 q - p \; \forall \; n, p, q \geq 1.$$

Proof of (1)

$$\sum_{k=1}^{n} \widehat{T}^{k} 1_A = \sum_{k=1}^{n} \widehat{T}_A^{k} 1_{[\varphi_k \leq n]}$$

$$\leq \sum_{k=1}^{n+p} \widehat{T}_A^{k} 1_{[\varphi_k \leq n]}$$

$$\leq p + \sum_{k=1}^{n} \widehat{T}_A^{p+k} 1_{[\varphi_{p+k} \leq n]}$$

$$\leq p + \sum_{k=1}^{n} \widehat{T}_A^{p+k} 1_{[\varphi_k \leq n]}$$

$$\leq p + \sum_{k=1}^{n} (1 + \epsilon_{p+1}) m([\varphi_k \leq n]) \quad \because [\varphi_k \leq n] \in \sigma(\alpha_0^{n-1})$$

$$= p + (1 + \epsilon_{p+1}) a(n).$$

Proof of (2)

$$\sum_{k=1}^{n} \widehat{T}^{k} 1_A \geq \sum_{k=1}^{n} \widehat{T}_A^{p+k} 1_{[\varphi_{p+k} \leq n]} - p$$

$$= \sum_{k=1}^{n} \widehat{T}_A^{p+k} 1_{[\varphi_k \leq n]} - \sum_{k=1}^{n} \widehat{T}_A^{p+k} 1_{[\varphi_k \leq n \leq \varphi_{p+k}]} - p$$

The main term is estimated as in the proof of (1),

$$\sum_{k=1}^{n} \widehat{T}_A^{p+k} 1_{[\varphi_k \leq n]} \geq (1 - \epsilon_{p+1}) a(n).$$

$$\sum_{k=1}^{n} \widehat{T}_A^{p+k} 1_{[\varphi_k \leq n \leq \varphi_{p+k}]} \leq (1 + \epsilon_1) \sum_{k=1}^{n} m([\varphi_k \leq n \leq \varphi_{p+k}])$$

$$= (1 + \epsilon_1) \sum_{k=1}^{n} \sum_{\ell=1}^{n} m([\varphi_k = \ell, \; \varphi_p \circ T_A^k > n - \ell])$$

$$\leq (1 + \epsilon_1)^2 \sum_{k=1}^{n} \sum_{\ell=1}^{n} m([\varphi_k = \ell]) m([\varphi_p > n - \ell])$$

$$= (1 + \epsilon_1)^2 \left(\sum_{k=1}^{n} \sum_{\ell=1}^{n-q} + \sum_{k=1}^{n} \sum_{\ell=n-q+1}^{n} \right)$$

$$= I + II.$$

$$I = \sum_{k=1}^{n} \sum_{\ell=1}^{n-q} m([\varphi_k = \ell]) m([\varphi_p > n - \ell])$$

$$\leq m([\varphi_p \circ T_A^k > q]) \sum_{k=1}^{n} m([\varphi_k \leq n - q])$$

$$\leq m([\varphi_p \circ T_A^k > q]) a(n).$$

$$II = \sum_{k=1}^{n} \sum_{\ell=n-q+1}^{n} m([\varphi_k = \ell]) m([\varphi_p > n - \ell])$$

$$\leq \sum_{k=1}^{n} m([n - q < \varphi_p \leq n])$$

$$\leq \sum_{k=1}^{n} m([\varphi_p \leq n]) - \sum_{k=1}^{n-q} m([\varphi_p \leq n - q])$$

$$= a(n) - a(n - q)$$

$$\leq q.$$

□

3.7.5 PROPOSITION [**A10**].

Suppose that T is a conservative, ergodic, measure preserving transformation of $(X, \mathcal{B}, m)$, and suppose that for some $A, B \in \mathcal{B}$, $0 < m(A), m(B) < \infty$, and $a(N) \uparrow \infty$,

$$\frac{1}{a(N)} \sum_{n=0}^{N} \widehat{T}^n 1_B \to m(B) \ on \ A,$$

then T is pointwise dual ergodic.

In particular, if T has a Darling- Kac set, then T is pointwise dual ergodic.

PROOF. For $A \in \mathcal{B}$, $0 < m(A) < \infty$, set

$$A_0 := A, \ A_n := A \setminus \bigcup_{j=1}^{n} T^{-j} A \quad (n \geq 1).$$

Check that

$$\sum_{n=0}^{\infty} \widehat{T}^n 1_{A_n} \equiv 1 \ \text{a.e.},$$

and $\forall \ B \in \mathcal{B}$, $0 < m(B) < \infty$:

$$\sum_{n=0}^{N} \widehat{T}^n 1_B = \sum_{k=0}^{N} \widehat{T}^k (1_{A_k} \sum_{n=0}^{N-k} \widehat{T}^n 1_B) + \sum_{n=0}^{N} \widehat{T}^n 1_{A \setminus \bigcup_{j=0}^{n} T^{-j} A}.$$

By possibly passing to a subset of A, we can assume (by Egorov's theorem) that

$$\frac{1}{a(N)} \sum_{n=0}^{N} \widehat{T}^n 1_B \to m(B) \ \text{uniformly on} \ A.$$

Let $M_k \uparrow 1$ be such that

$$\sum_{n=0}^{N} \widehat{T}^n 1_B \geq M_N m(B) a(N) \text{ on } A \;\; \forall \, N \geq 1,$$

then

$$\begin{aligned}
\sum_{n=0}^{N} \widehat{T}^n 1_B &= \sum_{k=0}^{N} \widehat{T}^k \big(1_{A_k} \sum_{n=0}^{N-k} \widehat{T}^n 1_B\big) + \sum_{n=0}^{N} \widehat{T}^n 1_{A \setminus \bigcup_{j=0}^{n} T^{-j} A} \\
&\geq \sum_{k=0}^{N} \widehat{T}^k \big(1_{A_k} \sum_{n=0}^{N-k} \widehat{T}^n 1_B\big) \\
&\geq a(N) \sum_{k=0}^{N} \frac{M_{N-k} a(N-k)}{a(N)} \widehat{T}^k 1_{A_k} \\
&\sim a(N) \text{ a.e. as } N \to \infty.
\end{aligned}$$

It now suffices show that

$$\limsup_{N \to \infty} \frac{1}{a(N)} \sum_{n=0}^{N} \widehat{T}^n 1_B \leq (1+\epsilon) m(B) \text{ a.e. } \forall \, \epsilon > 0$$

for then,

$$\frac{1}{a(N)} \sum_{n=0}^{N} \widehat{T}^n 1_B \to m(B) \text{ a.e.,}$$

and pointwise dual ergodicity follows from Hurewicz's ergodic theorem.

Let $\epsilon > 0$. By Hurewicz's ergodic theorem,

$$\frac{1}{a(N)} \sum_{n=0}^{N} \widehat{T}^n 1_A \to m(A) \text{ a.e. on } A,$$

and by Egorov's theorem, $\exists \, A' \in \mathcal{B}, \; A' \subset A$ such that $m(A') > \frac{1}{1+\epsilon} m(A)$, and

$$\frac{1}{a(N)} \sum_{n=0}^{N} \widehat{T}^n 1_A \to m(A) \text{ uniformly on } A'.$$

Let $\mathcal{M}_k \downarrow 1$ be such that

$$\sum_{n=0}^{N} \widehat{T}^n 1_A \leq M_N m(A) a(N) \text{ on } A'.$$

It follows that

$$\sum_{n=0}^{N} \widehat{T}^n 1_{A'} = \sum_{k=0}^{N} \widehat{T}^k (1_{A'_k} \sum_{n=0}^{N-k} \widehat{T}^n 1_{A'})$$

$$\leq \sum_{k=0}^{N} \widehat{T}^k (1_{A'_k} \sum_{n=0}^{N-k} \widehat{T}^n 1_A)$$

$$\leq \sum_{k=0}^{N} \widehat{T}^k M_{N-k} m(A) a(N-k) 1_{A'_k}$$

$$\leq m(A) a(N) \sum_{k=0}^{N} M_{N-k} \widehat{T}^k (1_{A'_k}$$

$$\sim m(A) a(N) \text{ a.e.}$$

$$\leq (1+\epsilon) m(A').$$

Using Hurewicz's theorem again, we see that

$$\limsup_{N \to \infty} \frac{1}{a(N)} \sum_{n=0}^{N} \widehat{T}^n 1_B \leq (1+\epsilon) m(B) \text{ a.e..}$$

$\square$

3.7.6 PROPOSITION. *A factor of a pointwise dual ergodic transformation is also pointwise dual ergodic.*

PROOF. Suppose that T is p.d.e, and that $\pi : T \xrightarrow{1} S$. It is easily checked that

$$\widehat{S}^n f = E(\widehat{T}^n (f \circ \pi)|\pi) \quad (f \in L^1(m_S)$$

and

$$E(\widehat{T}^n g|\pi) = \widehat{S}^n E(g|\pi)) \quad (g \in L^1(m_T)).$$

where for $g \in L^1(m_T)$, $E(g|\pi) \in L^1(m_S)$ is as in §3.1.

Let $a(n) \uparrow \infty$ be the return sequence of T, and suppose that $A \in \mathcal{B}$, $0 < m(A) < \infty$, then

$$\frac{1}{a(n)} \sum_{k=0}^{n-1} \widehat{T}^k 1_A \to m(A) \text{ a.e.},$$

and by Egorov's theorem, $\exists B \in \mathcal{B}$, $B \subset A$, $m(B) > 0$ so that this convergence is uniform on B. It follows that $\exists M > 0$ such that

$$\frac{1}{a(n)} \sum_{k=0}^{n-1} \widehat{T}^k 1_B \leq M \text{ on } B \ \forall \ n \geq 1,$$

whence

$$\frac{1}{a(n)} \sum_{k=0}^{n-1} \widehat{T}^k 1_B \leq M \text{ a.e. } B \ \forall \ n \geq 1.$$

By pointwise dual ergodicity,

$$\frac{1}{a(n)} \sum_{k=0}^{n-1} \widehat{T}^k 1_B \to m(B) \text{ a.e.}$$

and it follows from the bounded convergence theorem for conditional expectations that

$$\frac{1}{a(n)} \sum_{k=0}^{n-1} \widehat{S}^k E(1_B|\pi) = \frac{1}{a(n)} \sum_{k=0}^{n-1} E(\widehat{T}^k 1_B|\pi) = E\left(\frac{1}{a(n)} \sum_{k=0}^{n-1} \widehat{T}^k 1_B \Big|\pi\right) \to m(B) \text{ a.e.}$$

□

3.7.7 COROLLARY.

A pointwise dual ergodic transformation of an infinite measure space has no invertible factors.

The tail field of a pointwise dual ergodic transformation of an infinite measure space contains no sets of positive, finite measure.

PROOF. By theorem 2.4.2, no invertible measure preserving transformation of an infinite measure space can be pointwise dual ergodic, and by proposition 3.7.6, any factor of a pointwise dual ergodic transformation is itself pointwise dual ergodic. The conclusion is that no factor of a pointwise dual ergodic transformation of an infinite measure space can be invertible.

Let T be a conservative, ergodic, measure preserving transformation of $(X, \mathcal{B}, m)$. The *tail* of T is $\mathfrak{T} := \bigcap_{n=1}^{\infty} T^{-n}\mathcal{B}$ and is strictly T-invariant. In case $\exists\, A \in \mathfrak{T}$, $0 < m(A) < \infty$, then $\mathfrak{T}$ is σ-finite as

$$X = A \cup \bigcup_{n=1}^{\infty} T^{-n}A \setminus \bigcup_{k=0}^{n-1} T^{-k}A,$$

a countable union of sets in $\mathfrak{T}$ of finite measure. As a strictly T-invariant, sub-σ-algebra of $\mathcal{B}_T$, $\mathfrak{T}$ is associated by proposition 3.1.1 to an invertible factor of T which cannot exist!   □

EXERCISE 3.7.1: POINTWISE DUAL ERGODICITY WITHOUT INVARIANT MEASURE.

Let T be a conservative, ergodic, non-singular transformation of $(X, \mathcal{B}, \mu)$ with the property that there are constants a_n such that $\frac{1}{a_n} \sum_{k=0}^{n-1} \widehat{T}^k f$ converges a.e. for some $f \in L^1(m)_+$ to a non-zero limit.

1) Show that

$$\frac{1}{a_n} \sum_{k=0}^{n-1} \widehat{T}^k f \to h \int_X f d\mu \text{ a.e. } \forall\, f \in L^1(m)_+$$

where $h : X \to \mathbb{R}$ is measurable, $h > 0$ and $\widehat{T}h = h$.

2) If $dm = hd\mu$, then $(X, \mathcal{B}, m, T)$ ia a pointwise dual ergodic measure preserving transformation.

§3.8 Wandering Rates

Let T be a conservative, ergodic, measure preserving transformation of $(X, \mathcal{B}, m)$, and let $A \in \mathcal{B}$, $0 < m(A) < \infty$. The *wandering rate* of A is the sequence

$$L_A(n) := m\left(\bigcup_{k=0}^{n} T^{-k}A\right).$$

Note that $L_A(n) \uparrow m(X)$ as $n \to \infty$. The wandering rate of A measures the amount of X which visits X up to time n. A calculation shows that

$$L_A(n) = \int_A (\varphi_A \wedge n) dm.$$

There are connections between wandering rates and expected sojourn times.

3.8.1 THEOREM [**A7**].
Let T be a conservative, ergodic, measure preserving transformation of $(X, \mathcal{B}, m)$ and let $A \in \mathcal{B}$, $0 < m(A) < \infty$, then

$$\lim_{n \to \infty} \frac{L_A(n)}{n} \sum_{k=0}^{n-1} \int_X f \circ T^k dP \geq \frac{1}{2} \int_X f dm \quad \forall \, f \in L^1(m)_+, \, \& \, P \in \mathcal{P}(X), \, P << m.$$

In particular, if T is weakly homogeneous, then

$$\liminf_{n \to \infty} \frac{L_A(n) a_n(T)}{n} \geq \frac{1}{2}.$$

PROOF.
For $f \in L^1(m)_+$, $P \in \mathcal{P}(X)$, $P << m$, and $0 < y < 1$, set

$$\alpha_n(P, f, y) := \sup\{t \geq 0 : \; P([S_n(f) \geq t]) \geq y\}.$$

We claim first that for $f, g \in L^1(m)_+$, $P, Q \in \mathcal{P}(X)$, $P, Q << m$, and $0 < u < v < 1$;

(1)
$$\liminf_{n \to \infty} \frac{\alpha_n(P, f, u)}{\alpha_n(Q, g, v)} \geq \frac{\int_X f dm}{\int_X g dm}.$$

If not $\exists \, n_k \to \infty$ such that

$$\frac{\alpha_{n_k}(P, f, u)}{\alpha_{n_k}(Q, g, v)} \to \kappa < \frac{\int_X f dm}{\int_X g dm}.$$

Possibly passing to a subsequence, we may assume by corollary 3.6.2 that in addition

$$\frac{S_{n_k}^T}{\alpha_{n_k}(Q, g, v)} \xrightarrow{\partial} Y$$

for some random variable Y on $[0, \infty]$. Let $c(x) =$ Prob.$(Y \geq x)$, and let Λ be the (co-countable) set of continuity points of c.
On the one hand, we have that

$$Q([S_{n_k}(g) \geq x\alpha_{n_k}(Q, g, v)]) \to c(x/\int_X g dm)$$

as $k \to \infty$ whenever $x/\int_X g dm \in \Lambda$,

whence $c(x/\int_X g\,dm) \geq v\ \forall\ x < 1,\ x \in \Lambda$.
On the other hand,

$$P([S_{n_k}(f) \geq x\alpha_{n_k}(Q,g,v)]) \to c(x/\int_X f\,dm)$$

as $k \to \infty$ whenever $x/\int_X f\,dm \in \Lambda$,

whence $\forall\ \epsilon > 0$ such that $(1-\epsilon)\frac{\int_X f\,dm}{\int_X g\,dm} \in \Lambda$:

$$P([S_{n_k}(f) \geq (1-\epsilon)\frac{\int_X f\,dm}{\int_X g\,dm}\alpha_{n_k}(Q,g,v)]) \to c((1-\epsilon)/\int_X g\,dm) > u$$

hence for k large enough

$$\alpha_{n_k}(P,f,u) \geq (1-\epsilon)\frac{\int_X f\,dm}{\int_X g\,dm}\alpha_{n_k}(Q,g,v)$$

contradicting

$$\kappa < \frac{\int_X f\,dm}{\int_X g\,dm}.$$

Next, we claim that for $A \in \mathcal{B},\ 0 < m(A) < \infty$,

(2) $$\frac{L_A(n+1)(1+\alpha_n(m_A,1_A,y))}{n+1} \geq (1-y)m(A)\quad (n \geq 1,\ 0 < y < 1).$$

To see this, fix $0 < y < 1$, write $\alpha_n := \alpha_n(m_A,1_A,y)$, let T_A be the induced transformation on A, and let φ be the return time function of T to A.
    Set

$$\omega_n := \sup\{t \geq 0 :\ m([\varphi_n \geq t]|A) \geq (1-y)\}.$$

We have that

$$1 - y \leq \int_0^1 m_a([\varphi_n \geq t\omega_n])dt = \frac{1}{\omega_n}\int_A (\varphi_n \wedge \omega_n)dm_A$$

$$\leq \frac{1}{\omega_n}\int_A \sum_{k=0}^{n-1}(\varphi \circ T_A^k \wedge \omega_n)dm_A = \frac{n}{\omega_n}\int_A (\varphi \wedge \omega_n)dm_A$$

$$= \frac{nL_A(\omega_n)}{\omega_n m(A)}.$$

Now note that

$$m_A([\varphi_{1+\alpha_n} \leq n]) \leq m_A([S_n(1_A) \geq 1 + \alpha_n]) < y$$

whence $m_A([\varphi_{1+\alpha_n} \geq n+1]) > 1 - y$ and $\omega_{1+\alpha_n} \geq n+1$.
    Thus

$$(1-y)m(A) \leq (1+\alpha_n)\frac{L_A(\omega_{1+\alpha_n})}{\omega_{1+\alpha_n}} \leq (1+\alpha_n)\frac{L_A(n+1)}{n+1}.$$

To conclude: for $A \in \mathcal{B}$, $0 < m(A) < \infty$, set $a(n) := \frac{L_A(n)}{n}$. Let $f \in L^1(m)_+$ and $P \in \mathcal{P}(X)$, $P << m$, then

$$\frac{1}{a(n)} \sum_{k=0}^{n-1} \int_X f \circ T^k dP = \frac{1}{a(n)} \int_X S_n(f) dP$$

$$= \int_X f dm \int_0^\infty P([S_n(f) \geq t \int_X f dm a(n)]) dt$$

$$\geq \int_X f dm \int_0^1 P([S_n(f) \geq (1-y) \int_X f dm a(n)]) dy.$$

It follows from (1)and (2) that

$$\liminf_{n \to \infty} P([S_n(f) \geq (1-y) \int_X f dm a(n)]) \geq y \ \forall \ 0 < y < 1$$

whence by Fatou's lemma

$$\liminf_{n \to \infty} \int_0^1 P([S_n(f) \geq (1-y) \int_X f dm a(n)]) \geq \frac{1}{2}$$

and

$$\liminf_{n \to \infty} \frac{1}{a(n)} \sum_{k=0}^{n-1} \int_X f \circ T^k dP \geq \frac{\int_X f dm}{2}.$$

Supposing T to be weakly homogeneous, we have that $\forall \ n_k \to \infty$, $\exists \ m_\ell = n_{k_\ell} \to \infty$ and $A \in \mathcal{B}$, $m(A) = 1$ such that

$$\sum_{k=0}^{m_\ell-1} m(A \cap T^{-k}A) \sim a_{m_\ell}(T)$$

whence

$$\liminf_{\ell \to \infty} \frac{a_{m_\ell}(T)}{a(m_\ell)} \geq \frac{1}{2}.$$

$\square$

By theorem 3.8.1, for weakly homogeneous measure preserving transformations, there are minimal rates for the growth of wandering rates to ∞.

The following proposition shows that there are never maximal rates for their growth to ∞.

3.8.2 PROPOSITION.
Let T be a conservative, ergodic, measure preserving transformation of $(X, \mathcal{B}, m)$, and let $a_n \to \infty$, $a_n = o(n)$ as $n \to \infty$, then $\exists \ A \in \mathcal{B}$, $0 < m(A) < \infty$ such that

$$L_A(n) \geq a_n \ \forall \ n \geq 1.$$

PROOF. Set

$$c_n := \sup_{k \geq n} \frac{a_k}{k}, \ \& \ a(n) := \sum_{k=1}^{n} c_k,$$

then $c_n \downarrow 0$ and

$$a(n) \geq nc_n \geq a_n.$$

Let $f_n := c_n - c_{n+1}$. We claim that $\exists A_n \in \mathcal{B}$ $(n \geq 1)$ such that $\{T^{-k}A_n :\}_{k=0}^{n}$ are disjoint $\forall \ n \geq 1$, $m(A_n) = 2f_n$, and

$$m\left(\bigcup_{0 \leq k \leq n \leq N-1} T^{-k}A_n \cap \bigcup_{\ell=0}^{N} T^{-\ell}A_N \right) \leq \frac{f_N}{2} \ \forall \ N \geq 1.$$

To see this inductively, suppose that $\{A_n : 1 \leq n \leq N-1\}$ have been chosen. Use Rokhlin's tower theorem to choose $B_N \in \mathcal{B}$ such that $\{T^{-k}B_N\}_{k=0}^{N}$ are disjoint, and $m(B_N) = f_N$. It follows from the ergodic theorem that $\exists \ j \geq 1$ so that

$$m\left(\bigcup_{0 \leq k \leq n \leq N-1} T^{-k}A_n \cap T^{-j} \bigcup_{\ell=0}^{N} T^{-\ell}B_N \right) \leq \frac{f_N}{2}.$$

The set $A_N := T^{-j}B_N$ is as desired.

Set $A = \bigcup_{n=1}^{\infty} A_n$. It follows that

$$m(A) = \sum_{n=1}^{\infty} m(A_n) = c_1 < \infty,$$

and for fixed $N \geq 1$,

$$\bigcup_{k=1}^{N} T^{-k}A \supset \bigcup_{n=1}^{N} \bigcup_{k=1}^{n} T^{-k}A_n \cup \bigcup_{n=N+1}^{\infty} \bigcup_{k=1}^{N} T^{-k}A_n := \bigcup_{n=1}^{\infty} C_n$$

where

$$C_n = \begin{cases} \bigcup_{k=1}^{n} T^{-k}A_n & 1 \leq n \leq N, \\ \bigcup_{k=1}^{N} T^{-k}A_n & n > N. \end{cases}$$

We have that

$$L_A(N) = m\left(\bigcup_{k=1}^{N} T^{-k}A \right) \geq m\left(\bigcup_{n=1}^{\infty} C_n \right)$$

$$= \sum_{n=1}^{\infty} m\left(C_n \setminus \bigcup_{1 \leq k < n} C_k \right) \geq \sum_{n=1}^{\infty} \left(m(C_n) - \frac{f_n}{2} \right)$$

$$\geq \sum_{n=1}^{\infty} \frac{m(C_n)}{2} = \sum_{n=1}^{N} nf_n + \sum_{n=N+1}^{\infty} Nf_n$$

$$= \sum_{n=1}^{N} c_n = a(N) \geq a_N.$$

$\square$

For certain conservative, ergodic, measure preserving transformations T of infinite measure spaces $(X, \mathcal{B}, m)$, there are sets $A \in \mathcal{B}$, $0 < m(A) < \infty$ so that

$$L_B(n) \sim L_A(n) \quad \forall\, B \in \mathcal{B} \cap A,\ 0 < m(B) < \infty.$$

These give rise to *minimal wandering rates* in the sense that

$$\liminf_{n \to \infty} \frac{L_B(n)}{L_A(n)} \geq 1 \quad \forall\, B \in \mathcal{B},\ m(B) > 0.$$

To see this, note that by conservativity and ergodicity of T, $\exists\, C \in \mathcal{B}$, $m(C) > 0$ and $N \geq 1$ such that $C \subset A \cap T^{-N}B$, whence:

$$L_B(n) \sim L_{T^{-N}B}(n) \geq L_C(n) \sim L_A(n) \text{ as } n \to \infty.$$

The next result (theorem 3.8.3) shows that pointwise dual ergodic transformations have this property (i.e. have sets with minimal wandering rates).

Suppose that T is pointwise dual ergodic with return sequence $a(n) = \sum_{k=0}^{n} u_k$ where $u_k > 0$, then

$$\frac{1}{u(\lambda)} \sum_{n=0}^{\infty} e^{-\lambda n} \widehat{T}^n f \to \int_X f\,dm \text{ a.e. as } \lambda \to 0\ \forall\, f \in L^1(m)$$

where $u(\lambda) = \sum_{n=0}^{\infty} u_n e^{-\lambda n}$.

Call a set $A \in \mathcal{B}$, $0 < m(A) < \infty$ *uniform for $f \in L^1(m)_+$*, if

$$\frac{1}{a(n)} \sum_{k=0}^{n} \widehat{T}^k f \to \int_X f\,dm \text{ almost uniformly on } A \text{ as } n \to \infty.$$

If A is uniform for f, then

$$\frac{1}{u(\lambda)} \sum_{n=0}^{\infty} e^{-\lambda n} \widehat{T}^n f \to \int_X f\,dm \text{ almost uniformly on } A \text{ as } \lambda \to 0.$$

We denote the collection of sets which are uniform for some $f \in L^1(m)_+$ by $\mathcal{U}(T)$. This collection is heriditary in the sense that

$$A \in \mathcal{U}(T),\ B \in \mathcal{B}_+,\ B \subset A \implies B \in \mathcal{U}(T).$$

3.8.3 THEOREM. *Let T be pointwise dual ergodic. There is a sequence $L(n) \uparrow \infty$ such that*

$$L_A(n) \sim L(n) \quad \forall\, A \in \mathcal{U}(T).$$

Before proving theorem 3.8.3, we establish some auxilliary results.

3.8.4 LEMMA. *Let $A \in \mathcal{B}$, $0 < m(A) < \infty$ and $f \in L^1(m)$, then*

$$\int_A \left(\sum_{k=0}^{\infty} e^{-\lambda k} \widehat{T}^k f \right) (1 - e^{-\lambda \varphi_A})\,dm = \sum_{n=0}^{\infty} e^{-\lambda n} \int_{A_n} f\,dm \quad \forall\, \lambda > 0$$

where $A_0 = A$ and $A_n = T^{-n}A \setminus \bigcup_{k=0}^{n-1} T^{-k}A$ for $n \geq 1$.

PROOF. Define $\nu(B) := \int_B f\,dm \leq \infty$ for $B \in \mathcal{B}$.
For $A \in \mathcal{B}$, and $n \geq 1$, we have that

$$\int_A \widehat{T}^n f\,dm = \nu(T^{-n}A)$$

$$= \nu\left(T^{-n}A \setminus \bigcup_{k=0}^{n-1} T^{-k}A\right) + \sum_{k=0}^{n-1} \nu(T^{-k}(A \cap [\varphi_A = n-k]))$$

$$= \nu\left(T^{-n}A \setminus \bigcup_{k=0}^{n-1} T^{-k}A\right) + \int_A \sum_{k=0}^{n-1} \widehat{T}^k f 1_{[\varphi_A = n-k]}\,dm$$

whence for $n \geq 1$,

$$\int_A \sum_{n=0}^{\infty} e^{-\lambda n}\widehat{T}^n f\,dm = \sum_{n=0}^{\infty} e^{-\lambda n}\nu(A_n) + \sum_{n=1}^{\infty} e^{-\lambda n}\int_A \sum_{k=0}^{n-1} \widehat{T}^k f 1_{[\varphi_A = n-k]}\,dm$$

$$= \sum_{n=0}^{\infty} e^{-\lambda n}\nu(A_n) + \int_A e^{-\lambda \varphi_A} \sum_{n=0}^{\infty} e^{-\lambda n}\widehat{T}^n f\,dm.$$

$\square$

3.8.5 LEMMA.

$$1 \lesssim \frac{L_A(n)a(n)}{n} \lesssim 2 \quad \forall\, A \in \mathcal{U}(T).$$

PROOF. By lemma 3.8.4,

$$\int_A \left(\sum_{k=0}^{\infty} e^{-\lambda k}\widehat{T}^k f\right)(1 - e^{-\lambda \varphi_A})\,dm = \sum_{n=0}^{\infty} e^{-\lambda n}\int_{A_n} f\,dm.$$

Note that

$$\int_A (1 - e^{-\lambda \varphi_A})\sum_{k=0}^{\infty} e^{-\lambda n}\widehat{T}^k f\,dm = (1 - e^{-\lambda})\sum_{n=0}^{\infty} e^{-\lambda n}\int_A \sum_{k=0}^{n} \widehat{T}^k f 1_{[\varphi_A \geq n-k+1]}\,dm$$

and

$$\sum_{n=0}^{\infty} e^{-\lambda n}\int_{A_n} f\,dm = (1 - e^{-\lambda})\sum_{n=0}^{\infty} e^{-\lambda n}\int_{\bigcup_{k=0}^{n} T^{-k}A} f\,dm,$$

whence equating coefficients,

$$\int_A \left(\sum_{k=0}^{n} \widehat{T}^k f 1_{[\varphi_A \geq n-k+1]}\right)dm = \int_{\bigcup_{k=0}^{n} T^{-k}A} f\,dm.$$

Next

$$\sum_{n=0}^{N} \int_A \left(\sum_{k=0}^{n} \widehat{T}^k f 1_{[\varphi_A \geq n-k+1]}\right)dm = \sum_{n=0}^{N} \int_{\bigcup_{k=0}^{n} T^{-k}A} f\,dm \sim N\int_X f\,dm.$$

Now

$$\sum_{n=0}^{N}\sum_{k=0}^{n}\widehat{T}^{k}f1_{[\varphi_A\geq n-k+1]} = \sum_{k=0}^{N}\widehat{T}^{k}f\sum_{m=0}^{N-k}1_{[\varphi_A\geq m+1]}$$

$$= \sum_{k=0}^{N}\widehat{T}^{k}f\left(\varphi_A\wedge(N-k+1)\right)$$

where $a\wedge b := \max\{a,b\}$ for $a,b\in\mathbb{R}$.

We therefore have

$$\int_A\sum_{k=0}^{N}\widehat{T}^{k}f\left(\varphi_A\wedge(N-k+1)\right)\sim N\int_X fdm.$$

Now

$$\sum_{k=0}^{N}\widehat{T}^{k}f\left(\varphi_A\wedge(N-k+1)\right) \leq \sum_{k=0}^{N}\widehat{T}^{k}f(\varphi_A\wedge N)$$

$$\leq \sum_{k=0}^{2N}\widehat{T}^{k}f\left(\varphi_A\wedge(2N-k+1)\right) \text{ on } A,$$

and that

$$\sum_{k=0}^{N}\widehat{T}^{k}f(\varphi_A\wedge N) \sim \int_X fdm\, a(N)\,(\varphi_A\wedge N) \text{ as } N\to\infty \text{ uniformly on } A.$$

The lemma follows by integrating these on A. $\square$

3.8.6 ASYMPTOTIC RENEWAL EQUATION   [A7].

$$\int_A(1-e^{-\lambda\varphi_A})dm \sim \frac{1}{u(\lambda)} \text{ as } \lambda\to 0 \,\forall\, A\in\mathcal{U}(T).$$

PROOF. By lemma 3.8.4,

$$\int_A(1-e^{-\lambda\varphi_A})\sum_{n=0}^{\infty}e^{-\lambda n}\widehat{T}^{n}fdm = \sum_{n=0}^{\infty}e^{-\lambda n}\int_{A_n}fdm \to \int_X fdm \text{ as } \lambda\to 0$$

where $A_0 = A$ and $A_n = T^{-n}A\setminus\bigcup_{k=0}^{n-1}T^{-k}A$ for $n\geq 1$.

By assumption,

$$\int_A(1-e^{-\lambda\varphi_A})\sum_{n=0}^{\infty}e^{-\lambda n}\widehat{T}^{n}fdm \sim \int_X fdm u(\lambda)\int_A(1-e^{-\lambda\varphi_A})dm \text{ as } \lambda\to 0$$

and the proposition follows. $\square$

For $A\in\mathcal{B}$, $0<m(A)<\infty$ and $\lambda>0$, set

$$c_A(\lambda) := m(A) + \sum_{n=1}^{\infty}e^{-\lambda n}m(T^{-n}A\setminus\bigcup_{k=0}^{n-1}T^{-k}A).$$

PROOF OF THEOREM 3.8.3. We show that if $A \in \mathcal{U}(T)$, then

$$L_A(n) - L_B(n) = o(L_A(n)) \text{ as } n \to \infty \ \forall \ B \in \mathcal{B}_+, \ B \subset A.$$

We begin by noting that if $b_k \geq 0$ $(k \geq 0)$, and $b(\lambda) := \sum_{k \geq 0} b_k e^{-\lambda k}$ $(\lambda \geq 0)$, then for $n \geq 1$,

$$\sum_{k=0}^{n} b_k \leq e \sum_{k=0}^{n} b_k e^{-\frac{k}{n}} \leq e b(\frac{1}{n}).$$

Suppose that $A \in \mathcal{U}(T)$ and $B \in \mathcal{B}_+$, $B \subset A$, then

$$L_A(n) - L_B(n) = \sum_{k=0}^{n} (m(A_k) - m(B_k)) \leq e(c_A(\frac{1}{n}) - c_B(\frac{1}{n})) = o(\frac{n}{u(\frac{1}{n})})$$

by the asymptotic renewal equation.

To conclude,

$$\frac{n}{u(\frac{1}{n})} \leq \frac{en}{a(n)} \leq e L_A(n)$$

by lemma 3.8.5, and $L_A(n) - L_B(n) = o(L_A(n))$ as $n \to \infty$. $\qquad \square$

Sometimes, it is relatively easy to calculate wandering rates. We conclude with a method of finding return sequences given wandering rates.

3.8.7 PROPOSITION [**A7**], [**A10**]. *Let T be pointwise dual ergodic, and suppose that $\exists \ A \in \mathcal{U}(T)$ such that $L_A(n)$ is regularly varying at ∞ with index $\alpha \in [0, 1]$, then*

$$a(n) \sim \frac{1}{\Gamma(2-\alpha)\Gamma(1+\alpha)} \cdot \frac{n}{L_A(n)}.$$

PROOF. Let $A \in \mathcal{U}(T)$, then

$$c_A(\lambda) \sim \frac{1}{\lambda} \int_A (1 - e^{-\lambda \varphi_A}) dm$$

and by the asymptotic renewal equation,

$$c_A(\lambda) \sim \frac{1}{\lambda} \int_A (1 - e^{-\lambda \varphi_A}) dm \sim \frac{1}{\lambda u(\lambda)} \text{ as } \lambda \to 0.$$

The proposition follows from this using Karamata's Tauberian theorem (see [**Fe3**]). $\square$

ILLUSTRATION.

Let $X = \mathbb{R}$, $m =$ Lebesgue measure and $Tx := x - \frac{1}{x}$ (Boole's transformation). Set $A = [-1, 1]$, then

$$\bigcup_{k=0}^{n} T^{-k} A = [-v_n, v_n]$$

where

$$v_n = T v_{n+1}.$$

It can be easily shown that $v_n \sim \sqrt{2n}$ as $n \to \infty$, whence

$$L_A(n) \sim \sqrt{8n} \text{ as } n \to \infty.$$

On the other hand by exercise 1.3.4 T is pointwise dual ergodic, and indeed

$$\frac{1}{a(n)} \sum_{k=0}^{n-1} \widehat{T}^k \varphi_{ib} \to 1 = \int_{\mathbb{R}} \varphi_{ib} dm$$

as $n \to \infty$ uniformly on compact subsets of X where $a(n) \sim \frac{\sqrt{2n}}{\pi}$.

We reconstruct the return sequence $a(n)$ using proposition 3.8.7. Evidently $A \in \mathcal{U}(T)$ and hence by proposition 3.8.7,

$$a_n(T) \sim \frac{1}{\Gamma(3/2)^2} \frac{n}{L_A(n)} \sim \frac{\sqrt{2n}}{\pi}.$$

CHAPTER 4

Markov maps

In this chapter we consider Markov maps, concentrating mainly on Markov shifts and Markov maps of the interval. We give here a unified treatment which based on the distortion properties they share.

§4.1 Markov partitions

DEFINITION: MARKOV PARTITION, MARKOV MAP.

Let T be a locally invertible non-singular transformation of the standard space $(X, \mathcal{B}, m)$. A countable partition $\alpha \subset \mathcal{B}$ is called *basic* for T if $T : a \to Ta$ is invertible $\forall\, a \in \alpha$, and $\sigma(\{T^{-n}a : a \in \alpha, n \geq 0\}) = \mathcal{B}$ mod m.

The basic partition α for the locally invertible non-singular transformation T is called a *Markov partition* if $Ta \in \sigma(\alpha)$ mod m $\forall\, a \in \alpha$, and a locally invertible non-singular transformation with a Markov basic partition α is called a *Markov map* and denoted $(X, \mathcal{B}, m, T, \alpha)$.

Our first examples of Markov maps are Markov shifts.

DEFINITION: MARKOV SHIFT. Let S be a countable set. A *stochastic matrix* on S is a matrix $P : S \times S \to [0, 1)$ satisfying

$$\sum_{t \in S} p_{s,t} = 1 \ \forall\, s \in S.$$

For $q \in \mathcal{P}(S)$ such that $q_s > 0$ $\forall s$, define the measure $m_q \in \mathcal{P}(S^{\mathbb{N}})$ by

$$m_q([s_1, .., s_n]) = q_{s_1} \prod_{k=1}^{n-1} p_{s_k, s_{k+1}}$$

where $[s_1, .., s_n] := \{x = (x_1, x_2, \dots) \in S^{\mathbb{N}} : x_i = s_i \ 1 \leq i \leq n\}$.

Let $T : S^{\mathbb{N}} \to S^{\mathbb{N}}$ be the shift, then T is a measurable transformation of $(S^{\mathbb{N}}, \mathcal{B}(S^{\mathbb{N}}), m_q)$. It is called the *Markov shift of P with initial distribution q*. The Markov shift of P is called *non-singular* if $m_q \circ T^{-1} \sim m_q$ $\forall\, q \in \mathcal{P}(S)$ such that $q_s > 0$ $\forall s$.

Consider the partition $\alpha := \{[s] : s \in S\}$.

4.1.1 PROPOSITION.
The Markov shift T of P is non-singular iff $\forall\, t \in S \ \exists\, s \in S$ such that $p_{s,t} > 0$, and (in this case) α is a Markov partition for T.

PROOF. It is always the case that $m_q \circ T^{-1} << m_q$ and

$$\frac{dm_q \circ T^{-1}}{dm_q}(x) = \sum_{s \in S} \frac{q_s p_{s,x_1}}{q_{x_1}}.$$

The first part follows from this.

To see that α is a Markov partition for T, note that

$$T[s] = \bigcup_{t \in S,\ p_{s,t}>0} [t] \qquad \mod m_q.$$

$\square$

We continue with a study of some of the combinatorial and topological aspects of the theory of Markov maps.

§4.2 Graph shifts

Let X be a Polish space. A continuous map $T : X \to X$ is called
topologically recurrent if $\forall\ \emptyset \neq U \subset X$ open, $\exists\ n \geq 1$ such that $U \cap T^{-n}U \neq \emptyset$;
topologically transitive if $\forall\ \emptyset \neq U, V \subset X$ open, $\exists\ n \geq 1$ such that $U \cap T^{-n}V \neq \emptyset$;
and
topologically mixing if $\forall\ \emptyset \neq U, V \subset X$ open, $\exists\ n_0 \geq 1$ such that $U \cap T^{-n}V \neq \emptyset\ \forall\ n \geq n_0$.

Clearly topological mixing implies topological transitivity, which implies topological recurrence.

The topological properties are connected to the analogous metric properties. Indeed, it is not hard to check that if $\exists\ m \in \mathcal{P}(X)$, globally supported such that $m \circ T^{-1} \sim m$, then
T conservative implies T topologically recurrent;
T conservative and ergodic implies T topologically transitive,
and
if $m \circ T^{-1} = m$ and T is mixing, then T is topologically mixing.

Now let S be a countable set and let $T : S^{\mathbb{N}} \to S^{\mathbb{N}}$ be the *shift* defined by $T(s_1, s_2, \dots) := (s_2, s_3, \dots)$. The set $S^{\mathbb{N}}$, when equipped with the product discrete topology (generated by the *cylinder sets*

$$\{[s_1, \dots, s_n] := \{x \in S^{\mathbb{N}} : x_k = s_k\ 1 \leq k \leq n\} : n \geq 1, s_1, \dots, s_n \in S\})$$

becomes a Polish space, and $T : S^{\mathbb{N}} \to S^{\mathbb{N}}$ is continuous. It is easy to check that T is topologically mixing.

A closed, T-invariant $X \subset S^{\mathbb{N}}$ is called a *subshift*, and a subshift is a *graph shift* (or *topological Markov shift*) if there is a (directed) graph $G = (S, E_G)$ (called the *incidence graph*) such that

$$X = X_G := \{x = (x_1, x_2, ..) \in S^{\mathbb{N}} : (x_k, x_{k+1}) \in E_G\ \forall k \geq 1\}.$$

The vertices of G are called *states*, and the edges are called *transitions*.

For example if P is a stochastic matrix on S, $q \in \mathcal{P}(S), q_s > 0\ \forall s$ then the closed support of m_q is given by

$$X_P = \{x \in S^{\mathbb{N}} : p_{x_n, x_{n+1}} > 0\ \forall\ n \geq 1\}.$$

This is a subshift if and only if the Markov shift of P is non-singular, and in this case it is a graph shift with incidence graph $G_P := (S, E_P)$ where $E_P := \{(s, t) \in S \times S : p_{s,t} > 0\}$.

The definition of X_G can be made for any directed graph $G = (S, E_G)$ to obtain a closed, T-sub-invariant $X_G \subset S^{\mathbb{N}}$ (i.e. $TX_G \subseteq X_G$), but X_G is T-invariant iff $\forall\, x \in S\ \exists\, x \in S,\ (s, x) \in E$. We'll call X_G a *proper* graph shift (and G a proper incidence graph) if $TX_G = X_G$.

All graphs under consideration are directed.

A *path* in $G = (S, E)$ is a sequence $x_0, \ldots, x_n \in S$ such that $n \geq 1$ and $(x_k, x_{k+1}) \in E$ $(0 \leq k \leq n - 1)$. The *length* of the path $x_0, \ldots, x_n$ is n (the number of edges involved), the point x_0 is its *source*, and x_n is its *endpoint*. For $x, y \in S$, write $x \xrightarrow{n} y$ if there is a path of length n with source x and endpoint y, $x \rightarrow y$ if $x \xrightarrow{n} y$ for some $n \geq 1$.

A state $x \in S$ is called *persistent* if $x \rightarrow x$ and a graph is called *persistent* if all its states are persistent.

As can be easily checked, a graph shift is topologically recurrent iff its incidence graph is persistent.

Let $G = (S, E)$ be a graph. A subset $S' \subset S$ is called *G-closed* if $x \in S'$, $x \rightarrow y \implies y \in S'$. Let $S' \subset S$ be G-closed and set $X' := \bigcup_{s \in S'}[s] \subset X_G$, then $TX' \subset X'$ and $X' = X_{G'}$ where $G' = (S', E \cap S' \times S)$. We call X' the *subsystem* of X_G corresponding to S'. The subsystem X' is *proper* if $TX' = X'$.

A G-closed subset $S' \subset S$ called *irreducible* if it is minimal in the sense that it has no non-trivial G-closed subsets of states, equivalently: $x \rightarrow y\ \forall\, x, y \in S$. The graph G is called *irreducible* if S is irreducible.

Clearly, an irreducible graph is persistent, and a subsystem is topologically transitive iff the corresponding closed subset of states is irreducible.

EXAMPLE. Let $G = (\mathbb{Z}, E)$ where $E = \{(n, n), (n, n+1) :\ n \in \mathbb{Z}\}$, then G is persistent, but there is no closed, irreducible subset of states. Hence, (X_G, T) is topologically recurrent but has no topologically transitive subsystem.

This phenomenon is impossible when S is finite. The next result gives a sufficient condition for its elimination for infinite S.

Let $G = (S, E)$ be a graph, and for $x \in S$ set $E_x = \{y \in S :\ (x, y) \in E\}$. Note that $T[s] = \bigcup_{t \in E_s}[t]$.

4.2.1 PROPOSITION. *Let (X, T) be a graph shift with state space S, and assume that $|\{T[s] :\ s \in S\}| < \infty$, then $\overline{\bigcup_{n=0}^{\infty} T^{-n}X_0} = X$ where $X_0 := \bigcup_{k=1}^{N} X_k$, the union of the finitely many topologically transitive subsystems.*

PROOF.

The image TX' of a subsystem X' of X is nonempty, itself a subsystem and a union of elements of $\{T[s] : s \in S\}$. The assumption implies that there are only finitely many sets. Therefore, there are only finitely many proper subsystems and any proper subsystem contains a topologically transitive subsystem.

Also, if X' is a subsystem, then $\exists\, n \geq 1$ such that $T^n X' = \bigcap_{k=1}^{\infty} T^k X' \subset X$ is a proper subsystem.

Let the topologically transitive subsystems be $X_1, \ldots, X_N$ and let $X_0 := \bigcup_{k=1}^{N} X_k$.

We must show that $U \cap \bigcup_{n=0}^{\infty} T^{-n} X_0 \neq \emptyset \ \forall \ U \subset X$ open and nonempty. We claim first that it suffices to show that $\forall \ s \in S \ \exists \ n \geq 1$ such that $T^n[s] \cap X_0 \neq \emptyset$.

Indeed if $U \neq \emptyset$ is open then $\exists \ u_1, \ldots, u_k \in S$ such that $[u_1, \ldots, u_k] \subset U$. Assuming $\exists \ n \geq 1$ such that $T^n[u_k] \cap X_0 \neq \emptyset$, we have that

$$T^{n+k} U \cap X_0 \supset T^{n+k}[u_1, \ldots, u_k] \cap X_0 \supset T^n[u_k] \cap X_0 \neq \emptyset,$$

whence $U \cap \bigcup_{n=0}^{\infty} T^{-n} X_0 \neq \emptyset$.

Now fix $s \in S$, then $X_s := \bigcup_{n \geq 0} T^n[s]$ is a subsystem. As above, $\exists \ n \in \mathbb{N}$ such that $T^n X_s \cap X_0 \neq \emptyset$, equivalently $\exists \ n' \in \mathbb{N}$ such that $\emptyset \neq T^{n'}[s] \cap X_0 \neq \emptyset$. $\square$

Let $G = (S, E)$ be persistent and let $x \in S$. Consider the collection $K_x \subset \mathbb{N}$ of lengths of paths with x as both source and endpoint. This is a semigroup. The *period* of a state $x \in S$ (written $\mathrm{Per}\,(x)$) is greatest common divisor of K_x. Equivalently,

$$\mathrm{Per}\,(x) = \max \{d \geq 1 : \ K_x \subset d\mathbb{Z}\},$$

or

$$K_x - K_x = \mathrm{Per}\,(x)\mathbb{Z}.$$

Suppose that $G = (S, E)$ is irreducible, then

$$\mathrm{Per}\,(x) = \mathrm{Per}\,(y) \quad \forall \ x, y \in S.$$

To see this, suppose that $x, y \in S$. By irreducibility, $\exists \ p, q \in \mathbb{N}$ such that $x \xrightarrow{p} y$ and $y \xrightarrow{q} x$; whence $p + q \in K_x \cap K_y$ and

$$K_x \subset K_y + p + q, \quad K_y \subset K_x + p + q.$$

It follows that $K_x - K_x = K_y - K_y$ whence $\mathrm{Per}\,(x) = \mathrm{Per}\,(y)$.

The *period* of an irreducible graph (S, E) is the number $d \in \mathbb{N}$ with $\mathrm{Per}\,(x) = d \ \forall \ x \in S$. A graph is called *aperiodic* if its period is 1. It is easily seen that a topologically transitive graph shift is topologically mixing if and only if its incidence graph is aperiodic.

Now again let $G = (S, E)$ be irreducible with period d. Fix $x \in S$ and define for $0 \leq k \leq d - 1$

$$S_{k,x} := \{y \in S : \ \exists \ n \geq 0 \text{ such that } x \xrightarrow{nd+k} y\}.$$

We claim that $\{S_{k,x}\}_{k=0}^{d-1}$ are disjoint, since if not $\exists \ 0 \leq i < j \leq d - 1$ and $y \in S_{i,x} \cap S_{j,x}$, whence if $y \xrightarrow{N} x$, we have $x \xrightarrow{ad+i+N} x$ and $x \xrightarrow{a'd+j+N} x$ whence $j - i \in d\mathbb{Z}$ contradicting $0 < j - i < d$.

In view of this

$$y \in S_{k,x} \text{ and } y \xrightarrow{N} z \implies z \in S_{\ell,x} \text{ where } \ell = k + N \mod d.$$

This proves

4.2.2 PROPOSITION. *Let (X_G, T) be a topologically transitive graph shift whose graph G has period d, then $\exists \ \{X_k\}_{k=0}^{d-1}$ disjoint (e.g. $X_k = \bigcup_{s \in S_{k,x}}[s]$) such that $TX_k = X_{k+1} \ (0 \leq k \leq d - 2)$, $TX_{d-1} = X_0$, $X_G = \bigcup_{k=0}^{d-1} X_k$ and (X_k, T^d) is topologically mixing $(0 \leq k \leq d - 1)$.*

Lastly, we show that any Markov map $(X, \mathcal{B}, m, T, \alpha)$ is isomorphic to a graph shift on α equipped with a non-singular measure.

Write $\alpha = \{a_s : s \in S\}$ and define $\phi : X \to S^{\mathbb{N}}$ by $T^{n-1}x \in a_{\phi(x)_n} \in \alpha$. Clearly $\phi : X \to S^{\mathbb{N}}$ is measurable. Let $\tilde{T} : S^{\mathbb{N}} \to S^{\mathbb{N}}$ be the shift. We have

4.2.3 PROPOSITION.
1) The support of $m \circ \phi^{-1}$ is X_G where $G = (S, E)$ and

$$E = \{(s,t) : m(a_s \cap T^{-1}a_t) > 0 \}.$$

2) ϕ is a conjugacy of $(X, \mathcal{B}, m, T)$ with $(X_G, \mathcal{B}(X_G), m \circ \phi^{-1}, \tilde{T})$.

– the proof being left to the reader.

Thus there is no loss of generality in assuming that $X = X_G$, T is the shift, and $\alpha = \{[s] : s \in S\}$.

CYLINDER SET NOTATION.
For each $a \in \alpha_0^{n-1}$, $\exists\, s_1, \ldots, s_n \in S$ such that

$$a = \{x \in S^{\mathbb{N}} : x_i = s_i \,\forall\, 1 \le i \le n\} := [s_1, \ldots, s_n].$$

Let $a \in \alpha_0^{n-1}$ and $b \in \alpha_0^{q-1}$ and suppose that $a = [s_1, \ldots, s_n]$, $b = [t_1, \ldots, t_q]$. We'll write

$$[a,b] := [s_1, \ldots, s_n, t_1, \ldots, t_q] = a \cap T^{-n}b.$$

§4.3 Distortion properties

Let $(X, \mathcal{B}, m, T, \alpha)$ be a Markov map. For each $n \ge 1$, there are m-nonsingular inverse branches of T^n denoted $v_a : \mathcal{D}(v_a) := T^n a \to a$ $(a \in \alpha_0^{n-1})$ with Radon Nikodym derivatives

$$v_a' = v_{a,m}' := \frac{dm \circ v_a}{dm} = \left(\frac{dm \circ T^n}{dm} \circ v_a \right)^{-1}.$$

Here, and throughout, the derivative of a function f of a real variable is denoted by $Df(x) := \frac{df}{dx}(x)$. Higher derivatives are denoted by $D^{n+1}f := D(D^n f)$.

Note that if J and J' are subintervals of $\mathbb{R}$, m is Lebesgue measure on $\mathbb{R}$ and $f : J \to J'$ is a homeomorphism with both f and f^{-1} absolutely continuous, then in our notations,

$$f' : \frac{dm \circ f}{m} \equiv |Df|.$$

A *distortion property* is a property of the multiplicative variation of v_a' on $\mathcal{D}(v_a)$.

Let

$$\tilde{\alpha} = \bigcup_{n \in \mathbb{N}} \alpha_0^{n-1},$$

$$\tilde{\alpha}_+ = \{a \in \tilde{\alpha} : m(a) > 0\}$$

and for $C \in \mathbb{R}_+$ set

$$\mathfrak{g}(C, T) = \{a \in \tilde{\alpha}_+ : \frac{v_a'(x)}{v_a'(y)} \le C \text{ for } m \times m\text{-a.e. } (x, y) \in \mathcal{D}(v_a) \times \mathcal{D}(v_a)\},$$

and for $C \in \mathbb{R}_+$, $0 < r < 1$ set

$$\mathfrak{g}_r(C,T) :=$$
$$\left\{ a \in \tilde{\alpha}_+ : \left| \log \frac{v_a'(x)}{v_a'(y)} \right| \leq C r^{t(x,y)} \text{ for } m \times m\text{-a.e. } (x,y) \in \mathcal{D}(v_a) \times \mathcal{D}(v_a) \right\}$$

where $t(x,y) := \min\{n \geq 1 : x_n \neq y_n\}$.

Define the *distortion* of $a \in \tilde{\alpha}_+$ to be $\min\{C \geq 1 : a \in \mathfrak{g}(C,T)\}$.

4.3.1 DISTORTION PROPOSITION.
Let $a = [s_1, \ldots, s_n] \in \alpha_0^{n-1}$, $b \in \tilde{\alpha}$ and suppose that a, b, $a \cap T^{-n}b \in \mathfrak{g}(C,T)$, then

$$m(T^{-n}b|a) = C^{\pm 2} m(b|Ts_n).$$

PROOF. We have $\mathcal{D}(v_a) = T^n a = Ts_n$ and for m-a.e. $x \in a$,

$$m(a) = m(v_a Ts_n) = \int_{Ts_n} v_a' dm = C^{\pm 1} m(Ts_n) v_a'(x).$$

Similarly,

$$m(a \cap T^{-n}b) = m(v_a b) = \int_b v_a' dm = C^{\pm 1} m(b) v_a'(x).$$

Therefore,

$$m(a \cap T^{-n}b) = C^{\pm 1} m(b) v_a'(x) = C^{\pm 2} \frac{m(a)m(b)}{m(Ts_n)},$$

or, equivalently,

$$m(T^{-n}b|a) = C^{\pm 2} m(b|Ts_n).$$

$\square$

4.3.2 PROPOSITION.
Let S be a countable set, P be a stochastic matrix on S and let $(X_P, \mathcal{B}(X_P), m_q, T, \alpha)$ be the Markov shift of P with initial distribution q.

(i) For $C > 1$, $\mathfrak{g}(C,T) = \mathfrak{g}_r(C,T) \ \forall \ 0 < r < 1$ and

$$[s_1, \ldots, s_n] \in \mathfrak{g}(C,T) \quad \Leftrightarrow \quad [s_n] \in \mathfrak{g}(C,T).$$

(ii) $\forall \ s \in S \ \exists$ an initial distribution q with respect to which $[s] \in \mathfrak{g}(4,T)$.

PROOF.

(i) For $n \geq 1, a = [s_1, \ldots, s_n] \in \alpha_0^{n-1}$,

$$v_a' = m(a) \frac{p_{s_n, t}}{q_t} \quad \text{a.e. on } [t] \quad (t \in S, \ T[s_n] \supset [t]).$$

Hence, $\mathfrak{g}(C,T) = \mathfrak{g}_r(C,T) \ \forall \ 0 < r < 1$ and

$$[s_1, \ldots, s_n] \in \mathfrak{g}(C,T) \quad \Leftrightarrow \quad [s_n] \in \mathfrak{g}(C,T).$$

(ii) For $s \in S$, if $1/2 \leq q_t/p_{s,t} \leq 2$ whenever $p_{s,t} > 0$, then $[s] \in \mathfrak{g}(4,T)$. Clearly, for $s \in S$, there is $q \in \mathcal{P}(S)$ with this property. $\square$

DEFINITION: C^2 MARKOV INTERVAL MAP. Let $I \subset \mathbb{R}$ be a bounded interval. We consider I endowed with Lebesgue measure. A non-singular map $T : I \to I$ is called C^2-*Markov* if there is a Markov partition α of I into intervals such that $\forall a \in \alpha$, $T|_a$ is strictly monotonic and extends to a C^2 function on $\bar{a}$, $DT(x) \neq 0 \ \forall \ x \in I$ and

$$\sup \frac{|D^2 T|}{(DT)^2} < \infty.$$

The property $\sup \frac{|D^2 T|}{(DT)^2} < \infty$ of the C^2-Markov interval map $(I, \mathcal{B}(I), m, T, \alpha)$ implies a distortion property.

To establish this property, use the relationship $T \circ v_a = \mathrm{Id}$ on $\mathcal{D}(v_a) \ \forall \ a \in \alpha$, to obtain that

$$\frac{D^2 v_a}{D v_a} = -\frac{D^2 T \circ v_a}{(DT \circ v_a)^2} \quad \forall \ a \in \alpha$$

whence for $a \in \alpha$, $x < y \in \mathcal{D}(v_a)$,

$$\left| \log \frac{D v_a(x)}{D v_a(y)} \right| = \left| \int_x^y \frac{D^2 v_a}{D v_a} dm \right| \leq \int_x^y \left| \frac{D^2 v_a}{D v_a} \right| dm \leq M|y - x|$$

where $M = \sup \frac{|D^2 T|}{(DT)^2}$.

It follows from this that

$$v_a'(x) = |D v_a(x)| = e^{\pm M} D v_a(y)| = e^{\pm M} v_a'(y) \quad \forall \ a \in \alpha, \ x, y \in \mathcal{D}(v_a)$$

whence the distortion property:

$$v_a'(x) = e^{\pm M} \frac{\int_{\mathcal{D}(v_a)} v_a'(y) dy}{m(\mathcal{D}(v_a))} = e^{\pm M} \frac{m(a)}{m(\mathcal{D}(v_a))}.$$

4.3.3 PROPOSITION [Ad].

Let $(I, \mathcal{B}(I), m, T, \alpha)$ be an expanding C^2 interval map, then $\exists \, C > 1$, $0 < r < 1$ such that $\mathfrak{g}_r(C, T) = \tilde{\alpha}_+$.

PROOF. Let $\inf |DT| = \lambda > 1$, and let $\sup \frac{|D^2 T|}{(DT)^2} = M$.

We prove the proposition with $r = \frac{1}{\lambda}$.

We claim first that

$$\sup \frac{|D^2(T^n)|}{D(T^n)^2} \leq M' := \frac{\lambda M}{\lambda - 1}.$$

To see this

$$\frac{|D^2(T^n)(x)|}{|D(T^n)(x)|^2} \leq \frac{1}{|D(T^n)(x)|^2} \left| \sum_{k=0}^{n-1} |D(T^n)(x) \frac{D^2 T(T^k x) D(T^k)(x)}{DT(T^k x)} \right|$$

$$\leq M \sum_{k=0}^{n-1} \frac{1}{|D(T^{(n-k)})(T^k x)|}$$

$$\leq \frac{\lambda M}{\lambda - 1} := M.$$

It follows that

$$\left| \frac{D^2 v_a(x)}{D v_a(x)} \right| \leq M' \ \forall \ x \in I, \ a \in \alpha_0^{n-1}, \ n \geq 1,$$

whence, noting that Dv_a does not change sign on $\mathcal{D}(v_a)$, for $n \geq 1$, $x, y \in \mathcal{D}(v_a)$,

$$\left| \log \frac{v_a'(x)}{v_a'(y)} \right| = \left| \log \left| \frac{Dv_a(x)}{Dv_a(y)} \right| \right| \leq \int_x^y \left| \frac{D^2 v_a(t)}{Dv_a(t)} \right| dt \leq M'(y - x).$$

To finish, note that if $t(x, y) = n$ then $\exists a \in \alpha_0^{n-1}$ with $x, y \in \alpha$ whence

$$|x - y| \leq m(a) = m(v_a \mathcal{D}(v_a)) = \int_{\mathcal{D}(v_a)} v_a' dm \leq \frac{1}{\lambda^n},$$

and

$$e^{-M' r^{t(x,y)}} \leq \left| \frac{v_a'(x)}{v_a'(y)} \right| \leq e^{M' r^{t(x,y)}}$$

where $r = \frac{1}{\lambda}$.   □

DEFINITION: STRONG DISTORTION PROPERTY.

We say that the Markov map $(X, \mathcal{B}, m, T, \alpha)$ has the *strong distortion* (or *Renyi*) *property* if $\exists\, C > 1$ such that $\mathfrak{g}(C, T) = \tilde{\alpha}_+$.

DEFINITION: SCHWEIGER COLLECTION.

Let $(X, \mathcal{B}, m, T, \alpha)$ be a Markov map and let $C > 1$. A collection $\mathfrak{r} \subseteq \tilde{\alpha}_+$ is called a *Schweiger collection* for $(X, \mathcal{B}, m, T, \alpha)$ if $\mathfrak{r} \subseteq \mathfrak{g}(C, T)$ for some $C > 1$,

$$[b] \in \mathfrak{r}, [a] \in \tilde{\alpha}_+, [a, b] \in \tilde{\alpha}_+ \;\Rightarrow\; [a, b] \in \mathfrak{r},$$

and

$$\bigcup_{B \in \mathfrak{r}} B = X \quad \mathrm{mod}\ m.$$

DEFINITION: WEAK DISTORTION PROPERTY.

We say that the Markov map $(X, \mathcal{B}, m, T, \alpha)$ has the *weak distortion property* if $\exists$ a Schweiger collection for $(X, \mathcal{B}, m, T, \alpha)$.

Suppose that the Markov map $(X, \mathcal{B}, m, T, \alpha)$ has the weak distortion property and let $\mathfrak{r} \subset \mathfrak{g}(C, T)$ be a Schweiger collection for $(X, \mathcal{B}, m, T, \alpha)$.

It follows that $\mathfrak{r}$ generates $\mathcal{B}$, and also

$$\mathfrak{r} \subseteq \bar{\mathfrak{r}}(C, T)$$

where

$$\bar{\mathfrak{r}}(C, T) := \{[a] \in \mathfrak{g}(C, T) : [b] \in \tilde{\alpha}_+, [b, a] \neq \emptyset \;\Rightarrow\; [b, a] \in \mathfrak{g}(C, T)\}$$

and so T has the weak distortion property if and only if

$$\bigcup_{B \in \bar{\mathfrak{r}}(C, T)} B = X \quad \mathrm{mod}\ m$$

for some $C \geq 1$.

This is the case when T is conservative and ergodic, and $\exists\, c \in \bar{\mathfrak{r}}(C, T)$, for then

$$X = \bigcup_{n=1}^{\infty} T^{-n} c = \bigcup_{n=1}^{\infty} \bigcup_{[b_1, \dots b_n] \in \alpha_0^{n-1}} [b_1, \dots, b_n, c] \subset \bigcup_{B \in \bar{\mathfrak{r}}(C, T)} B \quad \mathrm{mod}\ m.$$

Now suppose that $(X, \mathcal{B}, m, T, \alpha)$ is a conservative ergodic Markov shift, then proposition 4.3.2, $\mathfrak{g}(C, T) = \bar{\mathfrak{r}}(C, T)$, and so $(X, \mathcal{B}, m, T, \alpha)$ has the weak distortion property if and only if $\mathfrak{g}(C,T) \neq \emptyset$ for some $C \geq 1$; which in turn can be fixed for some initial probability distribution. We have proved

4.3.4 PROPOSITION   [A-De-Ur].

A conservative ergodic Markov shift has the weak distortion property with respect to some initial probability distribution.

DEFINITION: FIXED POINTS.

Let $(I, \mathcal{B}(I), m, T, \alpha)$ a C^2-Markov interval map. A point $x \in I$ is called a *fixed point* if $Tx = x$. Note that if $x \in a \in \alpha$ is a fixed point, then necessarily $Ta \supset a$.

The fixed point $x \in a \in \alpha$ is called *attractive, repulsive* or *indifferent* according to whether $|DT(x)|$ is smaller, larger or equal (respectively) to 1. An indifferent $x \in a \in \alpha$ is a *regular source* if $DT \downarrow$ on $a_- := a \cap (-\infty, x_a)$, and $DT \uparrow$ on $a_+ := a \cap (x_a, \infty)$ strictly. In particular, if $x_a \in a$ is a regular source, then x_a is a point of inflection for T on a, and a minimum for DT on a.

DEFINITION: THALER'S ASSUMPTIONS   [Tha1] [Tha2].

We say that a C^2-Markov interval map $(I, \mathcal{B}(I), m, T, \alpha)$ satisfies the *Thaler assumptions* if

a) For each $a \in \alpha$, $\bar{a}$ contains at most one fixed point.

b) The set $\emptyset \neq \Lambda = \Lambda(T)$ of indifferent fixed points is finite, and each one is a regular source.

c) $\forall \epsilon > 0 \; \exists \rho(\epsilon) > 1$ such that $|DT(x)| \geq \rho(\epsilon) \; \forall x \in I \setminus B_{\mathbb{R}}(\Lambda, \epsilon)$, and

$$\inf_{x \in a \in \alpha \setminus \alpha_\Lambda} |DT(x)| > 1$$

where $\alpha_\Lambda = \{a \in \alpha : a \cap \Lambda \neq \emptyset\}$.

Note that a C^2-Markov interval map satisfying Thaler's assumptions cannot be expanding as $\Lambda \neq \emptyset$.

Evidently, if $(I, \mathcal{B}(I), m, T, \alpha)$ is a C^2-Markov interval map satisfying Thaler's assumptions, then so is $(I, \mathcal{B}(I), m, T^n, \alpha_0^{n-1})$. Here, $\Lambda(T^n) = \Lambda(T)$.

For a C^2-Markov interval map $(I, \mathcal{B}(I), m, T, \alpha)$ satisfying Thaler's assumptions, let

$$\mathfrak{r} = \mathfrak{r}(I, \mathcal{B}(I), m, T, \alpha) := \{[s_1, ..., s_n] : \text{ either } s_n \notin \alpha_\Lambda, \text{ or } s_{n-1} \neq s_n\}.$$

4.3.5 THEOREM   [Th2].

Let $(I, \mathcal{B}(I), m, T, \alpha)$ be a C^2-Markov interval map satisfying Thaler's assumptions, then

1) $\exists \, C > 1$ such that $\left| \dfrac{D^2 v_a(x)}{D v_a(x)} \right| \leq C \; \forall \, a \in \mathfrak{r}$,

and

2) $\mathfrak{r}$ is a Schweiger collection for T.

PROOF.
Set

$$\alpha^* := (\alpha \setminus \alpha_\Lambda) \cup \{ [\underbrace{b, \ldots, b}_{n \text{ times}}, a] : \; n \geq 1, \; b \in \alpha_\Lambda, \; a \in \alpha \; a \neq b \} \subset \mathfrak{r}.$$

We first prove that $\bigcup_{a \in \alpha^*} a = I \mod m$ to which end we show that if $x \in a \in \alpha_\Lambda$, then $\exists \, n \geq 1$ such that $T^n x \notin a$.

To see this, let $a \in \alpha_\Lambda$ then a contains a regular source x_a for T and v_a : $Ta \to a$ satisfies $v_a(x) > x$ on a_- :, and $v_a(x) < x$ on a_+, whence $v_a^n x \to x_a$ as $n \to \infty \; \forall \; x \in Ta$. Now $Ta_\pm \supset a_\pm$, and by assumption c), $Ta_\pm \neq a_\pm$ (when $a_\pm \neq \emptyset$). It follows that $\forall \; x \in a \in \alpha_\Lambda$, $x \neq x_a$, $\exists \; n \geq 1$ and $y \notin a$ such that $v_a^n(y) = x$ or in other words $T^n x \notin a$.

This shows that $\bigcup_{a \in \mathfrak{r}} a = I \mod m$.

To continue, we prove 1).

By assumption $\exists \; M > 1$ such that $\left| \frac{D^2 v_a(x)}{D v_a(x)} \right| \leq M \; \forall \; x \in Ta$, $a \in \alpha$.

We now show that

$$(4.3.1) \qquad |D^2 v_g| \leq C |D v_g| \text{ on } \mathcal{D}\mathcal{D}(v_g) \; \forall \; g \in \alpha^*$$

where $C := M + M \sup_{a \in \alpha_\Lambda, \; x \in Ta \setminus a} \left| \frac{x - x_a}{x - v_a(x)} \right|$.

This is clear for $a \in \alpha \setminus \alpha_\Lambda$ and it is sufficient to consider $g = \underbrace{[a, \ldots, a}_{n \text{ times}}, b] = v_a^n(b)$, $(n \geq 1, \; a \in \alpha_\Lambda, \; b \in \alpha, \; b \neq a)$. In this case

$$Dv_g(x) = D(v_a^n) \circ v_b(x) D v_b(x),$$

and

$$D^2 v_g(x) = D^2(v_a^n) \circ v_b(x)(Dv_b(x))^2 + D(v_a^n) \circ v_b(x) D^2 v_b(x)$$

whence

$$\left| \frac{D^2 v_g(x)}{D v_g(x)} \right| \leq \left| \frac{D^2(v_a^n) \circ v_b(x)}{D(v_a^n) \circ v_b(x)} \right| + M |D v_g(x)|$$

and (since $|D v_g| \leq 1 \; \forall \; g \in \alpha^*$) it suffices to show that

$$\left| \frac{D^2(v_a^n)(x)}{D(v_a^n)(x)} \right| \leq M \sup_{a \in \alpha_\Lambda, \; x \notin a} \left| \frac{x - x_a}{x - v_a(x)} \right| \; \forall \; x \notin a.$$

To this end fix $x \notin a$, then

$$D^2(v_a^n)(x) = D(v_a^n)(x) \sum_{k=0}^{n-1} \frac{D^2 v_a(v_a^k(x))}{D v_a(v_a^k(x))} D v_a^k(x),$$

$$\therefore \left| \frac{D^2(v_a^n)(x)}{D(v_a^n)(x)} \right| \leq M \sum_{k=0}^{n-1} |D(v_a^k)(x)|.$$

The proof of (4.3.1) is completed by the following claim:

$$(4.3.2) \qquad \sum_{k=0}^{\infty} D(v_a^k)(x) \leq \left| \frac{x - x_a}{v_a(x) - x} \right| \; \forall \; a \in \alpha_\Lambda, \; x \in \mathcal{D}(v_a), \; x \neq x_a.$$

To see this, we assume that $x < x_a$ (the case $x > x_a$ being analogous). Since x_a is a regular source, it follows that $v_a^k(x) < x \; \forall \; k \geq 1$, whence $D v_a^k \uparrow$ on $(x, v_a(x))$.

Thus, using the mean value theorem,

$$\frac{x_a - x}{v_a(x) - x} = \sum_{k=0}^{\infty} \frac{v_a^{k+1}(x) - v_a^k(x)}{v_a(x) - x}$$

$$= \sum_{k=0}^{\infty} D(v_a^k)(x_k) \text{ where } x_k \in (x, v_a(x))$$

$$\geq \sum_{k=0}^{\infty} D(v_a^k)(x).$$

whence (4.3.2).

It also follows from the assumptions that $\exists\, \theta < 1$ such that

$$|Dv_g| \leq \theta \ \ \forall\, g \in \alpha^*$$

because for each $g \in \alpha^*$ either $g \in \alpha \setminus \alpha_\Lambda$ in which case $|Dv_g| \leq \max_{x \in a \in \alpha \setminus \alpha_\Lambda} |Dv_a(x)| < 1$, or $g = [\underbrace{a, \dots, a}_{n \text{ times}}, b]$ for some $n \geq 1$, $a \in \alpha_\Lambda$, $b \in \alpha$, $b \neq a$ in which case $|Dv_g| \leq |v_{[a,b]}| \leq \frac{1}{\rho(\epsilon)} < 1$ where $\epsilon > 0$ satisfies $\bigcup_{a \in \alpha_\Lambda, \, b \in \alpha, \, b \neq a} [a, b] \subset I \setminus B(\Lambda, \epsilon)$.

To continue, let $a \in \mathfrak{r}$, then $\exists\, N \geq 1$ and $g_1, \dots, g_N \in \alpha^*$ such that $a = [g_1, \dots, g_N]$. It follows that

$$v_a = v_{g_1} \circ \cdots \circ v_{g_N}, \ \ Dv_a = \prod_{k=1}^{N} Dv_{g_k} \circ v_{[g_{k+1}, \dots v_{g_N}]}$$

whence

$$\frac{D^2 v_a}{Dv_a} = \sum_{k=1}^{N} \frac{D^2 v_{g_k} \circ v_{[g_{k+1}, \dots, g_N]}}{Dv_{g_k} \circ v_{[g_{k+1}, \dots, g_N]}} \cdot Dv_{[g_{k+1}, \dots, v_{g_N}]}$$

and

$$\left| \frac{D^2 v_a}{Dv_a} \right| \leq C \sum_{k=1}^{N} \theta^{N-k} \leq C' := \frac{C}{1 - \theta}.$$

This proves (1).

To see (2) note that, as before, for $a \in \mathfrak{r}$, we have that

$$|\log v_a'(x) - \log v_a'(y)| \leq \int_{[x,y]} \left| \frac{D^2 v_a}{Dv_a} \right| dm \leq C' \ \ \forall\, x < y \in \mathcal{D}(v_a)$$

whence $\mathfrak{r}$ is a Schweiger collection for T. $\qquad \square$

§4.4 Ergodic properties of Markov maps with distortion properties

In this section we show that for Markov maps with the weak distortion property, both the conservative and dissipative parts are open modulo sets of measure zero.

We also show that for conservative Markov maps with the weak distortion property:
there is always a σ-finite invariant measure,
metric and topological transitivity are equivalent, as are exactness and topological mixing.

The following is a version of a theorem of Renyi in [**Ren1**].

4.4.1 LEMMA [**A-De-Ur**].
Suppose $(X, \mathcal{B}, m, T, \alpha)$ is a Markov map such that

$$\mathfrak{g}(C, T) = \tilde{\alpha}_+, \ \& \ \inf\{m(B) : B \in T\alpha\} > 0,$$

then there is a T-invariant probability $q << m$ such that $\frac{dq}{dm} \in L^\infty(m)$, and such that $X' := [\frac{dq}{dm} > 0]$ is a proper subsystem.
In case $\#T\alpha < \infty$,

$$\log \frac{dq}{dm} \in L^\infty(m|_{X'}).$$

PROOF. Since $\mathfrak{g}(C, T) = \tilde{\alpha}$, it follows from the distortion proposition that $\forall n \geq 1$, $a \in \alpha_0^{n-1}, A \in \mathcal{B}$,

$$C^{-2} m(A|Tb_n) \leq m(T^{-n}A|a) \leq C^2 m(A|Tb_n).$$

Set $\beta = \{Tb : b \in \alpha\}$, and suppose $m(B) \geq \epsilon > 0$ $\forall B \in \beta$. For $A \in \mathcal{B}$, and $n \geq 1$, we have

$$m(T^{-n}A) = \sum_{\underline{b} \in \alpha^n} m(T^{-n}A|a)m(a)$$

$$\leq C^2 \sum_{\underline{b} \in \alpha^n} m(A|Tb_n)m(a)$$

$$\leq \frac{C^2}{\epsilon} m(A).$$

Hence

$$\sup_{n \geq 1}\left\{\left\|\frac{1}{n}\sum_{k=0}^{n-1}\frac{dm \circ T^{-k}}{dm}\right\|_\infty\right\}_{n \geq 1} < \infty,$$

and by proposition 1.4.3 (Renyi's theorem) $\exists$ a T-invariant probability $q << m$ on X and $n_k \to \infty$ such that $\frac{dq}{dm} \in L^\infty(m)$ and

$$\frac{1}{n_k}\sum_{j=0}^{n_k-1}\frac{dm \circ T^{-j}}{dm} \ \longrightarrow \ \frac{dq}{dm} \ \text{weak} * \text{ in } L^\infty(m).$$

Now set
$$\nu_n(A) = \sum_{B \in \beta} m(A|B)m(T^{-n}B) \quad (n \in \mathbb{N}, A \in \mathcal{B}),$$

then, by the above, $\exists D > 0$ such that

$$\frac{m(T^{-n}A)}{\nu_n(A)} \in [D^{-1}, D] \quad (n \in \mathbb{N}, A \in \mathcal{B})$$

and

$$\frac{d\nu_n}{dm} = \sum_{B \in \beta}\frac{m(T^{-n}B)}{m(B)}1_B.$$

By construction of q, there is a subsequence $n_k \to \infty$ such that

$$\lim_{k \to \infty}\frac{1}{n_k}\sum_{j=0}^{n_k-1} m(T^{-j}B) = q(B) \quad \forall B \in \beta$$

whence if

$$\lambda_n = \frac{1}{n} \sum_{j=0}^{n-1} \nu_j,$$

then

$$\frac{d\lambda_{n_k}}{dm} \xrightarrow{L^\infty(m)} \sum_{B \in \beta} \frac{q(B)}{m(B)} 1_B$$

$$= \frac{d\nu}{dm}$$

where

$$\nu(A) = \sum_{B \in \beta} q(B) m(A|B) \quad \forall A \in \mathcal{B}.$$

It follows that $\frac{dq}{d\nu} \in [D^{-1}, D]$.

Let

$$\alpha' = \{b \in \alpha : b \subseteq B \in \beta, q(B) > 0\}$$

and

$$X' = \bigcup_{b \in \alpha'} b.$$

Next, we claim that X' is a proper subsystem. To see this, first note that, for $B \in \beta$,

$$\nu(B) > 0 \Leftrightarrow \nu(b) > 0 \text{ for some } b \subseteq B \Leftrightarrow \nu(b) > 0 \forall b \subseteq B$$

whence,

$$q(B) > 0 \Leftrightarrow q(b) > 0 \text{ for some } b \subseteq B \Leftrightarrow q(b) > 0 \ \forall b \subseteq B.$$

For $b \in \alpha'$, by the T–invariance of q, $q(Tb) > 0$, hence $Tb \subseteq \bigcup_{B \in \alpha'} B$, and X' is a subsystem. As $T^{-1}(X' \setminus TX') \subseteq X \setminus X'$, it follows that $q(X' \setminus TX') = 0$, whence, by the above, $TX' = X'$ and X' is proper.

Under the assumption $\#\beta < \infty$, the non–zero values of $\frac{d\nu}{dm}$ are uniformly bounded below. Hence, the non–zero values of $\frac{dq}{dm}$ are also uniformly bounded below, whence $\log \frac{dq}{dm} \in L^\infty(m|_{X'})$. $\qquad \square$

4.4.2 LEMMA.
Suppose $(X, \mathcal{B}, m, T, \alpha)$ is a Markov map and that $\mathfrak{r} \subset \mathfrak{g}(C, T)$ is a Schweiger collection for $(X, \mathcal{B}, m, T, \alpha)$.

If $A \in \mathfrak{r}$ and $B \in \mathcal{B} \cap A$, then

$$m(a \cap T^{-n}B) = C^{\pm 6} m(a \cap T^{-n}A) m(B|A) \quad \forall B \ n \geq 1, \ a \in \alpha_0^{n-1}.$$

PROOF. Suppose $A = [\underline{a}] = [a_1, ..., a_n] \in \mathfrak{r}$, and fix $[\underline{c}] \in \mathfrak{r}$. For $k \geq 1$, $a \in \alpha_0^{k-1}$, it follows from the distortion proposition that

$$m([\underline{b}, \underline{a}, \underline{c}]) = \frac{C^{\pm 2}}{m(Ta_n)} m([\underline{b}, \underline{a}]) m([\underline{c}]) \quad \text{since } [\underline{b}, \underline{c}] \in \mathfrak{g}(C, T)$$

$$= C^{\pm 6} \frac{m([\underline{b}, \underline{a}])}{m([\underline{a}])} m([\underline{a}, \underline{c}]) \quad \text{since } [\underline{a}, \underline{c}], [\underline{c}] \in \mathfrak{g}(C, T)$$

$$= C^{\pm 6} m([\underline{b}, \underline{a}]) m([\underline{a}, \underline{c}]|A),$$

and the result follows because $\mathfrak{r}$ generates $\mathcal{B}$. $\qquad \square$

4.4.3 THEOREM.
Suppose $(X, \mathcal{B}, m, T, \alpha)$ *is a Markov map and that* $\mathfrak{r} \subset \mathfrak{g}(C, T)$ *is a Schweiger collection for* $(X, \mathcal{B}, m, T, \alpha)$.
   Let $A \in \mathfrak{r}$, *then*

$$\sum_{n=1}^{\infty} m(T^{-n}A) = \infty \quad \Rightarrow \quad A \subset \mathfrak{C} \quad mod\ m$$

and

$$\sum_{n=1}^{\infty} m(T^{-n}A) < \infty \quad \Rightarrow \quad A \subset X \setminus \mathfrak{C} \quad mod\ m.$$

In particular, $\mathfrak{C}$*, and* $\mathcal{D}$ *are both unions of sets in* $\mathfrak{r}$.

PROOF. The second implication is clear. To prove the first, suppose $A \in \mathfrak{r}$ and $m(A \setminus \mathfrak{C}) > 0$, then

$$\exists B \in \mathcal{B} \cap A, \ m(B) > 0, \ \ni \sum_{n=1}^{\infty} m(T^{-n}B) < \infty$$

whence, by lemma 4.4.2

$$\sum_{n=1}^{\infty} m(T^{-n}A) < \infty.$$

$\square$

4.4.4 THEOREM [A-De-Ur].
Suppose $(X, \mathcal{B}, m, T, \alpha)$ *is a topologically transitive Markov map which has the weak distortion property, then* T *is either conservative, or totally dissipative.*
   If T *is conservative, then,* T *is ergodic.*

PROOF. Assume that

$$\bigcup_{B \in \mathfrak{r}} B = X \quad mod\ m.$$

It follows from theorem 4.4.3 that

$$\mathfrak{C} = \bigcup_{B \in \mathfrak{r} \cap \mathfrak{C}} B \quad mod\ m.$$

Therefore, it follows from irreducibility that T is either conservative, or totally dissipative. Suppose T is conservative. Since $\mathfrak{r}$ generates $\mathcal{B}$, it follows from the distortion proposition that

$$\frac{m(T^{-n}A|a)}{m(A|Tb_n)} \in [C^{-2}, C^2] \quad \forall n \in \mathbb{N}, a \in \alpha_0^{n-1} \cap \mathfrak{r}, \quad A \in \mathcal{B}.$$

Now suppose that $A \in \mathcal{B}$, $T^{-1}A = A$, and $m(A) > 0$, then

$$\frac{m(A|a)}{m(A|Tb_n)} \in [C^{-2}, C^2] \quad \forall n \in \mathbb{N}, a \in \alpha_0^{n-1} \cap \mathfrak{r}.$$

By the martingale convergence theorem (see [**Doo**]), for m–a.e. $x \in X$,

$$m(A|[b_1(x), .., b_n(x)]) \to_{n\to\infty} 1_A(x)$$

where, for $n \geq 1$, $b_n(x)$ is defined by $T^{n-1}x \in b_n(x) \in \alpha$.

By conservativity of T, if $a = [b_1, .., b_n] \in \mathfrak{r}$, then, for m-a.e. $x \in a$, $T^k x \in a$ for infinitely many k, hence

$$\frac{1_A(x)}{m(A|Tb_n)} \in [C^{-2}, C^2].$$

It follows that

$$A = \bigcup_{B \in \mathfrak{r}, m(A\cap B)>0} B \quad \bmod m.$$

Since $m(A) > 0$,

$$\exists B \in \mathfrak{r} \; \ni \; B \subset A \quad \bmod m.$$

By irreducibility, if $B' \in \mathfrak{r}$, then

$$\exists k \geq 0 \ni \quad m(B \cap T^{-k}B') > 0$$

whence $B' \subset A$. Thus $A = X \bmod m$, and T is ergodic. $\qquad \square$

Suppose that $(X, \mathcal{B}, m, T, \alpha)$ is a conservative Markov map, and let $k \in \mathbb{N}$ and $A \in \alpha_0^{k-1}$. Define a partition of A by

$$\alpha_A = \bigcup_{n=1}^{\infty} [\varphi = n] \cap \alpha_0^{n+k-1},$$

where $\varphi : A \to \mathbb{N}$ is the first return time to A under T.

4.4.5 LEMMA. $(A, \mathcal{B} \cap A, m_A, T_A, \alpha_A)$ *is a Markov map.*

PROOF. By the Markov property for $(X, \mathcal{B}, m, T, \alpha)$, we have that $T_A(b) = A$, $\forall b \in \alpha_A$. $\qquad \square$

$(A, \mathcal{B} \cap A, m_A, T_A, \alpha_A)$ is called the *induced Markov map*.

4.4.6 THEOREM [**A-De-Ur**].
Suppose that $(X, \mathcal{B}, m, T, \alpha)$ is an topologically transitive Markov map having the weak distortion property with respect to $\mathfrak{r}$, and T is conservative, then there is a σ-finite, T-invariant measure $\mu \sim m$ such that

$$\log \frac{d\mu}{dm} \in L^{\infty}(B) \quad \forall B \in \mathfrak{r}.$$

PROOF. By theorem 4.4.4 T is ergodic. By §1.5, it suffices to show that $\forall \; A \in \mathfrak{r}$, there is a T_A-invariant probability $\mu \sim m|_A$ such that $\log \frac{d\mu}{dm} \in L^{\infty}(A)$.

Fix $A \in \mathfrak{r}$. By lemma 4.4.5, $(A, \mathcal{B} \cap A, m_A, T_A, \alpha_A)$ is a Markov map and $T_A(b) = A$, $\forall b \in \alpha_A$. Also $\mathfrak{g}(C, T_A) = (\alpha_A)_+$. By lemma 4.4.1, there is a T_A-invariant probability $\mu \sim m|_A$ such that $\log \frac{d\mu}{dm} \in L^{\infty}(A)$. $\qquad \square$

REMARKS.

1) The idea of inducing on $A \in \mathfrak{r}$ in the proof of theorem 4.4.6 can be traced back to [**Bow**].

2) If T has the weak distortion property, and $A \in \overline{\mathfrak{r}(C, T)}$, then $\mathfrak{g}(C, T_A) = (\alpha_A)_+$ and by [**Bra**] $(A, \mathcal{B} \cap A, m_A, T_A, \alpha_A)$ is continued fraction mixing, and by lemma 3.7.4, A is a Darling-Kac set for T.

It follows from proposition 4.7.8 (below) that if for some $0 < r < 1$, $\mathfrak{g}_r(C, T_A) = (\alpha_A)_+$, then $(A, \mathcal{B} \cap A, m_A, T_A, \alpha_A)$ is exponentially continued fraction mixing.

4.4.7 THEOREM [**A-De-Ur**] [**Tha2**].

Suppose that $(X, \mathcal{B}, m, T, \alpha)$ is a topologically mixing Markov map having the weak distortion property with respect to $\mathfrak{r}$, and T is conservative, then, T is exact.

PROOF. By theorem 4.4.4, T is ergodic. Suppose

$$A \in \bigcap_{n=1}^{\infty} T^{-n} \mathcal{B}, \quad m(A) > 0,$$

then

$$\forall n \geq 0, \quad \exists A_n \in \mathcal{B} \ni \quad A = T^{-n} A_n.$$

It follows that $A_n = T^n A$, whence $A = T^{-n} T^n A$, and

$$
\begin{aligned}
T^{-k} A_{n+k} &= T^{-k} T^{n+k} A \\
&= T^n (T^{-(k+n)} T^{(k+n)} A) \\
&= T^n A = A_n.
\end{aligned}
$$

Since $m(A) > 0$,

$$\exists b = [\beta_1, ..., \beta_\nu] \in \mathfrak{r} \ni \quad m(A \cap b) > 0.$$

Set

$$\varphi(x) = \min \{n \in \mathbb{N} : T^n x \in b\},$$

then $\varphi < \infty$ a.e. as T is conservative, ergodic. Define $T^* : X \to b$ by

$$T^* x = T^{\varphi(x)} x,$$

then $T^*|_b = T_b$, the induced transformation which is conservative, ergodic on b. For $k \geq 1$, let

$$\varphi_k(x) = \sum_{j=0}^{k-1} \varphi(T^{*j} x),$$

then $T^{*k} x = T^{\varphi_k(x)} x$, and

$$[b_{\varphi_k(x)+1}, ..., b_{\varphi_k(x)+\nu}] = b.$$

For every $x \in X$, and $k \geq 1$,

$$[b_1(x), ..., b_{\varphi_k(x)+\nu}(x)] \in \mathfrak{r}$$

whence, by the distortion proposition, and the weak distortion property

$$m(A|[b_1(x), ..., b_{\varphi_k(x)+\nu}(x)]) = C^{\pm 2} m(A_{\varphi_k(x)+\nu} | T\beta_\nu).$$

It follows from the martingale convergence theorem (see [**Doo**]), that for a.e. $x \in A$,

$$m(A|[b_1(x), ..., b_{\varphi_k(x)+\nu}(x)]) \to_{k \to \infty} 1_A(x)$$

whence

$$\liminf_{k \to \infty} m(A_{\varphi_k(x)+\nu}|T\beta_\nu) > 0,$$

and

$$\liminf_{k \to \infty} m(A_{\varphi_k(x)+\nu}) > 0.$$

Since $T^*|_b$ is conservative, ergodic on b, $T^{*-1}b = X$, and $m(A \cap b) > 0$, it follows that

$$\bigcup_{n=0}^{\infty} T^{*-n}A = X \quad \text{mod } m.$$

Thus, for a.e. $x \in X$, $\exists\, t \in \mathbb{N}$ such that

$$\liminf_{k \to \infty} m(A_{\varphi_k(T^{*t}x)+\nu}) > 0.$$

Suppose $x \in X$, $t \in \mathbb{N}$, are such, then

$$\exists \epsilon > 0 \,\ni\, m(A_{\varphi_k(T^{*t}x)+\nu}) \geq \epsilon \ \ \forall k \text{ large}.$$

Recall that

$$A_{\varphi_{k+t}(x)+\nu} = A_{\varphi_k(T^{*t}x)+\varphi_t(x)+\nu} = T^{\varphi_t(x)} A_{\varphi_k(T^{*t}x)+\nu}.$$

Next, we claim that

$$(4.4.1) \qquad \exists \epsilon_1 > 0 \,\ni\, B \in \mathcal{B}, \ m(B) \geq \epsilon \ \Rightarrow \ m(T^{\varphi_t(x)}B) \geq \epsilon_1.$$

To see this note first that

$$\exists \alpha_\epsilon \subseteq \alpha_0^{\varphi_t(x)-1} \,\ni\, |\alpha_\epsilon| < \infty, \text{ and } m\Big(\bigcup_{b \in \alpha_\epsilon} b\Big) \geq 1 - \epsilon/2.$$

The map $T^{\varphi_t(x)}$ is invertible, non-singular on every $b \in \alpha_\epsilon$ whence, since $|\alpha_\epsilon| < \infty$,

$$\forall \eta > 0, \ \exists \eta' > 0 \ \ni\, B \in \mathcal{B}, m(B \cap b) \geq \eta \text{ for some } b \in \alpha_\epsilon \ \Rightarrow m(T^{\varphi_t(x)}B) \geq \eta'.$$

It follows from the definition of α_ϵ that

$$m(B) \geq \epsilon \ \Rightarrow \ \exists b \in \alpha_\epsilon \,\ni\, m(B \cap b) \geq \frac{\epsilon}{2|\alpha_\epsilon|}$$

and the claim is established.

From (4.4.1), we obtain that

$$\liminf_{k \to \infty} m(A_{\varphi_k(x)+\nu}) > 0 \text{ a.e. on } X.$$

To establish the theorem, it is sufficient (using the martingale convergence theorem [**Doo**]) to prove that

$$\limsup_{k \to \infty} m(A_{\varphi_k(x)+\nu} \cap T\beta_\nu) > 0 \text{ a.e. on } X.$$

To see this, let $x \in X$ with $m(A_{\varphi_k(x)+\nu}) \geq \epsilon > 0$ for some $\epsilon > 0$, and all large k. By topological mixing of $(X, \mathcal{B}, m, T, \alpha)$

$$\exists p_0 \in \mathbb{N} \,\ni\, m(T^p\beta_\nu) \geq 1 - \epsilon/4, \quad \forall p \geq p_0.$$

By ergodicity of $T^*|_b$,

$$\exists p_1 \geq p_0, l \geq 1 \ni \sum_{k=1}^{\infty} 1_{[\varphi_l = p_1]} \circ T^{*k} = \infty \text{ a.e. on } X.$$

As before,

$$\exists \alpha_{p_1, \epsilon} \subseteq \alpha_0^{p_1 - 1} \cap T\beta_\nu \ni |\alpha_{p_1, \epsilon}| < \infty, \text{ and } m(\bigcup_{b \in \alpha_{p_1, \epsilon}} T^{p_1} b) \geq 1 - \epsilon/2,$$

whence

$$\exists b \in \alpha_{p_1, \epsilon} \ni m(A_{\varphi_k(x) + \nu} \cap T^{p_1} b) \geq \frac{\epsilon}{2|\alpha_{p_1, \epsilon}|}$$

for large k. Also as before, there exists $\epsilon_1 > 0$, such that

$$\text{If } B \in \mathcal{B}, \text{ and } b \in \alpha_{p_1, \epsilon} \ni m(B \cap T^{p_1} b) \geq \frac{\epsilon}{2|\alpha_{p_1, \epsilon}|} \quad \text{then}$$

$$m(T^{-p_1} B \cap b) \geq \epsilon_1, \text{ and hence } m(T^{-p_1} B \cap T\beta_\nu) \geq \epsilon_1.$$

This shows that

$$m(A_{\varphi_k(x) - p_1 + \nu} \cap T\beta_\nu) = m(T^{-p_1} A_{\varphi_k(x) + \nu} \cap T\beta_\nu) \geq \epsilon_1$$

for large k.

Because of the choice of p_1, we have

$$\epsilon_1 \leq m(A_{\varphi_{j+l}(x) - p_1 + \nu} \cap T\beta_\nu)$$
$$= m(A_{\varphi_j(x) + \varphi_l(T^{*j}x) - p_1 + \nu} \cap T\beta_\nu)$$
$$= m(A_{\varphi_j(x) + \nu} \cap T\beta_\nu)$$

for infinitely many j, and

$$\limsup_{k \to \infty} m(A_{\varphi_k(x) + \nu} \cap T\beta_\nu) \geq \epsilon_1.$$

This establishes the theorem. $\square$

§4.5 Markov shifts

Let S be a countable set, and P be a stochastic matrix on S.

Let $(X_P, \mathcal{B}(X_P), m_\pi, T)$ be the Markov shift of P (assumed proper) where $\pi \in \mathcal{P}(S)$, $\pi_s > 0 \, \forall \, s \in S$. In this case T is non-singular, and for $a = [s_1, \ldots, s_n]$ we have

$$v_a'(x) = m_\pi(a) \frac{p_{s_n, x_1}}{\pi_{x_1}} \quad (x \in T^n a = T[s_n]).$$

Recall the Frobenius-Perron operator $\widehat{T} := \widehat{T}_{m_\pi} : L^1(m_\pi) \to L^1(m_\pi)$ defined by

$$\int_X \widehat{T} f g \, dm = \int_X f g \circ T \, dm.$$

We have

$$\widehat{T}^n f(x) = \sum_{a \in \alpha_0^{n-1}} 1_{\mathcal{D}(v_a)}(x) v_a' f \circ v_a(x)$$

$$= \sum_{s_1, \ldots, s_n \in S} 1_{T[s_n]}(x) m_\pi([s_1, \ldots, s_n]) \frac{p_{s_n, x_1}}{\pi_{x_1}} f(s_1, \ldots, s_n, x)$$

and in particular

$$\widehat{T}^n 1_{[t]}(x) = \sum_{s_1,\ldots,s_n \in S} 1_{T[s_n]}(x) m_\pi([s_1,\ldots,s_n]) \frac{p_{s_n,x_1}}{\pi_{x_1}} 1_{[t]}(s_1,\ldots,s_n,x)$$

$$= \sum_{s_2,\ldots,s_n \in S} 1_{T[s_n]}(x) \pi_t p_{t,s_2} p_{s_2,s_3} \cdots p_{s_{n-1},s_n} p_{s_n,x_1} \frac{1}{\pi_{x_1}}$$

$$= \frac{p_{t,x_1}^{(n)}}{\pi_{x_1}}.$$

This fact implies that cylinder sets are Darling-Kac sets for T when T is conservative and ergodic.

DEFINITION: RECURRENT STATE. A state $s \in S$ is called *(probabilistically) recurrent* if

$$\sum_{n=0}^\infty p_{s,s}^{(n)} = \infty.$$

Note that recurrent states are G_P-persistent, but that there may be non-recurrent, G_P-persistent states.

We denote the collection of recurrent states by S_r. In case $S_r = S$, we call P *recurrent*.

4.5.1 THEOREM.
$$\mathfrak{C}(T) = \bigcup_{s \in S_r} [s].$$

In particular, if P is recurrent, then T is conservative.

PROOF.   If $s \notin S_r$, then

$$\sum_{n=1}^\infty 1_{[s]} \circ T^n < \infty \text{ a.e. on } [s],$$

and by the Halmos recurrence theorem,

$$[s] \subset \mathcal{D}(T) \qquad \mod m.$$

If $m([s] \setminus \mathfrak{C}(T)) > 0$, then $\exists W \in \mathcal{W}_+ \cap [s]$, and

$$m(W) \sum_{n=0}^\infty p_{s,s}^{(n)} = \int_W \sum_{n=0}^\infty \widehat{T}^n 1_{[s]}$$

$$= \int_{[s]} \sum_{n=0}^\infty 1_W \circ T^n dm$$

$$\leq \pi_s < \infty.$$

This implies that $s \notin S_r$.

Thus $s \in S_r \iff m([s] \setminus \mathfrak{C}(T)) = 0$, and the theorem is established.

$\square$

A stochastic matrix P is *irreducible* if its incidence graph G_P is irreducible, equivalently

$$\forall s,t \in S, \quad \exists n \in \mathbb{N} \ni p_{s,t}^{(n)} > 0.$$

If a stochastic matrix is irreducible, then

$$S_r \neq \emptyset \Rightarrow S_r = S.$$

To see this, let $s \in S_r, t \in S$, and $k, \ell \geq 1$ such that

$$p_{s,t}^{(k)}, \ p_{t,s}^{(\ell)} > 0,$$

then,

$$\sum_{n=1}^{\infty} p_{t,t}^{(n)} \geq \sum_{n=1}^{\infty} p_{t,t}^{(n+k+\ell)}$$

$$\geq \sum_{n=1}^{\infty} p_{t,s}^{(\ell)} p_{s,s}^{(n)} p_{s,t}^{(k)}$$

$$= \infty.$$

It follows from the theorem, that the shift T is either conservative, or totally dissipative.

4.5.2 LEMMA.
Fix $s \in S_r$, and let m_s be that constant multiple of $m|_{[s]}$ with $m_s([s]) = 1$, then $T_{[s]}$ is an exact measure preserving transformation of $([s], \mathcal{B} \cap [s], m_s)$.

PROOF.   Set

$$\alpha = \{[s, t_1, t_2, ..., t_{n-1}, s] : n \in \mathbb{N}, \ t_1, ..., t_{n-1} \in S \setminus \{s\}\}.$$

By theorem 2.1.4, $[s] \subset \mathfrak{C}(T)$, and it follows that α is a partition of $[s]$ mod m.
Let

$$\tilde{\alpha} = \{[s, t_1, t_2, ..., t_{n-1}, s] : n \in \mathbb{N}, \ t_1, ..., t_{n-1} \in S\}.$$

Clearly, the σ-algebra of subsets of $[s]$ generated by $\{T_{[s]}^{-n} A : n \geq 0, A \in \alpha\}$ is that generated by $\tilde{\alpha}$. It also follows from $[s] \subset \mathfrak{C}(T)$ that any cylinder which is a subset of $[s]$ is a countable union mod m of cylinders in $\tilde{\alpha}$, whence the σ-algebra of subsets of $[s]$ generated by $\{T_{[s]}^{-n} A : n \geq 0, A \in \alpha\}$ is $\mathcal{B} \cap [s]$ mod m.

For $A_1, ..., A_k \in \alpha$,

$$m_s\left(\bigcap_{j=1}^{k} T_{[s]}^{-(j-1)} A_j\right) = \prod_{j=1}^{k} m_s(A_j).$$

Thus

$$m_s \circ T_{[s]}^{-1} = m_s,$$

and, by the Kolmogorov $0 - 1$ law, $T_{[s]}$ is exact.

$\square$

4.5.3 THEOREM. *If P is irreducible, recurrent then T is conservative and ergodic with respect to m, there is a σ-finite T-invariant measure $\mu \sim m$*
    and
if G_P is aperiodic, then T is exact.

PROOF. By theorem 4.5.1, T is conservative.
Suppose that $T^{-1}A = A \in \mathcal{B}_+$. We'll show that

$$A \supset [s] \mod m \ \forall \ s \in S.$$

For every $s \in S$,

$$T_{[s]}^{-1}(A \cap [s]) = \bigcup_{n=1}^{\infty} [\varphi_{[s]} = n] \cap T^{-n}A = A \cap [s],$$

whence,

$$m(A \cap [s]) > 0 \ \Rightarrow \ A \supset [s] \mod m.$$

There is $t \in S$ such that $A \supset [t] \mod m$. Let $s \in S$, then, by irreducibility, there exists $n = n(s,t) \in \mathbb{N}$ such that $p_{s,t}^{(n)} > 0$, whence

$$m([s] \cap A) = m([s] \cap T^{-n}A)$$
$$\geq m([s] \cap T^{-n}[t]) > 0$$
$$\Rightarrow \ A \supset [s] \mod m.$$

This shows that T is ergodic.

It follows that there is an initial distribution π such that $(X_P, \mathcal{B}(X_P), m_\pi, T, \alpha)$ has the weak distortion property, and hence by theorem 4.4.6 there is a σ-finite T-invariant measure $\mu \sim m$. Exactness under the assumption of aperiodicity and recurrence follows from theorem 4.4.7. $\qquad\square$

Our next result is that the T-invariant measure μ (unique up to constant multiplication by theorem 1.5.6) is given by the so called *Chung taboo* distribution.
    Set for $s, t \in S$:

$$_sp_{s,t}^{(1)} = p_{s,t}, \quad _sp_{s,t}^{(n+1)} = \sum_{u \neq s} {_sp_{s,u}^{(n)}} p_{u,t} \quad (n \geq 1),$$

and

$$_sp_{s,t}^* = \sum_{n=1}^{\infty} {_sp_{s,t}^{(n)}}.$$

4.5.4 PROPOSITION [Chu].
If P is irreducible and recurrent, then

$$\sum_{u \in S} {_sp_{s,u}^*} p_{u,t} = {_sp_{s,t}^*},$$

$$\mu = m_\tau \ where \ \tau_t = c {_sp_{s,t}^*} \ \forall \ t \in S \ for \ some \ c > 0, \ s \in S;$$
and $[s]$ is a Darling- Kac set for $T_P \ \forall \ s \in S$.

PROOF. For $s \in S$, let

$$\mu_s(B) = \int_{[s]} \left(\sum_{k=0}^{\varphi_{[s]}-1} 1_B \circ T^k \right) dm_s,$$

then, by proposition 1.5.7, $\mu_s \circ T^{-1} = \mu_s \ll m$.

By unicity of invariant measure (theorem 1.5.6), $\mu_s = c\mu$ for some $c > 0$, and

$$\mu_t \equiv \mu_s([t])\mu_s \quad \forall \, t \in S,$$

and, necessarily, $0 < \mu_s([t]) < \infty$ for $s, t \in S$. Since $\mu_s|_{[s]} = m_s|_{[s]}$, it follows that

$$\mu_s([t_1, ..., t_n]) = \mu_s([t_1]) \prod_{k=1}^{n-1} p_{t_k, t_{k+1}}.$$

To calculate $\mu_s([t])$, note that

$$_s p_{s,t}^{(n)} = \mu_s([s] \cap T^{-n}[t] \setminus \bigcup_{k=1}^{n-1} T^{-k}[s]).$$

It follows that

$$_s p_{s,t}^* = \sum_{n=1}^{\infty} \mu_s([s] \cap T^{-n}[t] \setminus \bigcup_{k=1}^{n-1} T^{-k}[s])$$

$$= \sum_{n=1}^{\infty} \mu_s([s] \cap T^{-1} B_{n-1})$$

where

$$B_0 = [t], \text{ and } B_n = T^{-n}[t] \setminus \bigcup_{k=0}^{n-1} T^{-k}[s].$$

Clearly

$$T^{-1} B_n = ([s] \cap T^{-1} B_n) \cup B_{n+1}$$

this union being disjoint, whence

$$_s p_{s,t}^{(n+1)} \mu_s([s] \cap T^{-1} B_n) = \mu_s(B_n) - \mu_s(B_{n+1})$$

and

$$_s p_{s,t}^* = \mu_s(B_0) = \mu_s([t]).$$

The first two statements now follow.

We now consider the conservative, ergodic measure preserving transformation $(X_P, \mathcal{B}(X_P), m_\mu, T)$ where $\mu_t =_s p_{s,t}^*$. It follows that

$$\widehat{T}^k 1_{[t]}(x) = \frac{\mu_t}{\mu_{x_1}} p_{t,x_1}^{(k)} \quad \forall \, k \in \mathbb{N}, \, t \in S$$

and so each $[t]$ is a Darling-Kac set for T. $\square$

DEFINITION: POSITIVE, AND NULL RECURRENCE. A state $s \in S$ is called *positively recurrent* if $\varliminf_{n \to \infty} \frac{1}{n} \sum_{k=0}^{n-1} p_{s,s}^{(k)} > 0$, and *null recurrent* otherwise. For irreducible recurrent P, either all states are positively recurrent, or all states are null recurrent and P is called *positively-* or *null recurrent* accordingly.

4.5.5 PROPOSITION. *Suppose that P is irreducible recurrent, then P is positively recurrent $\Leftrightarrow$ for some, and hence all $s \in S$, $\sum_{t \in S} {}_s p_{s,t}^* < \infty$.*

PROOF.
$(X_P, \mathcal{B}(X_P), m_\mu, T)$ is a conservative, ergodic measure preserving transformation, whence by the ergodic theorem

$$\frac{1}{n} \sum_{k=0}^{n-1} p_{s,s}^{(k)} = \frac{1}{n\mu_s} \sum_{k=0}^{n-1} m_\mu([s] \cap T^{-k}[s]) \to \frac{\mu_s}{m_\mu(X_P)}.$$

It follows that P is positively recurrent if and only if $\mu(S) = m_\mu(X_P) < \infty$. $\square$

Lastly, we give a sufficient condition for positive recurrence based on Renyi's theorem.

4.5.6 PROPOSITION. *Suppose that P is a stochastic matrix on S, and suppose that $\exists\, M > 1$ and $q \in \mathcal{P}(S)$, $q_s > 0\ \forall\, s \in S$ such that*

$$p_{s,t} \leq M q_t \ \forall\ s, t \in S,$$

then P is positively recurrent and $\sup_{t \in S} \frac{{}_s p_{s,t}^*}{q_t} < \infty\ \forall\, s \in S$.

PROOF.
Consider $(X_P, \mathcal{B}(X_P), m_q, T, \alpha)$, the Markov shift of P with initial distribution q ($\alpha = \{[s] : s \in S\}$). For $f : X_P \to \mathbb{R}$ bounded, measurable,

$$|\widehat{T}^n f(x)| \leq \sum_{s_1, \dots, s_n \in S} 1_{T[s_n]}(x) m_q([s_1, \dots, s_n]) \frac{p_{s_n, x_1}}{q_{x_1}} |f(s_1, \dots, s_n, x)| \leq M \|f\|_\infty.$$

The conditions of lemma 4.4.1 are satisfied and so $\exists$ a T-invariant, m_q-absolutely continuous probability with bounded density. The proposition follows from this. $\square$

EXERCISE 4.5.1. Show by example that the condition of proposition 4.5.6 is not necessary for positive recurrence.

More generally,

§4.6 Schweiger's Jump transformation

Suppose that $\mathfrak{r}$ is a Schweiger collection for the Markov map $(X, \mathcal{B}, m, T, \alpha)$ and let $N_\mathfrak{r} : X \to \mathbb{N} \cup \infty$ be defined by

$$N_\mathfrak{r}(x) = \inf\{n \in \mathbb{N} : [a_1(x), \dots, a_n(x)] \in \mathfrak{r}\}$$

where $T^{k-1}x \in a_k(x) \in \alpha$ and $\inf \emptyset := \infty$, then $N_\mathfrak{r} < \infty$ a.e.

Schweiger's *jump transformation* ([**Schw1**], see also [**Schw3**]) $T^* : X \to X$ is defined by

$$T^*(x) = T^{N_\mathfrak{r}(x)}(x).$$

If $T : B \to X$ is onto $\forall B \in \alpha$, then $(X, \mathcal{B}, m, T^*)$ is a non–singular transformation.

In general, by the Markov property,

$$\forall n \geq 1, \ \exists \tau_n \subseteq \alpha \ \ni \ T^{*n} X = \bigcup_{b \in \tau_n} b.$$

Clearly, $T^{*n} X \supset T^{*n+1} X$, and $\tau_n \supset \tau_{n+1}$, $\forall n \geq 1$, whence

$$X_{T^*} := \bigcap_{n=1}^{\infty} T^{*n} X = \bigcup_{b \in \tau} b \quad \mathrm{mod} \ m$$

where $\tau = \bigcap_{n \in \mathbb{N}} \tau_n$.

In general (i.e. among Markov maps with the weak distortion property), it may be that $m(X_{T^*}) = 0$ (i.e. $\tau = \emptyset$).

4.6.1 PROPOSITION. *Suppose that* $\inf\{m(B) : B \in T\alpha\} > 0$, *then* $m(X_{T^*}) > 0$.

PROOF. Let $\beta = \{Tb : b \in \alpha\}$, then it follows from $\inf\{m(B) : B \in T\alpha\} > 0$ that

$$\exists \alpha_\Lambda \subseteq \alpha, \ |\alpha_\Lambda| < \infty \ \ni \ B \cap \alpha_\Lambda \neq \emptyset \ \forall B \in \beta.$$

Also,

$$\forall n \geq 1, \ \exists \emptyset \neq \alpha_n \subseteq \alpha \ \ni \ T^{*n} X = \bigcup_{B \in \alpha_n} B.$$

Since, for every $n \geq 1$, $T^{*n} X$ is a union of sets in β, it follows that $1 \leq |\alpha_n \cap \alpha_\Lambda| < \infty$ $\forall n \geq 1$, whence, since $\alpha_n \supset \alpha_{n+1}$,

$$\bigcap_{n \in \mathbb{N}} \alpha_n \supset \bigcap_{n \in \mathbb{N}} \alpha_n \cap \alpha_\Lambda \neq \emptyset$$

whence $m(X_{T^*}) > 0$. $\qquad\qquad\qquad\qquad\qquad\qquad\qquad\qquad\qquad\qquad \square$

Note that $N_{\mathfrak{r}}$ is a stopping time with respect to α in the sense that $[N_{\mathfrak{r}} = n] \in \sigma(\alpha_0^{n-1})$ $\forall n \in \mathbb{N}$. We have

4.6.2 PROPOSITION.
Suppose that $m(X_{T^*}) > 0$, *then* $(X_{T^*}, \mathcal{B} \cap X_{T^*}, m|_{X_{T^*}}, T^*, \alpha^*)$ *is a Markov map where* $\alpha^* = \bigcup_{n=1}^{\infty} X_{T^*} \cap [N_{\mathfrak{r}} = n] \cap \alpha_0^{n-1}$, $\mathfrak{g}(C, T^*) = \widetilde{\alpha^*}_+$ *and* $\{T^*b : b \in \alpha^*\} \subseteq \{Tb : b \in \alpha\}$.

The first result that we'll prove using the jump transformation is

4.6.3 THEOREM [A-De-Ur].
Suppose $(X, \mathcal{B}, m, T, \alpha)$ *is a topologically transitive Markov map with the weak distortion property, and such that* $\inf\{m(B) : B \in T\alpha\} > 0$, *then* $\mathfrak{C}(T) \neq \emptyset$ *mod* m.

Consequently, if $(X, \mathcal{B}, m, T, \alpha)$ *is topologically transitive, then* T *is conservative and ergodic.*

PROOF.

By propositions 4.6.1 and 4.6.2, the Markov map $(X_{T^*}, \mathcal{B} \cap X_{T^*}, m|_{X_{T^*}}, T^*, \alpha^*)$ satisfies the conditions of lemma 4.4.1 and so $\exists$ a m-absolutely continuous, T^*-invariant probability on X_{T^*}. It follows that the conservative part of T^*, $\mathfrak{C}(T^*)$ has positive measure, and, by theorem 4.4.3, is a union of sets in $\widetilde{\alpha^*}_+$.

If T is totally dissipative, then $\exists \, W \in \mathcal{B}$ a wandering set for T with $m(W) > 0$ such that $W \subset \mathfrak{C}(T^*)$. However,

$$\bigcup_{\nu=1}^{\infty} T^{*-\nu} W \subseteq \bigcup_{\nu=1}^{\infty} T^{-\nu} W \subseteq W^c$$

whence, by the conservativity of T^* on $\mathfrak{C}$, $m(W) = 0$, and $\mathfrak{C}(T) \supset \mathfrak{C}(T^*) \neq \emptyset$ mod m.

If $(X, \mathcal{B}, m, T, \alpha)$ is topologically transitive, then by theorem 4.4.4, T is conservative and ergodic. $\qquad \square$

To conclude this section, we consider the jump transformation of $(I, \mathcal{B}(I), m, T, \alpha)$, a C^2-Markov interval map satisfying Thaler's assumptions (as in §4.3) with respect to

$$\mathfrak{r} = \mathfrak{r}(I, \mathcal{B}(I), m, T, \alpha) := \{[s_1, ..., s_n] : \text{ either } s_n \notin \alpha_\Lambda, \text{ or } s_{n-1} \neq s_n\}.$$

As in §4.3, set

$$\alpha^* := \alpha \setminus \alpha_\Lambda \cup \{[\underbrace{b, \ldots, b}_{n \text{ times}}, a] : n \geq 1, \ b \in \alpha_\Lambda, \ a \in \alpha \ a \neq b\},$$

then (see the proof of proposition 4.3.5) α^* is a partition of I.

The jump transformation $T^* : I \to I$ is now defined by $T^* x = T^{N(x)} x$ where

$$N(x) = \begin{cases} 1 & x \in a \notin \alpha_\Lambda, \\ \min\{n \geq 1 : T^n x \notin a\} + 1 & x \in a \in \alpha_\Lambda. \end{cases}$$

If $m(X_{T^*}) > 0$ $(X_{T^*} := \bigcap_{n=1}^{\infty} T^{*n} I)$, then $(X_{T^*}, \mathcal{B}(I), m, T^*, \alpha^*)$ is an expanding C^2-Markov interval map; and

$$T^* \alpha^* = T\left((\alpha \setminus \alpha_\Lambda) \cup (\alpha \cap \bigcup_{a \in \alpha_\Lambda} Ta) \right) \subset T\alpha.$$

Recall from §4.3 that $\forall \, n \geq 1$, $(I, \mathcal{B}(I), m, T^n, \alpha_0^{n-1})$ also satisfies Thaler's assumptions. Here $\alpha_{\Lambda(T^n)} = \{[\underbrace{s, \ldots, s}_{n-\text{times}}] : s \in \alpha_{\Lambda(T)}\}$.

4.6.4 PROPOSITION. *If $(I, \mathcal{B}(I), m, T, \alpha)$ is topologically mixing and $|T\alpha| < \infty$, then $\exists \, n \geq 1$ such that $X_{(T^n)^*} = I$ mod m.*

PROOF. Firstly, we may assume that $\alpha_\Lambda \neq \alpha$, since this is the case for T^2 which is also topologically mixing.

Since $(I, \mathcal{B}(I), m, T, \alpha)$ is topologically mixing, $\forall \, s, t \in \alpha$, $\exists \, n_{s,t} \geq 1$ such that $T^n s \supset t \, \forall \, n \geq n_{s,t}$.

Let $s_1, \ldots, s_N \in \alpha$ be such that $T\alpha = \{Ts_1, \ldots, Ts_N\}$, then $T^n s = I$ mod m $\forall \, s \in S$, $n \geq \max_{1 \leq i,j \leq N} n_{s_i, s_j} + 1$.

Fixing such an n and $s \in \alpha \setminus \alpha_{\Lambda(T)}$, we have that

$$(T^n)^* I \supseteq \bigcup_{a \in \alpha_0^{n-1} \setminus \alpha_{\Lambda(T^n)}} T^n a \supseteq \bigcup_{a \in \alpha_0^{n-1} \setminus \alpha_{\Lambda(T^n)}, \ a \subset s} T^n a = T^n s = I.$$

$\square$

§4.7 Smooth Frobenius-Perron operators and the Gibbs property

Throughout this section, we fix $r \in (0,1)$ and consider Markov maps $(X, \mathcal{B}, m, T, \alpha)$ satisfying $\mathfrak{g}_r(C,T) = \tilde{\alpha}_+$. We call this property the *Gibbs property*.

It was shown in §4.3 that all Markov shifts with the strong distortion property, and expanding C^2-Markov interval maps have the (stronger) Gibbs property.

A topologically mixing Markov map $(X, \mathcal{B}, m, T, \alpha)$ with the Gibbs property satisfying

$\inf_{a \in \alpha} m(Ta) > 0$ has (by lemma 4.4.1) a bounded, invariant density (which is also bounded away from 0 in case $|T\alpha| < \infty$) and is exact (by theorem 4.4.7).

Define the metric $d = d_r$ on $X \subset S^{\mathbb{N}}$ by $d(x,y) = r^{t(x,y)}$ where $t(x,y) = \min\{n \geq 1 : x_n \neq y_n\} \leq \infty$, then (X,d) is a Polish space and $T : X \to X$ is Lipschitz continuous on each $a \in \alpha$.

In this section we show that

the invariant density is in fact Lipschitz continuous (theorem 4.7.2 below due to M. Halfant see [**Half**]),

and that

the Frobenius-Perron operator is quasi compact on the space of Lipschitz continuous functions (theorem 4.7.3 below).

In case $(X, \mathcal{B}, m, T, \alpha)$ is probability preserving then it is *exponentially continued fraction mixing* in the sense that $\exists \ K > 1$ and $0 < \theta < 1$ such that

$$|m(a \cap T^{-(n+k)}B) - m(a)m(B)| \leq K\theta^n m(a)m(B) \ \forall \ n, k \geq 1, \ a \in \alpha_0^{k-1}, \ B \in \mathcal{B}.$$

We'll use this last property to show that certain infinite ergodic interval maps have Darling - Kac sets.

The Frobenius-Perron operators $\widehat{T}^n : L^1(m) \to L^1(m)$ defined by

$$\int_X \widehat{T}^n f \cdot g \, dm = \int_X f \cdot g \circ T^n dm$$

have the form

$$\widehat{T}^n f = \sum_{b \in \beta} 1_b \sum_{a \in \alpha_0^{n-1}, \ T^n a \supset b} v_a' \cdot f \circ v_a$$

where β is the partition generated by $T\alpha$.

DEFINITION: LIPSCHITZ CONTINUITY.
1) A function $f : X \to \mathbb{R}$ is *Lipschitz continuous on* $A \subset X$ if

$$D_A f := \sup_{x,y \in A} \frac{|f(x) - f(y)|}{d(x,y)} < \infty,$$

and *Lipschitz continuous at* $x \in X$ if it is Lipschitz continuous on some neighbourhood of x.

2) A function is *locally Lipschitz continuous on* $A \subset X$ if it is Lipschitz continuous at every point of A.

3) Given a partition γ of X, a function $f : X \to \mathbb{R}$ is γ-*piecewise Lipschitz continuous* on X if it is Lipschitz continuous on each $A \in \gamma$ and $D_\gamma f := \sup_{A \in \gamma} D_A f < \infty$. Note that any bounded γ-piecewise Lipschitz continuous function is Lipschitz continuous on X.

4) The collection of β-piecewise Lipschitz continuous functions on X is denoted by L and equipped with the norm $\|f\|_L := \|f\|_{L^1(m)} + D_\beta f$.

Note that $\|f\|_{L^\infty(m)} \leq \|f\|_L$, and that in this notation,

$$\mathfrak{g}_r(C, T) := \{a \in \tilde{\alpha}_+ : D_{\mathcal{D}(v_a)}(\log v_a') \leq C\}.$$

4.7.1 Proposition. *Suppose that* $(X, \mathcal{B}, m, T, \alpha)$ *has the Gibbs property, that* $\mu \sim m$ *and that* $\log \frac{d\mu}{dm}$ *is Lipschitz continuous on* X,
then, $(X, \mathcal{B}, \mu, T, \alpha)$ *also has the Gibbs property.*

Proof.
Note first that $(X, \mathcal{B}, m, T, \alpha)$ has the Gibbs property iff

$$\sup_{a \in \bigcup_{n=1}^\infty \alpha_0^{n-1}} D_{\mathcal{D}(v_a)}(\log v_{a,m}') < \infty$$

where $v_{a,m}' := \frac{dm \circ v_a}{dm}$.
For each $n \geq 1$, $a \in \alpha_0^{n-1}$, we have that

$$v_{a,\mu}' = v_{a,m}' \frac{h \circ v_a}{h}$$

whence

$$D_{\mathcal{D}(v_a)}(\log v_{a,\mu}') \leq D_{\mathcal{D}(v_a)}(\log v_{a,m}') + D_{\mathcal{D}(v_a)}(\log h) + D_{\mathcal{D}(v_a)}(\log h \circ v_a)$$
$$\leq D_{\mathcal{D}(v_a)}(\log v_{a,m}') + 2D_X(\log h).$$

and the result follows. $\qquad\qquad\qquad\qquad\qquad\qquad\qquad\qquad\qquad\qquad\quad\square$

Definition: Adapted pair.
We'll call a pair of Banach spaces $(\mathcal{C}, \mathcal{L})$ *adapted* if $\mathcal{L} \subset \mathcal{C}$, $\|\cdot\|_{\mathcal{C}} \leq \|\cdot\|_{\mathcal{L}}$, $(\overline{\mathcal{L}})_{\mathcal{C}} = \mathcal{C}$,

$$x_n \in \mathcal{L} \ (n \geq 1), \ x_n \xrightarrow{\mathcal{C}} x \implies x \in \mathcal{L}, \|x\|_{\mathcal{L}} \leq \sup_n \|x_n\|_{\mathcal{L}},$$

and $\mathcal{L}$-bounded sets are precompact in $\mathcal{C}$.

Examples of adapted pairs.
1) If K is a compact metric space, $\mathcal{C}$ the collection of continuous $\mathbb{R}$-functions, and $\mathcal{L}$ the collection of Lipschitz continuous $\mathbb{R}$-valued functions then by the (classical) Arzela-Ascoli theorem, $(\mathcal{C}, \mathcal{L})$ is adapted.
2) If $\mathcal{C} = L^1([0, 1])$ and $\mathcal{L} = BV([0, 1])$ then by Helly's selection principle, the pair $(\mathcal{C}, \mathcal{L})$ is adapted.

It will follow from the version of the Arzela-Ascoli theorem below that if $(X, \mathcal{B}, m, T, \alpha)$ is a Markov map, and $\mathcal{C} = L^1(m)$, $\mathcal{L} = L$ then the pair $(\mathcal{C}, \mathcal{L})$ is adapted.

Let $(\mathcal{C}, \mathcal{L})$ be an adapted pair of Banach spaces. We consider a linear operator $P : \mathcal{L} \to \mathcal{L}$ satisfying

(4.7.1) $$\|P^n x\|_{\mathcal{C}} \le H\|x\|_{\mathcal{C}} \ \forall n \in \mathbb{N}, x \in \mathcal{L},$$

and

(4.7.2) $$\|Px\|_{\mathcal{L}} \le \theta\|x\|_{\mathcal{L}} + R\|x\|_{\mathcal{C}} \ \forall \ x \in \mathcal{L}$$

where $R, H \in \mathbb{R}_+$ and $\theta \in (0, 1)$.

The important result for operators of this type is the theorem of Ionescu-Tulcea and Marinescu (see [**Io-Mar**]) which asserts that such an operator P acts quasi-compactly on $\mathcal{L}$.

Let $(X, \mathcal{B}, m, T, \alpha)$ be a Markov map satisfying the Gibbs property and $\inf_{a \in \alpha} m(Ta) > 0$; $\mathcal{C} = L^1(m)$, $\mathcal{L} = L$ and $P = \widehat{T}$. We'll show that P satisfies (4.7.1), and that P^k satisfies (4.7.2) for k large, deducing our results from a special case of the Ionescu-Tulcea, Marinescu theorem.

REMARK.

Interval maps with Frobenius-Perron operators satisfying (4.7.1) and (4.7.2) for the adapted pair
$(L^1([0, 1]), BV([0, 1]))$ are considered in [**Hof-Kel**].

ARZELA-ASCOLI THEOREM. *If $f_n \in L$, and $\sup_{n \ge 1} \|f_n\|_L < \infty$, then $\exists \ n_k \to \infty$ and $g \in L$ such that*

$$v_{n_k}(x) \to g(x) \quad as \ k \to \infty \ \forall \ x \in X,$$

$$\|g\|_L \le \liminf_{n \to \infty} \|f_n\|_L,$$

and

$$\|v_{n_k} - g\|_1 \to 0 \quad as \ k \to \infty.$$

PROOF. Let $f_n \in L$, $\|f_n\|_L \le K$ $(n \ge 1)$ where $K > 0$. We have that $|f_n(x)| \le 2K$ $(x \in X, \ n \ge 1)$ and $|f_n(x) - f_n(y)| \le Kd(x, y) \ \forall \ x, y \in X, \ n \ge 1$.

Let $\Gamma \subset X$ be a countable dense set in X. There is a subsequence v_{n_k} so that $(v_{n_k}(\gamma))_{k \ge 1}$ is a Cauchy sequence $\forall \gamma \in \Gamma$

It follows that $(v_{n_k}(x))_{k \ge 1}$ is a Cauchy sequence $\forall x \in X$, and so $\exists \ g : X \to \mathbb{R}$ such that $v_{n_k}(x) \to g(x) \ \forall x \in X$. It follows from Fatou's lemma that $\|g\|_1 \le \liminf_{k \to \infty} \|v_{n_k}\|_1$. We have that for $x, y \in b \in \beta$, $x \ne y$

$$\frac{|g(x) - g(y)|}{d(x, y)} \leftarrow \frac{|v_{n_k}(x) - v_{n_k}(y)|}{d(x, y)} \le D_\beta v_{n_k}$$

whence $D_\beta g \le \liminf_{k \to \infty} D_\beta v_{n_k}$. Thus $g \in L$ and $\|g\|_L \le \liminf_{k \to \infty} \|v_{n_k}\|_L$.

Lastly, by the dominated convergence theorem,

$$\|v_{n_k} - g\|_1 \to 0 \ \text{ as } k \to \infty.$$

$\square$

We now consider $\widehat{T}$ acting on the space L.

4.7.2 LEMMA. *Suppose that $g \in L$ and $a \in \alpha_0^{n-1}$, then*

$$|v_a'(x)g(v_ax) - v_a'(y)g(v_ay)| \le M''d(x,y)\left(M\int_a |g|dm + (M+1)m(a)r^nD_ag\right).$$

PROOF.

$$|v_a'(x)g(v_a(x)) - v_a'(y)g(v_a(y))|$$
$$\le v_a'(x)|g(v_a(x))|\left|\frac{v_a'(y)}{v_a'(x)} - 1\right| + v_a'(y)|g(v_a(x)) - g(v_a(y))|$$
$$= I \ + \ II.$$

We have that

$$|g(v_ax)| \le \frac{1}{m(a)}\int_a |g|dm + r^nD_ag \ \ \forall \ x \in X.$$

Hence, by Renyi's property, and $(\mathcal{J})$,

$$I \le MM''d(x,y)m(a)|g(v_a(x))|$$
$$\le MM''d(x,y)\left(\int_a |g|dm + m(a)r^nD_ag\right).$$

Also by Renyi's property,

$$II \le M''m(a)d(v_a(x),v_a(y))D_ag = M''d(x,y)m(a)r^nD_ag.$$

The result is that

$$I + II \le M''d(x,y)\left(M\int_a |g|dm + (M+1)m(a)r^nD_ag\right).$$

$\square$

4.7.3 PROPOSITION. *For $f \in L$,*

$$\|\widehat{T}^nf\|_L \le M''\left((M+2)r^nD_\beta f + (M+1)\|f\|_1\right)$$

PROOF. Let $g \in L$, then

$$\widehat{T}^ng = \sum_{a\in\alpha_0^{n-1}} 1_{T^na}v_a' \cdot g \circ v_a.$$

For each $n \ge 1$, and $a \in \alpha_0^{n-1}$ we have

$$|g(v_ax) - \frac{1}{m(a)}\int_a gdm| \le D_agr^n \ \forall \ x \in T^na$$

whence, using Renyi's property and $D_a g \le D_\beta g$,

$$|\widehat{T}^n g(x)| \le \sum_{a \in \alpha_0^{n-1}} 1_{T^n a}(x) v_a'(x) \left(\frac{1}{m(a)} \int_a |g| dm + D_\beta g r^n\right)$$

$$\le M'' \sum_{a \in \alpha_0^{n-1}} m(a) \left(\frac{1}{m(a)} \int_a |g| dm + D_\beta g r^n\right)$$

$$= M'' D_\beta g r^n + M'' \|g\|_1.$$

It follows that

$$\|\widehat{T}^n g\|_\infty \le M'' D_\beta g r^n + M'' \|g\|_1.$$

For $g \in L$, $x, y \in b \in \beta$,

$$|\widehat{T}^n g(x) - \widehat{T}^n g(y)|$$

$$\le \sum_{a \in \alpha_0^{n-1}, \ T^n a \supset b} |(v_a'(x) g(v_a(x)) - v_a'(y) g(v_a(y)))|$$

$$= M'' d(x,y) \sum_{a \in \alpha_0^{n-1}, \ T^n a \supset b} \left(M \int_a |g| dm + (M+1) m(a) r^n D_\beta g \right) \text{ by lemma 4.7.2}$$

$$\le M'' d(x,y) (M \|g\|_1 + r^n (M+1) D_\beta g).$$

$\square$

The following is a version of a theorem of M.Halfant (see [**Half**]).

4.7.4 THEOREM.
Suppose that $(X, \mathcal{B}, m, T, \alpha)$ has the Gibbs property. If $|T\alpha| < \infty$, then $\exists \mu \sim m$ such that $\log \frac{d\mu}{dm} \in L$.

PROOF. By proposition 4.2.1, we may assume that $(X, \mathcal{B}, m, T, \alpha)$ is topologically transitive. By lemma 4.4.1, $\exists h : X \to \mathbb{R}_+$ measurable such that $\widehat{T} h = h$ and $\log h \in L^\infty(m)$. It suffices to show that $h \in L$.

By ergodicity of T, we have that $\frac{1}{n} \sum_{k=0}^{n-1} \widehat{T}^k 1 \to h$ in $L^1(m)$ as $n \to \infty$. We show that $\sup_{n \ge 1} \|\widehat{T}^n 1\|_L < \infty$.

By proposition 4.7.3, $\exists q \ge 1$, $\theta \in (0,1)$ and $H > 0$ such that

$$\|\widehat{T}^q f\|_L \le \theta \|f\|_L + H \|f\|_1 \ \ \forall f \in H.$$

Iterating, we see that

$$\|\widehat{T}^{qn} f\|_L \le \theta^n \|f\|_L + \frac{H}{1-\theta} \|f\|_1 \le H' \|f\|_L \ \ \forall f \in H, \ n \ge 1$$

whence, writing $n \ge 1$ as $n = a_n q + b_n$ where $a_n \ge 0$ and $0 \le b_n < q$, we have

$$\|\widehat{T}^n 1\|_L = \|\widehat{T}^{a_n q + b_n} 1\|_L \le H' M^q$$

where $M = \sup_{0 \le b < q} \|T^b 1\|_L$ $(< \infty$ by proposition 4.7.3).

By the Arzela-Ascoli theorem, the Cesaro averages $(\frac{1}{n} \sum_{k=0}^{n-1} \widehat{T}^k 1)_{n \ge 1}$ have a $L^1(m)$-convergent subsequence with a limit in L. This limit must be h and therefore $h \in L$.

$\square$

HALFANT'S THEOREM FOR INTERVAL MAPS.

Suppose that $(I, \mathcal{B}, m, T, \alpha)$ is an expanding C^2 Markov interval map and $|T\alpha| < \infty$, and let β be the partition of I (into intervals) generated by $T\alpha$.

By proposition 4.3.3, $(I, \mathcal{B}, m, T, \alpha)$ has the Gibbs property, and by theorem 4.7.4 the invariant density h has logarithm $\log h \in L$.

In fact, h is Lipschitz continuous on each $b \in \beta$. To see this,

$$\widehat{T}^n 1 = \sum_{b \in \beta} 1_b \sum_{a \in \alpha_0^{n-1}, \, T^n a \supset b} v_a' \leq M' \sum_{b \in \beta} 1_b \sum_{a \in \alpha_0^{n-1}, \, T^n a \supset b} m(a) = M'$$

(where M' is as in the proof of proposition 4.3.3); and for $x < y \in b \in \beta$,

$$|\widehat{T}^n 1(x) - \widehat{T}^n 1(y)| \leq \sum_{a \in \alpha_0^{n-1}, \, T^n a \supset b} |v_a'(x) - v_a'(y)|$$

$$= \sum_{a \in \alpha_0^{n-1}, \, T^n a \supset b} |v_a'(x)| \left| \frac{Dv_a(y)}{Dv_a(x)} - 1 \right|$$

$$= \sum_{a \in \alpha_0^{n-1}, \, T^n a \supset b} |v_a'(x)| \left| e^{\int_x^y \frac{D^2 v_a}{Dv_a} dm} - 1 \right|$$

$$\leq M' \sum_{a \in \alpha_0^{n-1}, \, T^n a \supset b} m(a)(e^{M'|y-x|} - 1)$$

$$\leq M''|x - y|.$$

By the (classical) Arzela-Ascoli theorem, $\frac{1}{n} \sum_{k=1}^{n} \widehat{T}^k 1 \to h$ uniformly on each $b \in \beta$ and h (hence $\log h$) is Lipschitz continuous on each $b \in \beta$.

Recall that the *resolvent* set of the bounded linear operator $P : \mathcal{L} \to \mathcal{L}$,

$$\rho(P) = \{z \in \mathbb{C} : \exists \, (zI - P)^{-1} : \mathcal{L} \to \mathcal{L} \text{ bounded }\}$$

is open,

$$z \mapsto \|(zI - P)^{-1}\|_{\mathcal{L}} := \sup_{\|x\|_{\mathcal{L}}=1} \|(zI - P)^{-1}x\|_{\mathcal{L}}$$

is continuous on $\rho(P)$, and for $x \in \mathcal{L}$, $x^* \in \mathcal{L}^*$ (the dual space of $\mathcal{L}$), $z \mapsto \langle (zI - P)^{-1}x, x^* \rangle$ is analytic on $\rho(P)$. Recall also the *spectral radius theorem* which states that

$$\|P^n\|_{\mathcal{L}}^{\frac{1}{n}} \to \sup\{|z| : z \in \mathcal{C} \setminus \rho(P)\}.$$

4.7.5 PROPOSITION.

Suppose that $(\mathcal{C}, \mathcal{L})$ is adapted, and that $P : \mathcal{L} \to \mathcal{L}$ satisfies (4.7.1) and (4.7.2), then

$$\overline{B_{\mathbb{C}}(0,1)}^{\,c} \subset \rho(P).$$

PROOF.

It follows from (4.7.1), and (4.7.2) that

$$\|P^{n+1}x\|_{\mathcal{L}} \leq r\|P^n x\|_{\mathcal{L}} + RH\|x\|_C \; \forall x \in \mathcal{L}, n \in \mathbb{N},$$

whence

$$\|P^n x\|_{\mathcal{L}} \leq r^n \|x\|_{\mathcal{L}} + \frac{RH}{1-r}\|x\|_C \; \forall x \in \mathcal{L}, \; n \in \mathbb{N},$$

and

$$\|P^n x\|_{\mathcal{L}} \leq M \|x\|_{\mathcal{L}} \; \forall x \in \mathcal{L}, \; n \in \mathbb{N},$$

where $M = \left(1 + \frac{RH}{1-r}\right)$. Thus, for $|z| > 1$, the series

$$S(z) := \sum_{n=0}^{\infty} z^{-(n+1)} P^n$$

converges (in $\mathrm{Hom}\,(\mathcal{L},\mathcal{L})$), and satisfies $S(z) = (zI - P)^{-1}$. $\qquad\square$

An $\mathcal{L}$-eigenvalue of P is a complex number $\lambda \in \mathbb{C}$ such that $\exists\, 0 \neq x \in \mathcal{L} \ni Px = \lambda x$. Clearly no $\mathcal{L}$-eigenvalue is in $\rho(P)$. It follows that if P satisfies (4.7.1) and (4.7.2) then no $\mathcal{L}$-eigenvalue of P is outside the (closed) unit disc.

4.7.6 LEMMA [Io-Mar].
Suppose that $(\mathcal{C}, \mathcal{L})$ is adapted, and that $P : \mathcal{L} \to \mathcal{L}$ satisfies (4.7.1) and (4.7.2).
Suppose also that there are no $\mathcal{L}$-eigenvalues of P on $\partial B_{\mathbb{C}}(0,1)$ (the unit circle), then $\exists\, K \in \mathbb{R}_+$, and $\rho \in (0,1)$ such that

$$\|P^n f\|_{\mathcal{L}} \leq K\rho^n \|f\|_{\mathcal{L}} \quad \forall\, n \geq 1, \; f \in \mathcal{L}.$$

PROOF.
We prove that $\partial B_{\mathbb{C}}(0,1) \subset \rho(P)$. This will follow from

(4.7.3a) $\qquad\qquad (zI - P)\mathcal{L} = \mathcal{L} \; \forall\; z \in \partial B_{\mathbb{C}}(0,1),$

and

(4.7.3b) $\qquad\qquad \exists \epsilon > 0 \ni \; \|(zI - P)x\|_{\mathcal{L}} \geq \epsilon\|x\|_{\mathcal{L}} \; \forall x \in \mathcal{L}, \; z \in \mathbb{T}.$

To prove (4.7.3a), let $z \in \partial B_{\mathbb{C}}(0,1)$, $g \in \mathcal{L}$, and suppose $z_n \to z$, $|z_n| > 1$. Since $z_n \in \rho(P)$, $\exists\, f_n \in \mathcal{L}$ such that $(z_n I - P)f_n = g$. It follows that

$$|z_n|\|f_n\|_{\mathcal{L}} \leq \|Pf_n\|_{\mathcal{L}} + \|g\|_{\mathcal{L}} \leq r\|f_n\|_{\mathcal{L}} + \|f_n\|_{\mathcal{C}} + \|g\|_{\mathcal{L}},$$

whence

(4.7.4) $\qquad\qquad (1-r)\|f_n\|_{\mathcal{L}} \leq \|f_n\|_{\mathcal{C}} + \|g\|_{\mathcal{L}}.$

Let $a_n = \|f_n\|_{\mathcal{L}}$. If $\liminf_{n\to\infty} a_n < \infty$, then $\exists\, n_j \to \infty$, $f \in \mathcal{L}$ such that $v_{n_j} \xrightarrow{\mathcal{C}} f$, whence $(zI - P)f = g$.

If $a_n \to \infty$, set $f_n' = a_n^{-1} f_n$, then by (4.7.4), $\liminf_{n\to\infty} \|f_n'\|_{\mathcal{C}} \geq 1 - r > 0$ and $\exists\, n_j \to \infty$, $0 \neq f \in \mathcal{L}$ such that $f_{n_j}' \xrightarrow{\mathcal{C}} f$, whence $(zI - P)f' = 0$, contradicting the assumption that z is not an eigenvalue.

To prove (4.7.3b), suppose that no such $\epsilon > 0$ exists, then

$$\exists x_n \in \mathcal{L}, \|x_n\|_{\mathcal{L}} = 1, z_n \in \mathbb{T} \ni \; \|(z_n I - P)x_n\|_{\mathcal{L}} \to 0.$$

Without loss of generality,

$$x_n \xrightarrow{\mathcal{C}} x \in \mathcal{L}, \; z_n \to z \in \partial B_{\mathbb{C}}(0,1),$$

as $n \to \infty$. It follows that $Px = zx$, and we need to show that $x \neq 0$.

Suppose otherwise, that $x = 0$, then $\|x_n\|_C \to 0$, whence, by (4.7.2), for any $r' \in (r, 1)$ and n large,

$$\|Px_n\|_{\mathcal{L}} \leq r' < 1,$$

and

$$\|(z_n I - P)x_n\|_{\mathcal{L}} \geq 1 - \|Px_n\|_{\mathcal{L}} \geq 1 - r' > 0,$$

contradicting $\|(z_n I - P)x_n\|_{\mathcal{L}} \to 0$ and establishing (4.7.3b).

This shows that $\partial B_{\mathbb{C}}(0, 1) \subset \rho(P)$, whence, by proposition 4.7.5, $\{z \in \mathbb{C} : |z| \geq 1\} \subset \rho(P)$. By openness of $\rho(P)$, there exists $r \in (0, 1)$ such that $\{z \in \mathbb{C} : |z| > r\} \subset \rho(P)$, and the lemma follows from the spectral radius theorem. $\qquad\square$

4.7.7 THEOREM   ([Rue]).

Suppose that $(X, \mathcal{B}, m, T, \alpha)$ is a topologically mixing, probability preserving Markov map having the Gibbs property and such that $\inf\limits_{a \in \alpha} m(Ta) > 0$, then $\exists\, K > 1$, $0 < \theta < 1$ such that

$$\left\|\widehat{T}^n f - \int_X f dm\right\|_L \leq K\theta^n \|f\|_L \quad \forall\, f \in L.$$

PROOF.

As mentioned above, $(\mathcal{C}, \mathcal{L}) = (L^1(m), L)$ is adapted, and $\exists\, k \geq 1$ such that $P = \widehat{T}^k$ satisfies (4.7.1) and (4.7.2). Let $Q = P - E$, then $Q^n = P^n - E$. By the exactness of T, $\|Q^n f\|_1 \to 0 \; \forall\, f \in L^1(m)$, and so Q has no $\mathcal{L}$-eigenvalues on $\partial B_{\mathbb{C}}(0, 1)$.

Clearly Q satisfies (4.7.1). To see that Q satisfies (4.7.2), let $f \in \mathcal{L}$, then

$$\|Qf\|_{\mathcal{L}} \leq \|Pf\|_{\mathcal{L}} + \|Ef\|_{\mathcal{L}}$$
$$\leq r\|f\|_{\mathcal{L}} + R\|f\|_C + \int_I |f| dm \|h\|_{\mathcal{L}}$$
$$\leq r\|f\|_{\mathcal{L}} + (R + \|h\|_{\mathcal{L}})\|f\|_C.$$

By lemma 4.7.6, $\exists\, K \in \mathbb{R}_+$, and $r \in (0, 1)$ such that

$$\|Q^n f\|_{\mathcal{L}} \leq Kr^n \|f\|_{\mathcal{L}} \quad \forall\, n \geq 1, \; f \in \mathcal{L},$$

whence

$$\left\|\widehat{T}^{kn} f - \int_I f dm h\right\|_L \leq Kr^n \|f\|_L \; \forall\, f \in L,$$

whence

$$\left\|\widehat{T}^n f - \int_I f dm h\right\|_L \leq K'(r^{\frac{1}{k}})^n \|f\|_L \; \forall\, f \in L.$$

$\qquad\square$

4.7.8 COROLLARY.

If $(X, \mathcal{B}, m, T, \alpha)$ is a topologically mixing, probability preserving Markov map having the Gibbs property and such that $\inf\limits_{a \in \alpha} m(Ta) > 0$ then it is exponentially continued fraction mixing.

PROOF. Suppose that $\mathfrak{g}_r(C,T) = \tilde{\alpha}_+$, then $\forall \ a \in \tilde{\alpha}_+$, we have that $v'_a = e^{\pm C} m(a)$, and for $x, y \in \mathcal{D}(v_a)$,

$$|v'_a(x) - v'_a(y)| \leq e^C m(a) \left| \frac{v'_a(x)}{v'_a(y)} - 1 \right| \leq e^C m(a)(e^{Cr^{t(x,y)}} - 1) \leq m(a) M r^{t(x,y)}$$

for some constant M, and

$$\left\| \frac{v'_a}{m(a)} \right\|_L \leq C + M.$$

It follows from theorem 4.7.7 that $\exists \ K > 1$, $0 < \theta < 1$ such that

$$\|\widehat{T}^n v'_a - m(a)\|_L \leq K\theta^n (C + M) m(a) \quad \forall \ a \in \tilde{\alpha}_+.$$

The corollary follows from this. $\square$

§4.8 Non-expanding interval maps

Let $(I, \mathcal{B}(I), m, T, \alpha)$ be a C^2-Markov interval map satisfying Thaler's assumptions, let $\alpha_\Lambda = \{a \in \alpha : a \cap \Lambda \neq \emptyset\}$ and let

$$\mathfrak{r} = \mathfrak{r}(I, \mathcal{B}(I), m, T, \alpha) := \{[s_1, ..., s_n] : \text{ either } s_n \notin \alpha_\Lambda, \text{ or } s_{n-1} \neq s_n\}.$$

As shown in proposition 4.3.5, $\mathfrak{r}$ is a Schweiger collection for T.

4.8.1 THEOREM   [A10], SEE ALSO [Tha3].
If T is topologically transitive, and conservative, then $\exists$ a σ-finite, T-invariant measure $\mu \sim m$ with respect to which T is pointwise dual ergodic.
Indeed any set $A \in \mathfrak{r}$ is a Darling - Kac set for T.

PROOF.
Ergodicity and the existence of the T-invariant measure follow from the weak distortion property (via theorems 4.4.4 and 4.4.6 respectively). By proposition 3.7.5, it is sufficient to show that any set $A \in \mathfrak{r}$ is a Darling - Kac set for T. Let $A \in \mathfrak{r}$.

By proposition 4.3.5, the induced Markov map $(A, \mathcal{B}(A), m|_A, T_A, \alpha_A)$ is an expanding C^2-Markov interval map. By proposition 4.3.3, $\exists \ C > 1$, $0 < r < 1$ such that $\mathfrak{g}_r(C, T_A) = \tilde{\alpha}_{A+}$. Evidently $T_A b = A \ \forall \ b \in \alpha_A$ and so $(A, \mathcal{B}(A), m|_A, T_A, \alpha_A)$ is topologically mixing and has the Gibbs property.

By corollary 4.7.8, $(A, \mathcal{B}(A), m|_A, T_A, \alpha_A)$ is exponentially continued fraction mixing, and by lemma 3.7.4, A is a Darling-Kac set for T. $\square$

It now follows from theorem 3.8.3 that T has minimal wandering rates in the sense that

$$\exists \ L(n) \uparrow \infty \text{ such that } L_A(n) \sim L(n) \text{ as } n \to \infty \ \forall \ A \in \mathcal{U}(T).$$

The rest if this section is devoted to a method for calculating $L(n)$, and then $a_n(T)$ using the asymptotic renewal equation (see §3.8). This method works when T is piecewise onto.

Accordingly we assume for the rest of this chapter that $(I, \mathcal{B}(I), m, T, \alpha)$ is a piecewise onto, C^2-Markov interval map satisfying Thaler's assumptions.

We begin with an investigation of the invariant density $\frac{d\mu}{dm}$ using Schweiger's jump transformation (see §4.6).

As in §4.3, set

$$\alpha^* := \alpha \setminus \alpha_\Lambda \cup \{[\underbrace{b, \ldots, b}_{n \text{ times}}, a] : n \geq 1, \ b \in \alpha_\Lambda, \ a \in \alpha \ a \neq b\},$$

then (see the proof of proposition 4.3.5) α^* is a partition of I.

Recall from §4.6 that Schweiger's jump transformation $T^* : I \to I$ is defined by

$$T^* x = T^{N(x)} x$$

where

$$N(x) = \left\{ \begin{array}{ll} 1 & x \in a \notin \alpha_\Lambda, \\ \min\{n \geq 1 : T^n x \notin a\} + 1 & x \in a \in \alpha_\Lambda. \end{array} \right.$$

Recall statement (4.3.2) (established in the proof of theorem 4.3.5):

$$(4.3.2) \qquad \sum_{k=0}^{\infty} D(v_a^k)(x) \leq \left| \frac{x - x_a}{v_a(x) - x} \right| \ \forall \, a \in \alpha_\Lambda, \ x \neq x_a.$$

In addition to (4.3.2), we shall need

$$(4.8.1) \qquad \sum_{k=n}^{\infty} D(v_a^k)(x) \leq \left| \frac{v_a^n(x) - x_a}{v_a(x) - x} \right| \ \forall \, a \in \alpha_0, \ x \neq x_a, \ n \geq 1.$$

This is seen as follows

$$\sum_{k=n}^{\infty} D(v_a^k)(x) \leq \sum_{k=0}^{\infty} D(v_a^k) \circ v_a^n(x) D(v_a^n)(x)$$

$$\leq \left| \frac{D(v_a^n)(x)(v_a^n(x) - x_a)}{v_a^{n+1}(x) - v_a^n(x)} \right| \text{ by } (4.8.1)$$

$$\leq \left| \frac{v_a^n(x) - x_a}{v_a(x) - x} \right| \ (\because \ |v_a^n(x) - v_a^{n+1}(x)| \geq D(v_a^n)(x)|v_a(x) - x|).$$

4.8.2 THEOREM   [Th2].

$$\frac{d\mu}{dm}(x) = g(x) \prod_{a \in \alpha_\Lambda} \frac{x - x_a}{x - v_a(x)} \ \text{where } g > 0 \text{ is continuous on } I.$$

PROOF.   By Halfant's theorem there is a T^*-invariant probability $q \sim m$ with density $h = \frac{dq}{dm}$ Lipschitz continuous on I. Set

$$\mu(A) := \int_I \left(\sum_{k=0}^{N-1} 1_A \circ T^k \right) dq,$$

then μ is T-invariant:

$$\mu(T^{-1}A) = \int_I \left(\sum_{k=1}^{N} 1_A \circ T^k \right) dq$$

$$= \int_I \left(\sum_{k=0}^{N-1} 1_A \circ T^k + 1_A \circ T^N - 1_A \right) dq$$

$$= \int_I \left(\sum_{k=0}^{N-1} 1_A \circ T^k + 1_A \circ T^* - 1_A \right) dq$$

$$= \mu(A) + q(T^{*-1}A) - q(A) = \mu(A),$$

and

$$\frac{d\mu}{dm} = \sum_{k=0}^{\infty} \widehat{T}^k (h 1_{[N \geq k+1]}).$$

Now, for $k \geq 1$,

$$[N \geq k+1] = \bigcup_{a \in \alpha_\Lambda} \bigcup_{j=k}^{\infty} v_a^j(I \setminus a),$$

the union being disjoint, and so

$$\widehat{T}^k (h 1_{[N \geq k+1]}) = \sum_{a \in \alpha_\Lambda^{k-1}} v_a' h \circ v_a 1_{[N \geq k+1]} \circ v_a$$

$$= \sum_{a \in \alpha_\Lambda^{k-1}} v_a' h \circ v_a \left(\sum_{a \in \alpha_\Lambda} \sum_{j=k}^{\infty} 1_{v_a^j(I \setminus a)} \circ v_a \right)$$

$$= \sum_{a \in \alpha_\Lambda} v_a^{k\prime} h \circ v_a^k \left(\sum_{j=k}^{\infty} 1_{v_a^j(I \setminus a)} \circ v_a^k \right)$$

$$= \sum_{a \in \alpha_\Lambda} v_a^{k\prime} h \circ v_a^k,$$

since

$$\sum_{j=k}^{\infty} 1_{v_a^j(I \setminus a)} \circ v_a^k = \sum_{j=0}^{\infty} 1_{v_a^j(I \setminus a)} = 1_{\bigcup_{j=0}^{\infty} v_a^j(I \setminus a)} \equiv 1.$$

$$\therefore \quad \frac{d\mu}{dm} = \sum_{k=0}^{\infty} \widehat{T}^k (h 1_{[N \geq k+1]}) = h + \sum_{a \in \alpha_\Lambda} \sum_{k=1}^{\infty} v_a^{k\prime} h \circ v_a^k.$$

Now set, for $a \in \alpha_\Lambda$, $x \neq x_a$,

$$H_a(x) := \sum_{k=1}^{\infty} v_a^{k\prime}(x) h \circ v_a^k(x).$$

To complete the proof of the theorem, we show that H_a is continuous on $I \setminus \{x_a\}$, and that

(4.8.2) $$H_a(x) \left| \frac{v_a(x) - x}{x - x_a} \right| \to h(x_a) \text{ as } x \to x_a.$$

To establish continuity of H_a on $I \setminus \{x_a\}$, it suffices to show that $\forall \ x \neq x_a \ \exists$ an open interval $J \ni x$ such that $\sum_{k=1}^{\infty} v_a^{k\prime} h \circ v_a^k$ converges uniformly on J. This is a consequence of (4.8.1) as follows. Given $x \neq x_a \; \exists$ an open interval $J \ni x$, and $\epsilon_n \to 0$ as $n \to \infty$ such that

$$\left| \frac{v_a^n(y) - x_a}{v_a(y) - y} \right| \leq \epsilon_n \; \forall \; y \in J.$$

Thus (using (4.8.1)),

$$\sum_{k=n}^{\infty} v_a^{k\prime} h \circ v_a^k \leq \|h\|_\infty \sum_{k=n}^{\infty} v_a^{k\prime} \leq \|h\|_\infty \epsilon_n \to 0$$

as $n \to \infty$ uniformly on J.

To prove (4.8.2), note that for $x \neq x_a$,

$$H_a(x) = \sum_{k=1}^{\infty} v_a^{k\prime}(x) h(x_a) + \sum_{k=1}^{\infty} v_a^{k\prime}(x)(h \circ v_a^k(x) - h(x_a)).$$

By (4.8.1),

$$h(x_a) v_a'(x) \leq \left| \frac{v_a(x) - x}{x - x_a} \right| \sum_{k=1}^{\infty} v_a^{k\prime}(x) h(x_a) \leq h(x_a),$$

and

$$\sum_{k=1}^{\infty} v_a^{k\prime}(x)(h \circ v_a^k(x) - h(x_a)) \leq \|h\|_L \sum_{k=1}^{\infty} v_a^{k\prime}(x) |v_a^k(x) - x_a|$$

$$= o\left(\left| \frac{x - x_a}{x - v_a(x)} \right| \right) \text{ as } x \to x_a.$$

$$\therefore \quad H_a(x) \left| \frac{v_a(x) - x}{x - x_a} \right| \to h(x_a) \text{ as } x \to x_a.$$

$\square$

Let $\mathcal{A}(\mathfrak{r})$ denote the ring generated by $\mathfrak{r}$ and let $\mathfrak{H}(\mathfrak{r}) = \{B \in \mathcal{B} : \; B \subset A \in \mathcal{A}\}$.

4.8.3 THEOREM   [Th2].

$\exists L(n) \uparrow$ such that $L_B(n) \sim L(n)$ as $n \to \infty \; \forall B \in \mathfrak{H}(\mathfrak{r})$, $\mu(B) > 0$.

PROOF. It is sufficient to show that if $A \in \mathcal{B}$ is a finite union of sets in $\mathfrak{r}$, then

$$L_A(n) \sim L_B(n) \text{ as } n \to \infty, \; \forall B \in \mathcal{B}, \; m(B) > 0, \; B \subset A.$$

The proof of this uses

4.8.4 LEMMA   [Th2].
Suppose that $A \in \mathcal{A}(\mathfrak{r})$, then

$$\exists M \geq 1 \quad \text{such that} \quad \forall k \geq 1, \; B \in \mathcal{B}, \; B \subset A,$$

$$\mu\left(T^{-k}B \setminus \bigcup_{j=0}^{k-1} T^{-j}A\right) \leq Mm(B)\mu\left(T^{-k}A \setminus \bigcup_{j=0}^{k-1} T^{-j}A\right).$$

PROOF. Firstly, since $A \in \mathcal{A}(\mathfrak{r})$, by theorem 4.8.2, $\exists M \geq 1$ such that $\frac{d\mu}{dm} \in [1/M, M]$ on A. Also, since T is conservative and ergodic, for $B \in \mathcal{B}$, we have by Kac's formula,

$$\mu(B) = \sum_{\ell=0}^{\infty} \mu(A \cap T^{-\ell}B \setminus \bigcup_{j=1}^{\ell} T^{-j}A).$$

Thus, for $B \in \mathcal{B}$, $B \subset A$, and $k \geq 1$,

$$\mu(T^{-k}B \setminus \bigcup_{j=0}^{k-1} T^{-j}A) = \sum_{\ell=1}^{\infty} \mu(T^{-(k+\ell)}B \cap A \setminus \bigcup_{j=1}^{k+\ell-1} T^{-j}A)$$

$$\leq M \sum_{\ell=1}^{\infty} m(T^{-(k+\ell)}B \cap A \setminus \bigcup_{j=1}^{k+\ell-1} T^{-j}A).$$

Using the distortion proposition, it is not hard to see that $\exists K \geq 1$ such that $\forall B \in \mathcal{B}$, $B \subset A$, $k, \ell \geq 1$,

$$m(T^{-(k+\ell)}B \cap A \setminus \bigcup_{j=1}^{k+\ell-1} T^{-j}A) \leq Km(B)m(T^{-(k+\ell)}A \cap A \setminus \bigcup_{j=1}^{k+\ell-1} T^{-j}A)$$

whence

$$\mu(T^{-k}B \setminus \bigcup_{j=0}^{k-1} T^{-j}A) \leq MK \sum_{\ell=1}^{\infty} m(B)m(T^{-(k+\ell)}A \cap A \setminus \bigcup_{j=1}^{k+\ell-1} T^{-j}A)$$

$$\leq M^2 K \sum_{\ell=1}^{\infty} m(B)\mu(T^{-(k+\ell)}A \cap A \setminus \bigcup_{j=1}^{k+\ell-1} T^{-j}A)$$

$$= M^2 Km(B)\mu(T^{-k}A \setminus \bigcup_{j=1}^{k-1} T^{-j}A).$$

$\square$

Continuing the proof of theorem 4.8.3, let $B \in \mathcal{B}$, $B \subset A$, then

$$\bigcup_{k=0}^{n} T^{-k}A \setminus \bigcup_{k=0}^{n} T^{-k}B \subset (A \setminus B) \cup \bigcup_{k=1}^{n} T^{-k}(A \setminus B) \setminus \bigcup_{j=0}^{k-1} T^{-j}A$$

whence

$$L_A(n) - L_B(n) \leq \mu(A \setminus B) + \sum_{k=1}^{n} \mu(T^{-k}(A \setminus B) \setminus \bigcup_{j=0}^{k-1} T^{-j}A)$$

$$\leq Mm(A \setminus B)L_A(n)$$

by lemma 4.8.4. Using this, we obtain for any fixed $d \geq 1$,

$$L_B(n) \sim L_B(n+d) \text{ as } n \to \infty$$

$$\geq L_{A \cap \bigcup_{j=0}^{d} T^{-j}B}(n)$$

$$\geq (1 - Mm(A \setminus \bigcup_{j=0}^{d} T^{-j}B))L_A(n).$$

This establishes theorem 4.8.3 as

$$m(A \setminus \bigcup_{j=0}^{d} T^{-j}B) \to 0 \text{ as } d \to \infty.$$

$\square$

We next identify the growth rate of $L(n)$.

4.8.5 THEOREM [**Th2**].
If α is finite, then

$$L(n) \sim \sum_{a \in \alpha_\Lambda} h(x_a) \sum_{k=0}^{n-1} (v_a^k(1) - v_a^k(0)).$$

PROOF.
We first suppose that $\alpha_\Lambda \neq \alpha$ and set $B = \bigcup_{a \in \alpha \setminus \alpha_\Lambda} a$. Since α is finite, $B \in \mathcal{A}(\mathfrak{r})$ whence by theorem 4.8.3, $L_B(n) \sim L(n)$.

$$\therefore \ L(n) \sim \mu(\bigcup_{k=0}^{n} T^{-k}B)$$

$$\sim \sum_{a \in \alpha_\Lambda} h(x_a) \int_{I \setminus v_a^n(I)} \frac{x - x_a}{x - v_a(x)} dx$$

$$\sim \sum_{a \in \alpha_\Lambda} h(x_a) \int_{I \setminus v_a^n(I)} \left(\sum_{k=0}^{\infty} v_a^{k\prime}(x) \right) dx$$

$$= \sum_{a \in \alpha_\Lambda} h(x_a) \sum_{k=0}^{n-1} (v_a^k(1) - v_a^k(0)).$$

In case $\alpha_\Lambda = \alpha$, we must allow for the possibility that $x_a \in \partial a$ (i.e. $x_a = 0, 1$).
Let $J_a \ni x_a$ be an interval, open in $\overline{a}$, and with endpoints c_a, d_a (i.e. $J_a^o = (c_a, d_a)$), set $B = I \setminus \bigcup_{a \in \alpha} J_a$, and note that $B \in \mathfrak{H}(\mathfrak{r})$.
It follows again from theorem 4.8.3 that

$$L(n) \sim \mu(\bigcup_{k=0}^{n} T^{-k}B)$$

$$\sim \sum_{a \in \alpha_\Lambda} h(x_a) \int_{I \setminus v_a^n(J_a)} \frac{x - x_a}{x - v_a(x)} dx$$

$$\sim \sum_{a \in \alpha_\Lambda} h(x_a) \int_{I \setminus v_a^n(J_a)} \left(\sum_{k=0}^{\infty} v_a^{k\prime}(x) \right) dx$$

$$\sim \sum_{a \in \alpha_\Lambda} h(x_a) \sum_{k=0}^{n-1} (v_a^k(d_a) - v_a^k(c_a))$$

$$\sim \sum_{a \in \alpha_\Lambda} h(x_a) \sum_{k=0}^{n-1} (v_a^k(1) - v_a^k(0)).$$

$\square$

The return sequence $a_n(T)$ of T can now calculated using proposition 3.8.7 in case $L(n)$ is regularly varying since $\mathfrak{r} \subset \mathcal{U}(T)$.

In this case, $a_n(T)$ is also regularly varying and by corollary 3.7.3,

$$\frac{S_n^T}{a_n(T)} \xrightarrow{\mathfrak{d}} Y_\alpha$$

where α is the index of regular variation of $a_n(T)$, and Y_α has the normalised Mittag-Leffler distribution of order α.

We conclude with some examples.

4.8.6 LEMMA.
Suppose that $u : \mathbb{R}_+ \to (0,1)$ is regularly varying at 0 with index $\alpha > 0$ and let $v_0 = 1$, $v_{n+1} = v_n(1 - u(v_n))$,
then

$$v_n \sim \frac{1}{\alpha^{\frac{1}{\alpha}} a^{-1}(n)}$$

as $n \to \infty$ where $a(x) := \frac{1}{u(\frac{1}{x})}$.

PROOF.
Write $u(x) = x^\alpha L(x)$ where L is slowly varying at 0, then

$$a\left(\frac{1}{v_{n+1}}\right) = \frac{1}{v_n^\alpha} \frac{1}{(1 - u(v_n))^\alpha} \frac{1}{L(v_{n+1})}$$

$$= a\left(\frac{1}{v_n}\right)\left(1 + \alpha u(v_n) + o(u(v_n))\right)\frac{L(v_n)}{L(v_{n+1})}$$

$$= \left(a\left(\frac{1}{v_n}\right) + \alpha + o(1)\right)\frac{L(v_n)}{L(v_{n+1})}.$$

By the representation theorem for slowly varying functions, $\exists\, M > 1$ and $k : [M, \infty) \to \mathbb{R}_+$, $\epsilon : [M, \infty) \to \mathbb{R}$ measurable, such that $|\epsilon(t)| < \frac{\alpha}{2}\ \forall\, t \geq M$, $\epsilon(t) \to 0$ as $t \to \infty$, $k(y) \to k \in \mathbb{R}_+$ as $y \to 0$ and such that

$$L(y) = k(y)e^{\int_M^{\frac{1}{y}} \frac{\epsilon(t)}{t} dt} \quad (y \geq M).$$

It follows that

$$\log\left|\frac{L(v_n)}{L(v_{n+1})}\right| \leq \int_{\frac{1}{v_n}}^{\frac{1}{v_{n+1}}} \frac{|\epsilon(t)|}{t} dt$$

$$= o\left(\log\left|\frac{v_n}{v_{n+1}}\right|\right)$$

$$= o\left(\log\left|\frac{1}{1 - u(v_n)}\right|\right)$$

$$= o(u(v_n))$$

whence

$$\frac{L(v_n)}{L(v_{n+1})} = 1 + o(u(v_n)),$$

$$a\left(\frac{1}{v_{n+1}}\right) = a\left(\frac{1}{v_n}\right) + \alpha + o(1),$$

$$a\left(\frac{1}{v_n}\right) \sim \alpha n$$

and

$$v_n \sim \frac{1}{\alpha^{\frac{1}{\alpha}} a^{-1}(n)}.$$

$\square$

4.8.7 THEOREM.

Suppose that $b(n)$ is regularly varying with index in $(0,1)$, then $\exists$ a piecewise onto, C^2 Markov interval map T with a 2 set basic partition such that $a_n(T) \propto b(n)$.

PROOF.

Consider piecewise increasing maps where $\alpha = \{[0,a],[a,1]\}$ and $T : [a,1] \rightarrow [0,1]$ a linear bijection ($T' = \frac{1}{1-a}$ on $[a,1]$), and 0 is a regular source for T. By theorem 4.8.5,

$$L(n) \sim \sum_{k=0}^{n} v_0^n(1).$$

Define $u(x)$ by $v_0(x) = x(1 - u(x))$, and let $v_0^n(1) := v_n$, then $v_{n+1} = v_n(1 - u(v_n))$.

Suppose that u is regularly varying with index $\alpha > 1$ at 0. By lemma 4.8.6, $v_n \propto \frac{1}{a^{-1}(n)}$ as $n \rightarrow \infty$ which is regularly varying at ∞ with index $-\alpha^{\frac{1}{\alpha}}$. By Karamata's Tauberian theorem,

$$L(n) \propto \sum_{k=1}^{n} \frac{1}{a^{-1}(k)} \propto \frac{n}{a^{-1}(n)},$$

and by proposition 3.8.7, $a_n(T) \propto a^{-1}(n)$.

To prove the theorem, find $C : \mathbb{R}_+ \rightarrow \mathbb{R}_+$ increasing such that $\frac{x}{b(x)} \sim C(x)$ and set $u(x) := \frac{1}{C^{-1}(\frac{1}{x})}$.

Since u is regularly varying with index $\alpha > 1$ at 0, $\exists v_0$ C^2 near 0 so that $x - v_0(x) \sim x u(x)$ as $x \rightarrow 0$. To finish, define T so that $T \circ v_0 = \text{Id}$ near 0. $\square$

REMARK.

Using lemma 4.8.6 and the methods of theorem 4.8.7, one can obtain a piecewise onto, C^2 Markov interval map with a 2 set basic partition and $a_n(T)$ regularly varying at ∞ with index 1. If L is C^2 on $(0,1)$ and slowly varying at 0, then $x \rightarrow x^2 L(x)$ is C^2 near 0 if and only if $\exists \lim_{x \rightarrow 0} L(x) < \infty$. Consequently if T is a piecewise onto, C^2 Markov interval map as above with $x - v_0(x) \sim x^2 L(x)$ as $x \rightarrow 0$, then v_n is regularly varying at ∞ with index -1 and $\exists \lim_{n \rightarrow \infty} n v_n > 0$, and any such sequence v_n can be achieved in this way.

For example L constant, in which case $v_n \propto \frac{1}{n}$ and $a_n(T) \propto \frac{n}{\log n}$; or $L(x) = |\log x|^{-\alpha}$ (where $\alpha > 0$), in which case $v_n \propto \frac{(\log n)^\alpha}{n}$ and $a_n(T) \propto \frac{n}{(\log n)^{\alpha+1}}$.

EXAMPLE.

As in [**A10**], let $T : (0,1) \to (0,1)$ be defined by $Tx := \left\{\frac{x}{1-x}\right\}$ where $\{x\} :=$ $\min_{n \in \mathbb{Z},\ n \leq x} x - n$ denotes the fractional part of $x \in \mathbb{R}$, then T is a piecewise onto C^2 interval map with Markov partition $\alpha := \{I_n : n \in \mathbb{N}\}$ where $I_n := (1 - \frac{1}{n}, 1 - \frac{1}{n+1})$. It can be checked that T satisfies Thaler's assumptions with $\alpha_\Lambda = I_1$ and that the measure $d\mu(x) := \frac{dx}{x}$ is T-invariant.

We claim that $B := [0,1] \backslash I_1$ is a Darling-Kac set for T. To see this note that the induced transformation $T_B : B \to B$ is a piecewise onto, expanding Markov interval map and so by proposition 4.3.3 $\exists\, C > 1$, $0 < r < 1$ such that $\mathfrak{g}_r(C, T_B) = (\tilde{\alpha}_B)_+$.

By corollary 4.7.8, $(B, \mathcal{B} \cap B, \mu|_B, T_B, \alpha_B)$ is continued fraction mixing, whence by lemma 3.7.4 B is a Darling-Kac set for T.

Writing $v_n = v_{I_1}^n(1)$ we have $v_0 = 1$, $v_{n+1} = v_n(1 - u(v_n))$ where $u(x) \sim x$ as $x \to 0$. It follows from lemma 4.8.6 that $v_n \sim \frac{1}{n}$.

Next, $B \in \mathcal{U}(T)$ (being a Darling-Kac set) whence by theorem 3.8.3, $L(n) \sim L_B(n)$. The proof of theorem 4.8.5 shows that $L(n) \sim \log n$ whence by proposition 3.8.7 $a_n(T) \sim \frac{n}{\log n}$.

By corollary 3.7.3,

$$\frac{S_n^T(f)}{a_n(T)} \to \int_X f d\mu \text{ in measure, on sets of finite measure } \forall\, f \in L^1(\mu).$$

Define the functions $n_k : (0,1) \to \mathbb{N} \cap [2, \infty)$ $(k \geq 1)$ by $x = 1/n_1 - 1/n_2 - 1/n_3 - 1/\ldots$ then $n_1(x) = n(x) := \left[\frac{1}{x}\right] + 1$ and $n_k(x) = n(S^{k-1}x)$ where $Sx = 1 - \left\{\frac{1}{x}\right\}$.

Note that $1 - S(1-x) = Tx$ as above, and it is shown in [**A10**] using the above convergence in measure that

$$\frac{1}{K} \sum_{k=1}^{K} n_k \to 3 \text{ in measure.}$$

§4.9 Additional reading

1) Other examples of Markov maps having the weak distortion property, see [**A-De-Ur**] (parabolic rational maps), [**Kel**] (interval maps with negative Schwarzian derivative) and [**Yu**] (multidimensional number theoretical transformations).

2) A "law of the iterated logarithm" for certain infinite measure preserving transformations, including Markov maps with the weak distortion property can be found in [**A-De1**].

3) For more information on C^2 Markov interval maps of $\mathbb{R}$, see [**A-De2**].

CHAPTER 5

Recurrent events and
Similarity of Markov shifts

§5.1 Renewal sequences

DEFINITION: RENEWAL SEQUENCE.
A bounded sequence $\{u_n : n \geq 0\}$ is called a *renewal sequence* if $u_0 = 1$, and $f_n \geq 0$, where

$$f_1 = u_1, \ f_n := u_n - \sum_{k=1}^{n-1} f_k u_{n-k} \ (n \geq 2).$$

Writing for $x \leq 0$, $u(x) = \sum_{n=0}^{\infty} u_n x^n$ and $f(x) = \sum_{k=1}^{\infty} f_k x^k$, we have for $0 < x < 1$ that $u(x) - 1 = u(x)f(x)$, whence $f(x) = 1 - \frac{1}{u(x)}$ and $\sum_{k=1}^{\infty} f_k = f(1) = 1 - \frac{1}{u(1)} \leq 1$.

DEFINITION: RECURRENT RENEWAL SEQUENCE, LIFETIME DISTRIBUTION.
The renewal sequence $\{u_n : n \geq 0\}$ is called *recurrent* if

$$\sum_{n=0}^{\infty} u_n = \infty, \text{ or equivalently, } \sum_{k=1}^{\infty} f_k = 1.$$

In this case $f \in \mathcal{P}(\mathbb{N})$ is called the *lifetime distribution* corresponding to u and denoted $f = f^{(u)}$.

5.1.1 PROPOSITION [Chu]. *If P is a stochastic matrix on S, and $s \in S$, then $u_n := p_{s,s}^{(n)}$ is a renewal sequence.*

PROOF. Consider the probability p on $\subset S^{\mathbb{N}}$ defined by

$$p([t_1, \ldots, t_n]) := p_{s,t_1} p_{t_1,t_2} \cdots p_{t_{n-1},t_n}.$$

For $n \geq 1$ consider the event $A_n := \{x \in [s] : x_n = s, \ x_k \neq s \ \forall \ 1 \leq k < n\}$. Evidently, $F_n := p(A_n) \geq 0$ and it is sufficient to show that

$$\sum_{k=1}^{n} F_k p_{s,s}^{(n-k)} = p_{s,s}^{(n)} \ \forall \ n \geq 1.$$

To see this, note that $T^{-n}[s] = \bigcup_{k=1}^{n} A_k \cap T^{-n}[s]$, whence

$$p_{s,s}^{(n)} = p(T^{-n}[s] = \sum_{k=1}^{n} p(A_k \cap T^{-n}[s]).$$

For each $k \geq 1$ $A_k = \bigcup_{a \in \alpha_k} a$ where $\alpha_k = \{[t_1, \ldots, t_{k-1}, s] : t_1, \ldots, t_{k-1} \neq s\}$. It follows that for each $a = [t_1, \ldots, t_{k-1}, s]$,

$$
\begin{aligned}
p(a \cap T^{-n}[s]) &= p([t_1, \ldots, t_{k-1}, s] \cap T^{-n}[s]) \\
&= \sum_{x_1, \ldots, x_{n-k} \in S} p([t_1, \ldots, t_{k-1}, s] \cap T^{-k}[x_1, \ldots, x_{n-k}, s]) \\
&= \sum_{x_1, \ldots, x_{n-k-1} \in S} p([t_1, \ldots, t_{k-1}, s]) p([x_1, \ldots, x_{n-k-1}, s]) \\
&= p(a) p_{s,s}^{(n-k)},
\end{aligned}
$$

whence

$$
p(A_k \cap T^{-n}[s]) = \sum_{a \in \alpha_k} p(a \cap T^{-n}[s]) = \sum_{a \in \alpha_k} p(a) p_{s,s}^{(n-k)} = F_k p_{s,s}^{(n-k)}.
$$

$\square$

On the other hand, given the renewal sequence $\{u_n : n \geq 0\}$, one can define a stochastic matrix $P = P_u$ with state space $\mathbb{N} \cup \{\infty\}$ by

$$
p_{j,k} = \begin{cases}
f_k & \text{if } j = 1, k \in \mathbb{N} \\
1 - \sum_{n=1}^{\infty} f_n & \text{if } j = 1, k = \infty \\
1 & \text{if } j - k = 1, \\
0 & \text{else.}
\end{cases}
$$

In case u is recurrent, we consider P_u restricted to $\mathbb{N}$.

5.1.2 PROPOSITION   [**Chu**]. *Let $u = (u_0, u_1, \ldots)$ be a renewal sequence and let $P = P_u$, then*

(i)
$$
u_n = p_{1,1}^{(n)} \quad (n \geq 0).
$$

(ii)   *If u is recurrent, then P_u is irreducible, recurrent on $[1, N_f) \subseteq \mathbb{N}$ where $N_f := 1 + \sup_{k \in \mathbb{N}, \, f_k > 0} k \leq \infty$; has a stationary distribution on $[1, N_f)$ defined by*

$$
c_k = \sum_{k \leq j < N_f} f_j \quad \text{if } k \in [1, N_f);
$$

and is positively recurrent iff

$$
\sum_{n=1}^{\infty} n f_n < \infty.
$$

PROOF.
(i) Since

$$
[\varphi_{[1]} = n] := [1] \cap T^{-n}[1] \cap \bigcup_{k=1}^{n-1} T^{-k}[1] = [1, n, n-1, \ldots, 2, 1] = [1, n],
$$

we have

$$
f_{1,1}^{(n)} = m([\varphi_{[1]} = n] | [1]) = f_n
$$

whence by the renewal equation $p_{1,1}^{(n)} = u_n$.

(ii) Irreducibility and recurrence are evident. To check the stationary distribution,

$$\sum_{1 \le s < N_f} c_s p_{s,t} = \begin{cases} c_1 p_{1,t} + c_{t+1} p_{t+1,t} = c_t & t + 1 < N_f \\ c_1 p_{1,t} = f_t = c_t & t + 1 = N_f. \end{cases}$$

By definition, P_u is positively recurrent if and only if

$$\infty > \sum_{s=1}^{\infty} c_s = \sum_{n=1}^{\infty} n f_n.$$

$\square$

DEFINITION: RENEWAL PROCESS.
In case u is recurrent, the 2-sided Markov chain $T_u = T_{P_u}$ is called the *renewal process* with renewal sequence u.

EXERCISE 5.1.1.
Suppose that $a(n)$ is regularly varying at ∞ and that $\frac{a(n)}{n} \downarrow$ as $n \uparrow$. Show that $\exists$ a renewal process T_u with $a_n(T_u) \sim a(n)$.

5.1.3 PROPOSITION [**Ken**].
(i) If u, and v are renewal sequences, then so is uv, where $(uv)_n := u_n v_n$.
(ii) If $u^{(n)}$ $(n \ge 1)$ are renewal sequences and $u_k^{(n)} \to v_k$ as $n \to \infty \ \forall \ k \ge 1$, then $v = (v_0, v_1, \dots)$ is a renewal sequence.

PROOF.
(i) By proposition 5.1.2, there are stochastic matrices P and P' on S and S' respectively, $s \in S$ and $s' \in S'$ such that $u_n = p_{s,s}$, $u'_n = p'_{s',s'}$.
It follows that $Q = P \times P'$ defined by

$$q_{(a,b),(a',b')} := p_{a,b} p'_{a',b'} \quad (a, b \in S, \ a', b' \in S')$$

is a stochastic matrix on $S \times S'$ satisfying $q_{(a,a'),(b,b')}^{(n)} := p_{a,b}^{(n)} p_{a',b'}'^{(n)}$.
In particular,

$$q_{(s,s'),(s,s')}^{(n)} := p_{s,s}^{(n)} p_{s',s'}'^{(n)} = u_n v_n$$

whence by proposition 5.1.1 uv is a renewal sequence.
(ii) For each n, $k \ge 1$,

$$f_k^{(n)} = u_k^{(n)} - \sum_{j=1}^{k} f_j^{(n)} u_{k-j}^{(n)}.$$

An induction on $k \ge 1$ shows that $\exists \lim_{n \to \infty} f_k^{(n)} := g_k \ \forall \ k \ge 1$ whence $g_k = v_k - \sum_{j=1}^{k} g_j v_{k-j} \ge 0 \ \forall \ k \ge 1$ and v is a renewal sequence. $\square$

§5.2 Markov towers and recurrent events

DEFINITION: MARKOV TOWER, SIMPLE MARKOV TOWER.
A *Markov tower* is a tower equipped with an independent generator for the base transformation, with respect to which, the height function is measurable (i.e. the independent generator refines the partition generated by the height function).

A Markov tower T with base A is called *simple* if the partition $\{[\varphi_A = k] : k \in \mathbb{N}\}$ is a generator for T_A.

Let $f \in \mathcal{P}(\mathbb{N})$ and let $S : \Omega := \mathbb{N}^{\mathbb{Z}} \to \Omega$ be the shift $(Sx)_n = x_{n+1}$ and let $p_f \in \mathcal{P}(\Omega)$ be product measure. The simple Markov tower T_f with height function distribution f is the tower over $(\Omega, \mathcal{B}(\Omega), p_f, S)$ with height function $\varphi(x) := x_0$:

$$X_{T_f} := \bigcup_{n=1}^{\infty} [\varphi \geq n] \times \{n\} \subset X \times \mathbb{N}, \quad m_{T_f}(A \times \{n\}) := p_f(A)$$

and

$$T_f(x, n) = \left\{ \begin{array}{ll} n+1 & n \leq \varphi(x) - 1, \\ (Sx, 1) & n = \varphi(x). \end{array} \right.$$

Let T be a Markov tower over A with height function distribution f. Define $\pi : A \to \Omega$ by $\pi(x)_n := \varphi_A(T_A^n x)$, then by assumption $\pi : (A, \mathcal{B}_T \cap A, m|_A, T_A) \to (\Omega, \mathcal{B}(\Omega), p_f, S)$ and $\varphi(\pi x) = \varphi_A(x)$. By exercise 1.5.2, up to isomorphism, T is the tower over $(A, \mathcal{B}_T \cap A, m|_A, T_A)$ with height function φ_A so if $\tilde{\pi} : X_T \to X_{T_f}$ is defined by $\tilde{\pi}(x, n) := (\pi x, n)$ then $\tilde{\pi} : T \to T_f$.

Thus, any Markov tower T is a (canonical) extension of a simple Markov tower with the same base and height function and a simple Markov tower is determined up to isomorphism by the distribution of its height function.

5.2.1 PROPOSITION [A-Kea2].

(i) If P is an irreducible, recurrent stochastic matrix on S, and $s \in S$, $p_{s,s}^{(n)} = u_n$, then T_P is a Markov tower over $[s]$ whose height function is distributed according to $f^{(u)}$.

(ii) If u is a recurrent renewal sequence, then the renewal process T_u is the simple Markov tower whose height function has distribution $f^{(u)}$.

(iii) Any simple Markov tower is a renewal process.

PROOF.
(i) Suppose $m([s]) = 1$. The partition

$$\beta = \{[s, t_1, \ldots, t_n, s] : n \geq 0, \ t_1, \ldots, t_n \in S \setminus \{s\}\}$$

is an independent generator for $T_{[s]}$ with respect to which $\varphi_{[s]}$ is measurable. The distribution of the height function $\varphi_{[s]}$ is given by

$$m([\varphi_{[s]} = n]) = \sum_{t_1, \ldots, t_{n-1} \notin S} m([s, t_1, \ldots, , t_{n-1}, s]) = f_{s,s}^{(n)} = f_n^{(u)}.$$

(ii) By (i), T_u is a Markov tower over $[1]$ with height function distributed according to $f^{(u)}$. Evidently, the partition $\{[\varphi_{[1]} : k \in \mathbb{N}\}$ is a generator for $T_{[1]}$, so this Markov tower is simple.

(iii)   Let T be a simple Markov tower, and let the distribution of its height function be f. $\exists$ a recurrent renewal sequence u such that $f = f^{(u)}$. By (ii), T_u is the simple Markov tower whose height function has distribution f, and is thus isomorphic to T. $\square$

If S is a Markov tower over the base set $A \in \mathcal{B}_S$, then $S \to T$ where T is the simple Markov tower over the same base set with the same height function. It follows from the above that the sequence $u = u(A)$ defined by

$$u_n = m_T(T^{-n}A|A) = m_S(S^{-n}A|A) \quad (n \geq 0)$$

is a recurrent renewal sequence and $T = T_{u(A)}$.

Accordingly, we call u the *renewal sequence of the Markov tower T over A.*
Suppose T is a measure preserving transformation.

DEFINITION: RECURRENT EVENT.
A *recurrent event* for T is a set $A \in \mathcal{B}_T$ such that $0 < m(A) < \infty$, and

$$m\left(\bigcap_{k=0}^{N} T^{-n_k}A\right) = m(A) \prod_{k=1}^{N} u_{n_k - n_{k-1}}$$

whenever $0 = n_0 \leq n_1 \leq \ldots n_N$, and where $u_n = m_T(T^{-n}A|A)$. Let $M(T)$ denote
the collection of recurrent events for T.

5.2.2 THEOREM [A1].
If T is conservative, then

$$A \in M(T) \Leftrightarrow \{\varphi_A \circ T_A^n : n \geq 0\} \text{ are independent on A with respect to } m(\cdot|A).$$

In this case, $u = u(A) = \{u_n : n \geq 0\}$ is a recurrent renewal sequence and
$T \to T_{u(A)}$.

PROOF. Without loss of generality, we assume that $A \in \mathcal{B}$, $m(A) = 1$, and
that $\mathcal{B} = \sigma(\{T^n A : n \in \mathbb{Z}\})$.

The only part which does not follow from proposition 5.2.1 is
$A \in M(T) \implies \{\varphi_A \circ T_A^n : n \geq 0\}$ are independent, equivalently T is a simple
Markov tower over A.

To prove this, consider the shift $S : \Omega := \{0,1\}^{\mathbb{N}} \to \Omega$, let $\mathcal{A}$ denote the algebra
generated by cylinder sets and let $\mu : \mathcal{A} \to [0,\infty)$ be the additive set function
defined by

$$\mu([0^{r_1}, 1^{s_1}, 0^{r_2}, 1^{s_2}, \ldots]) :=$$

$$m_T\left(\bigcap_{i_1=0}^{r_1-1} T^{-i_1}A^c \cap \bigcap_{j_1=r_1}^{r_1+s_1-1} T^{-j_1}A \cap \bigcap_{i_2=r_1+s_1}^{r_1+s_1+r_2-1} T^{-i_2}A^c \ldots\right)$$

for $r_1, s_1, r_2, s_2, \cdots \geq 0$. We prove that
(i) u is a renewal sequence with lifetime distribution given by $\mu([1, 0^k, 1]) = f_{k+1}$ $(k \geq 0)$;
and

((ii)) $\mu([1, 0^{r_1}, 1, 0^{r_2}, 1 \ldots 1, 0^{r_k}, 1, *^{n_1}, 1, *^{n_2}, 1 \ldots 1, *^{n_\ell}, 1]) = \prod_{i=1}^{k} f_{r_i+1} \prod_{j=1}^{\ell} u_{n_j+1}$

(where $[*] := \Omega$) $\forall \ell \geq 1$, $n_1, \ldots, n_\ell \geq 0$ and $k \geq 0$, $r_1, \ldots, r_k \geq 1$.

To establish (i), we prove the statements $P(r)$ for $r = 0, 1 \ldots$ where $P(r)$ is:

$$\mu([1, 0^r, 1, *^{n_1}, 1, *^{n_2}, 1 \ldots 1, *^{n_\ell}, 1]) = f_{r+1} \prod_{j=1}^{\ell} u_{n_j+1} \ \forall \ell \geq 1, \ n_1, \ldots, n_\ell \geq 0.$$

This is true for $r = 0$ as

$$\mu([1, 1, *^{n_1}, 1, *^{n_2}, 1 \ldots 1, *^{n_\ell}, 1]) = u_1 \prod_{j=1}^{\ell} u_{n_j+1} = f_1 \prod_{j=1}^{\ell} u_{n_j+1}.$$

To see that $P(r) \implies P(r+1)$, assume $P(r)$; then

$$\mu([1, 0^{r+1}, 1, *^{n_1}, 1, *^{n_2}, 1 \ldots 1, *^{n_\ell}, 1])$$
$$= \mu([1, *^{r+1}, 1, *^{n_1}, 1, *^{n_2}, 1 \ldots 1, *^{n_\ell}, 1])$$
$$- \sum_{j=0}^{r} \mu([1, 0^j, 1, *^{r-j}, 1, *^{n_1}, 1, *^{n_2}, 1 \ldots 1, *^{n_\ell}, 1])$$
$$= \prod_{j=1}^{\ell} u_{n_j+1} \left(u_{r+2} - \sum_{j=0}^{r} f_{j+1} u_{r-j+1} \right)$$
$$= \prod_{j=1}^{\ell} u_{n_j+1} f_{r+2}$$

which is $P(r+1)$. Thus by induction, $P(r)$ is true $\forall\, r \geq 0$ and (i) is established.

Statement (ii) is also established by induction, and we consider the statements $Q(k, r)$ (for $k \geq 0$, $r \geq 1$):

$$\mu([1, 0^{r_1}, 1, 0^{r_2}, 1 \ldots 1, 0^{r_k}, 1, 0^r, 1, *^{n_1}, 1, *^{n_2}, 1 \ldots 1, *^{n_\ell}, 1])$$
$$= \prod_{i=1}^{k} f_{r_i+1} \prod_{j=1}^{\ell} u_{n_j+1} \forall\, \ell \geq 1,\ n_1, \ldots, n_\ell \geq 0,\ r_1, \ldots, r_k \geq 0.$$

Note that $Q(0, r) \equiv P(r)$ and $Q(k+1, 0) \equiv Q(k, r) \,\forall\, r \geq 0$. Thus, $Q(1, 0)$ has already been established, and we must show $Q(k, 0) \,\forall\, k \geq 0$.

To see $Q(k, 0) \implies Q(k+1, 0)$, we show $Q(k, 0) \implies Q(k, r) \,\forall\, r \geq 0$. Indeed, assuming $Q(k, r)$,

$$\mu([1, 0^{r_1}, 1, 0^{r_2}, 1 \ldots 1, 0^{r_k}, 1, 0^{r+1}, 1, *^{n_1}, 1, *^{n_2}, 1 \ldots 1, *^{n_\ell}, 1])$$
$$= \mu([1, 0^{r_1}, 1, 0^{r_2}, 1 \ldots 1, 0^{r_k}, 1, *^{r+1}, 1, *^{n_1}, 1, *^{n_2}, 1 \ldots 1, *^{n_\ell}, 1])$$
$$- \sum_{j=0}^{r} \mu([1, 0^{r_1}, 1, 0^{r_2}, 1 \ldots 1, 0^{r_k}, 1, 0^j, 1, *^{r-j}, 1, *^{n_1}, 1, *^{n_2}, 1 \ldots 1, *^{n_\ell}, 1])$$
$$= \prod_{i=1}^{k} f_{r_i+1} \prod_{j=1}^{\ell} u_{n_j+1} \left(u_{r+2} - \sum_{j=0}^{r} f_{j+1} u_{r-j+1} \right)$$
$$= \prod_{i=1}^{k} f_{r_i+1} \prod_{j=1}^{\ell} u_{n_j+1} f_{r+2}$$

which is $Q(k, r+1)$.

Thus by induction on $r \geq 0$ $Q(k, r)$ holds $\forall\, r \geq 0$ and hence $Q(k+1, 0)$ holds. By induction on $k \geq 0$ $Q(k, 0)$ holds $\forall\, k \geq 0$ and (ii) is established.   □

REMARK.

For non-invertible T, the proof of theorem 5.2.2 can be adapted to show that if $A \in \mathcal{B}$, $m(A) = 1$ and $\widehat{T}^n 1_A$ is constant on $A \ \forall \ n \geq 1$, then φ_A is independent of $T_A^{-1}(\mathcal{B} \cap A)$.

5.2.3 PROPOSITION [A-Lin-W]. *Let u be a recurrent renewal sequence, and let S be a conservative, ergodic, measure preserving transformation of $(X, \mathcal{B}, m)$, then $S \times T_u$ is conservative if and only if*

$$\sum_{n=1}^{\infty} u_n h \circ S^n = \infty \ a.e. \ \forall \ h \in L^1(m), \ h > 0.$$

PROOF. Evidently $S \times T_u$ conservative $\implies \sum_{n=1}^{\infty} u_n h \circ S^n = \infty$ a.e. $\forall \ h \in L^1(m)$, $h > 0$.

To prove the converse, let T be the one-sided shift of P_u, then T_u is the natural extension of T. There is no loss of generality in assuming S invertible.

Assume $\sum_{n=1}^{\infty} u_n h \circ S^n = \infty$ a.e. $\forall \ h \in L^1(m)$, $h > 0$.

Since $\widehat{T}^n 1_{[1]} = u_n$ a.e. on $[1]$, we have

$$\sum_{n=1}^{\infty} \widehat{T}^n 1_{[1]} h \circ S^n = \infty \ \text{a.e. on} [1] \times X \ \ \forall \ h \in L^1(m), \ h > 0$$

and by propositions 1.2.4 and 1.3.1, $S \times T$ is conservative.

Since Let $S \times T_u$ is the natural extension of $S \times T$, we have that $S \times T_u$ is conservative. $\square$

EXERCISE 5.2.1.

Let Y be a random variable on $\mathbb{N}$ and suppose that

$$L(x) := E(Y \wedge x)$$

is regularly varying at ∞ with index $\alpha \in (0, 1)$.

Let $S : \Omega := \mathbb{N}^{\mathbb{Z}} \to \Omega$ be the shift $(Sx)_n = x_{n+1}$, $p_f \in \mathcal{P}(\Omega)$ be product measure and define $\varphi : \Omega \to \mathbb{N}$ by $\varphi(x) = x_1$.

1) Show (as in [Fe3]) that if $b(n) = L(b(n))$, then

$$E(e^{-t \frac{\varphi_n}{b(n)}}) \to e^{-\Gamma(1+\alpha)t^{1-\alpha}}$$

as $n \to \infty \ \forall \ t > 0$.

Now let T be the tower over $(\Omega, \mathcal{B}(\Omega), p_f, S)$ with height function φ.

2) Show that $a_n(T) \propto \frac{n}{L(n)}$.

3) Show using the Darling-Kac theorem that $\exists \ c_\alpha \in \mathbb{R}_+$ such that $E(e^{-\frac{t}{Y_\alpha}}) = e^{-c_\alpha t^{1-\alpha}}$ where Y_α has the Mittag-Leffler distribution of order α.

REMARK.

The limiting distribution in exercise 5.2.1, part 1 is called (positive) *stable of order* α. For distributional convergence to stable laws as a consequence of the Darling-Kac theorem, see [A10].

§5.3 Kaluza sequences

DEFINITION: KALUZA SEQUENCE. A bounded sequence $u = \{u_n : n \geq 0\}$ is called a *Kaluza sequence* if $u_n > 0$, $u_0 = 1$, and

$$v_n := \frac{u_{n+1}}{u_n} \uparrow v \leq 1 \text{ as } n \uparrow \infty.$$

5.3.1 PROPOSITION [Ki].

If $\{u_n : n \geq 0\}$ is a Kaluza sequence, then $\exists\ v$, $p_k \in (0,1]$ $(k \geq 1)$ such that $\prod_{k=1}^{\infty} p_k > 0$, and

$$u_n = v^n \prod_{k=1}^{\infty} p_k^{k \wedge n}.$$

5.3.2 THEOREM [Kal]. *Kaluza sequences are renewal sequences.*

PROOF.
Let $\Omega = \{0,1\}^{\mathbb{N}}$ and let $T : \Omega \to \Omega$ be the shift. For $0 < p \leq 1$ let $\mu_p = \prod(1 - p, p)$ be the product measure, and consider the measure preserving transformation $(\Omega, \mathcal{B}(\Omega), \mu_p, T)$.

As can easily be checked, for $k \geq 1$, $A_k := \underbrace{[1, \ldots, 1]}_{k\text{-times}} \in M(T)$ whence by theorem 5.2.2, $u_n := \frac{\mu_p(A_k \cap T^{-n} A_k)}{\mu_p(A_k)} = p^{k \wedge n}$ defines a renewal sequence. The proposition now follows from proposition 5.1.3 which states that products and limits of renewal sequences are renewal sequences. $\qquad\square$

DEFINITION: MOMENT SEQUENCE.
A *moment sequence* is a sequence u of form $u_0 = 1$ and

$$u_n = \int_0^{\infty} e^{-tn} dp(t) \quad (n \geq 1)$$

where p is a probability on $[0, \infty)$.

5.3.3 PROPOSITION [Ki].

Any moment sequence is a Kaluza sequence.
The sequence $u_n = \int_0^{\infty} e^{-tn} dp(t)$ is recurrent if and only if $\int_0^{\epsilon} \frac{dp(t)}{t} = \infty\ \forall\ \epsilon > 0$.

PROOF.
Suppose that $u_n = \int_0^{\infty} e^{-tn} dp(t)$ where p is a probability on $[0, \infty)$, then by the Cauchy-Schwarz inequality,

$$
\begin{aligned}
u_n &= \int_0^{\infty} e^{-tn} dp(t) \\
&= \int_0^{\infty} e^{-(n-1)\frac{t}{2}} e^{-(n+1)\frac{t}{2}} dp(t) \\
&\leq \sqrt{\int_0^{\infty} e^{-(n-1)\frac{t}{2}} dp(t) \int_0^{\infty} e^{-(n+1)\frac{t}{2}} dp(t)} \\
&= \sqrt{u_{n-1} u_{n+1}} \ \ \forall\ n \geq 0
\end{aligned}
$$

and u is a Kaluza sequence.

The sequence $u_n = \int_0^\infty e^{-tn} dp(t)$ is recurrent if and only if $\sum_{n=0}^\infty u_n = \infty$, equivalently

$$\infty = \int_0^\infty \sum_{n=0}^\infty e^{-tn} dp(t) = \int_0^\infty \frac{dp(t)}{1 - e^{-t}}.$$

Evidently $\forall\, 0 < \epsilon < 1$, $\int_\epsilon^\infty \frac{dp(t)}{1-e^{-t}} \le \frac{1}{1-e^{-\epsilon}}$ and $\frac{1}{1-e^{-t}} = \frac{e^{\pm 1}}{t}$ as $t \to 0$, so

$$\int_0^\infty \frac{dp(t)}{1 - e^{-t}} = \infty \;\Leftrightarrow\; \int_0^\epsilon \frac{dp(t)}{t} = \infty.$$

$\square$

5.3.4 COROLLARY [Bru].

If S is a conservative, ergodic non-singular transformation of $(X, \mathcal{B}, m)$ with $\mathfrak{P}(S) = \emptyset \bmod m$, then $\exists$ a conservative, exact Markov chain T such that $S \times T$ is totally dissipative.

PROOF.

Since $\mathfrak{P}(S) = \emptyset \bmod m$, by theorem 1.4.4 $\exists\, A \in \mathcal{B}$ such that

$$\frac{1}{n} \sum_{k=1}^n m(A \cap S^{-k}A) \to 0 \text{ as } n \to \infty.$$

It follows that $t\epsilon(t) \to 0$ as $t \to 0$ where

$$\epsilon(t) := \sum_{k=1}^\infty m(A \cap S^{-k}A)e^{-tk}.$$

Let $y_k > y_{k+1} \downarrow 0$ $(k \ge 1)$ be such that $y_k < \frac{1}{2^k}$ and

$$t\epsilon(t) < \frac{1}{2^k} \;\forall\, 0 < t < y_k.$$

Define $f : (0, \infty) \to [0, \infty)$ by

$$f(t) = \left\{ \begin{array}{ll} \frac{2^k}{k(k+1)(y_k - y_{k+1})} & t \in [y_{k+1}, y_k) \ (k \ge 1) \\ 0 & t \ge y_1. \end{array} \right.$$

Evidently

$$\int_0^\infty f(t)dt = \sum_{k=1}^\infty \frac{2^k}{k(k+1)} = \infty,$$

$$\int_0^\infty tf(t)dt = \sum_{k=1}^\infty \frac{2^k(y_k + y_{k+1})}{k(k+1)} \le 2$$

and

$$\int_0^\infty t\epsilon(t)f(t)dt = \sum_{k=1}^\infty \frac{2^k}{k(k+1)(y_k - y_{k+1})} \int_{[y_{k+1}, y_k)} t\epsilon(t)dt$$

$$< \sum_{k=1}^\infty \frac{1}{k(k+1)} = 1.$$

Defining $dp(t) = \frac{tf(t)dt}{\int_0^\infty tf(t)dt}$, we obtain $p \in \mathcal{P}([0,\infty))$ and a moment sequence $u_n := \int_0^\infty e^{-tn} dp(t)$ which is recurrent since

$$\int_0^\infty \frac{dp(t)}{t} \propto \int_0^\infty f(t)dt = \infty.$$

Also

$$\sum_{k=0}^\infty u_k m(A \cap S^{-k}A) = \int_0^\infty \sum_{k=0}^\infty e^{-tk} m(A \cap S^{-k}A) dp(t)$$

$$= \int_0^\infty \epsilon(t) dp(t)$$

$$\propto \int_0^\infty t\epsilon(t) f(t) dt < \infty,$$

whence

$$\sum_{k=0}^\infty 1_{A \times [1]} \circ (S \times T_u)^k < \infty \text{ a.e. on } A \times [1],$$

$S \times T_u$ is not conservative, hence totally dissipative. □

5.3.5 COROLLARY [A-Lin-W].

If S is a conservative, ergodic measure preserving transformation of $(X, \mathcal{B}, m)$, then $\exists$ a conservative, exact Markov chain T with an infinite invariant measure such that $S \times T$ is ergodic.

PROOF. Fix $h \in L^1(m)$, $h > 0$, then

$$\sum_{n=1}^\infty e^{-tn} h \circ S^n = \infty \text{ as } t \to 0 \text{ a.e. on } X.$$

By Egorov's theorem, $\exists B \in \mathcal{B}_+$ and $\alpha(t) \to \infty$ as $t \to 0$ such that

$$\sum_{n=1}^\infty e^{-tn} h \circ S^n \geq \alpha(t) \text{ on } B \ \forall \ t > 0.$$

Since $\alpha(t) \uparrow \infty$ as $t \to 0$, $\exists f : (0,\infty) \to [0,\infty)$ such that $\int_{(0,\infty)} f(t)dt = 1$ and $\int_{(0,\infty)} \alpha(t)f(t)dt = \infty$. Consider the moment sequence

$$u_n := \int_{(0,\infty)} e^{-tn} f(t)dt,$$

which is null-recurrent, and which satisfies

$$\sum_{n=1}^\infty u_n h \circ S^n(x) = \int_{(0,\infty)} f(t) \sum_{n=1}^\infty e^{-tn} h \circ S^n(x) dt$$

$$\geq \int_{(0,\infty)} f(t)\alpha(t)f(t)dt$$

$$= \infty \ \forall \ x \in B$$

and hence for a.e. $x \in X$. By proposition 5.2.3, $S \times T_u$ is conservative. □

§5.4 Similarity of Markov towers

In this section, we study the structure of null recurrent Markov chains. There is a simple sufficient condition for similarity in terms of the renewal sequences attached to the chains.

DEFINITION: EQUIVALENT RENEWAL SEQUENCES.
Two renewal sequences u, and u' are *equivalent* (denoted $u \sim u'$) if there are positively recurrent, aperiodic renewal sequences v, and v' such that $uv = u'v'$.

5.4.1 THEOREM [**A-Kea2**].
Markov towers with equivalent renewal sequences are similar.

PROOF. Let S and S' be infinite Markov towers over the sets $A \in \mathcal{B}_S$ and $A' \in \mathcal{B}_{S'}$ respectively. By assumption, there are positively recurrent, aperiodic renewal sequences v and v' such that $u(A)v = u(A')v' := w$.

By proposition 5.2.1, $S \to T_{u(A)}$ and $S' \to T_{u(A')}$.

Now, $T_{u(A)} \times T_v$ is the Markov chain of the stochastic matrix $P_{u(A)} \times P_v$ on $\mathbb{N} \times \mathbb{N}$.

By proposition 5.2.1, $T_{u(A)} \times T_v$ is a Markov tower over $[1] \times [1]$ with renewal sequence w, whence $T_{u(A)} \times T_v \to T_w$, and

$$S \leftarrow S \times T_v \to T_{u(A)} \times T_v \to T_w.$$

An analogous proof shows that

$$S' \leftarrow S' \times T_{v'} \to T_{u(A')} \times T_{v'} \to T_w.$$

Therefore $S \sim S'$. $\square$

Showing that renewal sequences are equivalent involves constructing positively recurrent, aperiodic renewal sequences; and Kaluza sequences are instrumental.

Let u be any non-negative sequence, with $u_n > 0$ for all n large (for example an aperiodic renewal sequence), then

$$v_n(u) = \frac{u_{n+1}}{u_n}, \text{ and } p_n(u) = \frac{v_{n-1}}{v_n}$$

are defined for large enough n. Indeed, if

$$n_u := \min \{n \geq 1, \ u_k > 0 \ \forall \ k \geq n - 1\}$$

then $p_n(u)$ is defined $\forall \ n \geq n_u - 1$.

Suppose $u_n > 0$ for every $n \geq 0$, then

$$\frac{v_n}{v_{n+k+1}} = \prod_{j=1}^{k} p_{n+j}.$$

Thus, if $u_n \sim u_{n+1}$, and $p_n \leq 1 \ \forall \ n \geq 1$, then

$$0 < v_n \xleftarrow[k\to\infty]{} \frac{v_n}{v_{n+k+1}} = \prod_{j=1}^{k} p_{n+j} \xrightarrow[k\to\infty]{} \prod_{j=1}^{\infty} p_{n+j},$$

whence u is a Kaluza sequence, and

$$\sum_{k=1}^{\infty} \log\left(\frac{1}{p_k(u)}\right) < \infty.$$

On the other hand, if

$$\sum_{n=1}^{\infty} |\log p_n(u)| < \infty,$$

we have

$$u_n = \prod_{k=1}^{\infty} p_k^{k \wedge n}.$$

5.4.2 PROPOSITION [A-Kea2]. *Let u be an aperiodic renewal sequence satisfying $u_n \sim u_{n+1}$ as $n \to \infty$. If*

$$\sum_{n=n_u}^{\infty} n(\log p_n(u))_+ < \infty,$$

then u is equivalent to a Kaluza sequence w such that

$$\sum_{n=n_u}^{\infty} n|\log p_n(u) - \log p_n(w)| < \infty.$$

PROOF. By aperiodicity, there exists $n_0 \geq 0$ such that the semigroup in $\mathbb{N}$ generated by $\{1 \leq k < n_0 : u_k > 0\}$ contains every $n \in \mathbb{N} \cap [n_0, \infty)$. If $\{g_n : n \in \mathbb{N}\}$ is the probability distribution on $\mathbb{N}$ satisfying

$$u_n = \sum_{k=1}^{n} g_k u_{n-k},$$

set

$$h_n = \begin{cases} g_n \text{ for } 1 \leq n < n_0 \\ \sum_{k=n_0}^{\infty} g_k \text{ for } n = n_0 \\ 0 \text{ else }, \end{cases}$$

and let v be the renewal sequence defined by

$$v_n = \sum_{k=1}^{n} h_k v_{n-k}.$$

Since $v_n = u_n$ for $1 \leq n < n_0$, we have that $v_n > 0$ for $n \geq n_0$. Since P_v is defined on finitely many states, we have that v is positively recurrent, and, moreover

$$\exists v_\infty > 0, \ r \in (0,1) \ \ni \ v_n = v_\infty(1 + O(r^n)).$$

Define a by

$$a_n = \begin{cases} 1 \text{ for } 1 \leq n < n_0 \\ \frac{u_n}{v_n} \text{ for } n \geq n_0, \end{cases}$$

then $a_n > 0$ for every $n \geq 1$, $a_n \sim a_{n+1}$, and (as $u_n = v_n \ \forall \ 1 \leq n < n_0$)

$$u_n = a_n v_n \ \forall \ n \geq 1.$$

Moreover,

$$\log p_n(a) - \log p_n(u) = O(r^n) \text{ as } n \to \infty,$$

whence

$$\sum_{n=1}^{\infty} n(\log p_n(a))_+ < \infty.$$

Define sequences v', and w by

$$v'_n := \prod_{k \ni \, p_k(a) > 1} \left(\frac{1}{p_k(a)} \right)^{k \wedge n}, \quad w_n = a_n v'_n.$$

This product converges, and v' is a positively recurrent Kaluza sequence, since

$$\sum_{k=1}^{\infty} k \log \left(\frac{1}{p_k(v')} \right) < \infty.$$

Since $av' = w$, we have $w_n \sim w_{n+1}$, and by inspection,

$$1 \geq p_n(w) = \begin{cases} p_n(a) \text{ for } p_n(a) \leq 1, \\ 1 \text{ for } p_n(a) > 1, \end{cases}$$

whence, by the remarks preceding this proposition, w is a Kaluza sequence.

Moreover,

$$uv' = avv' = wv$$

and

$$\sum_{n=n_u}^{\infty} n|\log p_n(u) - \log p_n(w)| < \infty.$$

$\square$

5.4.3 PROPOSITION [A-Kea2]. *If u, u' are Kaluza sequences, and*

$$\sum_{n=1}^{\infty} n|\log p_n(u) - \log p_n(u')| < \infty,$$

then u and u' are equivalent.

PROOF. Define another Kaluza sequence w by

$$p_n(w) := p_n(u) \vee p_n(u').$$

It follows easily from the definitions that

$$u = vw, \quad u' = v'w$$

where v, v' are Kaluza sequences defined by

$$p_n(v) = \frac{p_n(u)}{p_n(w)}, \quad p_n(v') = \frac{p_n(u')}{p_n(w)},$$

indeed, it follows that v, and v' are positively recurrent, whence u and u' are equivalent. $\square$

5.4.4 PROPOSITION   [**A-Kea2**].

Let u be an aperiodic renewal sequence, and let w be a Kaluza sequence such that

$$\sum_{n=n_u}^{\infty} n|\log p_n(u) - \log p_n(w)| < \infty,$$

then

$$u \sim w.$$

PROOF.   This follows from propositions 5.4.1 and 5.4.2.   □

For $\beta \geq 0$, $u^{(\beta)}$ defined by

$$u_n^{(\beta)} = \frac{1}{(n+1)^{\beta}}$$

is a Kaluza sequence, and

$$\log\left(\frac{1}{p_n(u^{(\beta)})}\right) = \frac{\beta}{n^2} + \delta_n \text{ where } \sum_{n=1}^{\infty} n|\delta_n| < \infty.$$

The renewal sequence $u^{(\beta)}$ is recurrent for $0 < \beta \leq 1$.

The following exhibits an uncountable family of pairwise strongly disjoint, conservative, ergodic Markov shifts.

5.4.5 PROPOSITION.

$T_{u^{(\beta)}}$ and $T_{u^{(\beta')}}$ are strongly disjoint when $0 < \beta \neq \beta' \leq 1$.

PROOF.

$$a_n(T_{u^{(\beta)}}) \propto n^{1-\beta}$$

$$\therefore T_{u^{(\beta)}} \sim T_{u^{(\beta')}} \implies a_n(T_{u^{(\beta)}}) \propto a_n(T_{u^{(\beta')}}) \implies \beta = \beta'.$$

                □

5.4.6 COROLLARY   [**A-Kea2**]. *Let u be an aperiodic renewal sequence such that*

$$\sum_{n=n_u}^{\infty} n|\log\left(\frac{1}{p_n(u)}\right) - \frac{\beta}{n^2}| < \infty,$$

then

$$u \sim u^{(\beta)}.$$

§5.5 Random walks

In this section, we consider random walks on $\mathbb{Z}^d$ $(d \geq 1)$. Let f be a probability on $\mathbb{Z}^d$ and define a stochastic matrix $P = P_f$ on $S = \mathbb{Z}^d$ by $p_{s,t} := f_{t-s}$. It is evident that $m_s = 1$ is a stationary distribution for P_f, and the shift T_f of (P_f, m) is known as the *random walk* on $\mathbb{Z}^d$ with *jump distribution* f. The stochastic matrix P_f is irreducible iff $S_f := \{t \in \mathbb{Z}^d : f_t > 0\}$ is contained in no proper subgroup of $\mathbb{Z}^d$, and P_f is aperiodic if S_f is contained in no coset of any proper subgroup of $\mathbb{Z}^d$.

A random walk on $\mathbb{Z}$ is a skew product, and therefore conservative if, for example its jump distribution has first moment and is centred (see §8.1)

A random walk with irreducible and recurrent stochastic matrix is by §4.5 pointwise dual ergodic.

LOCAL LIMIT THEOREM   [**Bre**].

Let $f \in \mathcal{P}(\mathbb{Z}^d)$ be aperiodic, and suppose that $\sigma_f^2 := \sum_{n \in \mathbb{Z}^d} \|n\|^2 f_n < \infty$ and $\sum_{n \in \mathbb{Z}^d} n f_n = 0$, then

$$p_{0,0}^{(n)}(f) \sim \frac{1}{\sqrt{\det \Gamma (2\pi n)^d}} \quad as \ n \to \infty$$

where $\Gamma_{i,j} := \sum_{x \in \mathbb{Z}^d} x_i x_j f_x$.

PROOF.

For $s \in [-\pi, \pi)^d$,

$$\phi(s) := \sum_{x \in \mathbb{Z}^d} f_x e^{i\langle s, x \rangle} = 1 - \frac{s^t \Gamma s}{2} + o(\|s\|^2) \quad as \ s \to 0.$$

By irreducibility of f Γ is non-singular and so $\exists \ \epsilon > 0$ such that $s^t \Gamma s \geq \epsilon \|s\|^2 \ \forall \ s \in \mathbb{T}^d$. It follows that $\exists \ \delta > 0$ such that

$$|\phi(s)| \leq 1 - \frac{\epsilon \|s\|^2}{4} \quad \forall \ \|s\| \leq \delta.$$

By aperiodicity of f, $\exists \ 0 < r < 1$ such that

$$|\phi(s)| \leq r \quad \forall \ s \in \mathbb{T}, \ \|s\| \geq \delta,$$

whence

$$p_{0,0}^{(n)}(f) = \int_{[-\pi,\pi)^d} \phi(s)^n ds$$

$$= \int_{B(0,\delta)} \phi(s)^n ds + \int_{[-\pi,\pi)^d \setminus B(0,\delta)} \phi(s)^n ds$$

$$= I + II.$$

Changing variables,

$$I = \frac{1}{n^{\frac{d}{2}}} \int_{B(0,\delta \sqrt{n})} \phi(\frac{y}{\sqrt{n}})^n dy.$$

Evidently $\phi(\frac{y}{\sqrt{n}})^n \to e^{-\frac{y^t \Gamma y}{2}}$ as $n \to \infty$ whence by dominated convergence,

$$I \sim \frac{1}{n^{\frac{d}{2}}} \int_{\mathbb{R}^d} e^{-\frac{y^t \Gamma y}{2}} dy = \frac{1}{\sqrt{\det \Gamma (2\pi n)^d}}.$$

On the other hand

$$II \leq 2\pi r^n = o\left(\frac{1}{n^{\frac{d}{2}}}\right) \quad as \ n \to \infty.$$

$\square$

It now follows from §4.5 that irreducible random walks on $\mathbb{Z}^d$ $(d = 1, 2)$ with centred jump distributions f with finite second moment have recurrent stochastic matrices, and are pointwise dual ergodic with return sequences of form

$$a_n(T_f) \sim \begin{cases} c_f^{(1)} \sqrt{n} & d = 1, \\ c_f^{(2)} \log n & d = 2 \end{cases}$$

where

$$c_f^{(1)} = \sqrt{\frac{2}{\pi \sigma_f^2}}, \quad \sigma_f^2 = \sum_{n \in \mathbb{Z}} n^2 f_n,$$

and

$$c_f^{(2)} = \frac{1}{2\pi \sqrt{\det \Gamma}}.$$

In a similar manner, it can be shown that for $f \in \mathcal{P}(\mathbb{Z}^d)$ not supported on a proper subgroup,

$$p_{0,0}^{(n)}(f) = O\left(\frac{1}{n^{\frac{d}{2}}}\right)$$

as $n \to \infty$, whence there is no ergodic random walk on $\mathbb{Z}^d$ for $d \geq 3$.

The following exhibits an uncountable family of pairwise strongly disjoint, conservative, ergodic random walks.

5.5.1 PROPOSITION   [**A1**].
For each $\alpha \in (0, \frac{1}{2})$ there is a conservative, ergodic random walk T_α on $\mathbb{Z}$ with $a_n(T_\alpha) \propto n^\alpha$.

PROOF.
Given $\alpha \in (0, \frac{1}{2})$, set $a = a(\alpha) := \frac{1}{1-\alpha} \in (1, 2)$ and define $f^{(\alpha)} \in \mathcal{P}(\mathbb{Z})$ by

$$f_n^{(\alpha)} := \begin{cases} \frac{1}{2\zeta(1+a)|n|^{1+a}} & n \neq 0, \\ 0 & n = 0. \end{cases}$$

It follows (as in [**Bar-Nin**] p. 141 ff.) that if

$$\phi_\alpha(s) := \sum_{x \in \mathbb{Z}} f_x e^{ixs} = \frac{1}{\zeta(1+a)} \sum_{x=1}^{\infty} f_x \cos xs,$$

then

$$1 - \phi_\alpha(s) \sim K_\alpha |s|^a \text{ as } s \to 0,$$

where $K_\alpha \in \mathbb{R}_+$, whence (as in the proof of the local limit theorem),

$$p_{0,0}^{(n)}(f^{(\alpha)}) \sim \frac{1}{n^{\frac{1}{a}}} \int_{\mathbb{R}} e^{-K_\alpha |s|^a} ds \propto \frac{1}{n^{1-\alpha}}.$$

It follows that $T_{f^{(\alpha)}}$ is conservative and ergodic with $a_n(T_\alpha) \propto n^\alpha$.   □

Here, we consider random walks on $\mathbb{Z}^d$ $(d \geq 1)$ with centred jump distributions f satisfying

(7)
$$\sum_{x \in \mathbb{Z}^d} \|x\|^7 f_x < \infty,$$

for example

$$\pi_k^{(d)} = \begin{cases} \frac{1}{3^d} & k \in \{0, \pm 1\}^d \\ 0 & \text{el se.} \end{cases}$$

5.5.2 THEOREM   [**A-Kea2**].

If f is an aperiodic probability on $\mathbb{Z}^d$ satisfying (7), and $u_n(f) = p_{0,0}^{(n)}(f)$, then

$$u(f) \sim u^{(\frac{d}{2})}.$$

5.5.3 LEMMA   [**A-Kea2**].

Suppose that u is a bounded, non-negative sequence, $\beta > 0$, $\Gamma \subset (0,2]$ is a finite set, $\epsilon > 0$, and that

$$u_n = K u_n^{(\beta)} \left(1 + \sum_{\gamma \in \Gamma} \frac{a_\gamma}{n^\gamma} + O\left(\frac{1}{n^{2+\epsilon}}\right) \right) \text{ as } n \to \infty$$

where $K > 0$, and $a_\gamma \in \mathbb{R}$ $(\gamma \in \Gamma)$, then

$$\log\left(\frac{1}{p_n(u)}\right) = \frac{\beta}{n^2} + \delta_n \text{ where } \sum_{n=1}^{\infty} n|\delta_n| < \infty.$$

PROOF.   We'll show that

$$\log\left(\frac{1}{p_n}\right) = \log\left(\frac{1}{p_n(u^{(\beta)})}\right) + O\left(\frac{1}{n^{2+\epsilon'}}\right) \text{ as } n \to \infty$$

where $\epsilon' > 0$. Set

$$A_n := \frac{u_n}{u_n^{(\beta)}} = K\left(1 + \sum_{\gamma \in \Gamma} \frac{a_\gamma}{n^\gamma} + O\left(\frac{1}{n^{2+\epsilon}}\right) \right) \text{ as } n \to \infty.$$

Using the Taylor expansion of $\log(1 + x)$ near 0, it follows that

$$B_n := \log\left(\frac{1}{A_n}\right) = -\log K + \sum_{\gamma \in \Gamma'} \frac{a_\gamma'}{n^\gamma} + O\left(\frac{1}{n^{2+\epsilon'}}\right) \text{ as } n \to \infty,$$

where $\Gamma' \subset (0,2]$ is a finite set, $a_\gamma' \in \mathbb{R}$ $(\gamma \in \Gamma')$, and $\epsilon' > 0$. If $B_n' = B_{n+1} - B_n$, and $B_n'' = B_{n+1}' - B_n'$, then

$$\log\left(\frac{1}{p_n}\right) = \log\left(\frac{1}{p_n(u^{(\beta)})}\right) + B_{n-1}''.$$

Using the Taylor expansion of $(1 + x)^{-\gamma}$, $(\gamma > 0)$ near 0, we see that

$$\frac{1}{n^\gamma} - \frac{1}{(n+1)^\gamma} = \frac{\gamma}{n^{\gamma+1}} + O\left(\frac{1}{n^{\gamma+2}}\right),$$

whence

$$B_n'' = O\left(\frac{1}{n^{2+\epsilon''}}\right) \text{ as } n \to \infty$$

where $\epsilon'' > 0$. □

PROOF OF THEOREM 5.5.2.   We'll establish the precondition of lemma 5.5.3. Let X be a random variable on $\mathbb{Z}^d$ distributed as f, and let

$$\varphi(x) = E(e^{ix \cdot X}) = \sum_{n \in \mathbb{Z}^d} f_n e^{ix \cdot n}, \quad (x \in \mathbb{R}^d).$$

Choose $\delta > 0$ such that $|1 - \varphi(x)| \le r < 1$ for $|x| \le \delta$, and set

$$\psi(x) := \log \varphi(x) + \frac{\sigma(x)}{2},$$

where $\sigma(x) = E((x \cdot X)^2)$, and log is defined in a small neighbourhood of 1 as that holomorphic inverse of the exponential function with $\log 1 = 0$. Since f is aperiodic, we have that

$$|\varphi(x)| \le r' < 1 \ \forall \ \delta \le |x| \le \pi.$$

It follows that

$$u_n = \frac{1}{(2\pi)^d} \int_{N(0,\delta)} (\varphi(t))^n dt + O(r'^n)$$

$$= \frac{1}{(2\pi)^d n^{\frac{d}{2}}} \int_{N(0,\delta\sqrt{n})} \left(\varphi\left(\frac{t}{\sqrt{n}}\right) \right)^n dt + O(r'^n)$$

$$= \frac{A_n}{(2\pi)^d n^{\frac{d}{2}}} + O(r'^n)$$

where $N(0,r)^{\cdot} = \{x \in \mathbb{R}^d : |x| < r\}$, and

$$A_n = \int_{N(0,\delta\sqrt{n})} \left(\varphi\left(\frac{t}{\sqrt{n}}\right) \right)^n dt$$

$$= \int_{N(0,\delta\sqrt{n})} \left(\varphi\left(\frac{t}{\sqrt{n}}\right) \right)^n dt$$

$$= \int_{N(0,\delta\sqrt{n})} e^{-\frac{\sigma(t)}{2}} \exp\left(n\psi\left(\frac{t}{\sqrt{n}}\right) \right) dt.$$

Clearly

$$A_n \to A = \int_{\mathbb{R}^d} e^{-\frac{\sigma(t)}{2}} dt,$$

and

$$A - A_n = \int_{N(0,\delta\sqrt{n})} e^{-\frac{\sigma(t)}{2}} \left(1 - \exp\left(n\psi\left(\frac{t}{\sqrt{n}}\right) \right) \right) dt + O(r''^n)$$

where $r'' \in (0,1)$.

Using (7) and Taylor's theorem, for $|x| \le \delta$,

$$\psi(x) = \sum_{k=3}^{6} \sum_{1 \le i_1,\dots,i_k \le d} a(i_1,\dots,i_k) \prod_{\nu=1}^{k} x_{i_\nu} + O(|x|^7)$$

as $x \to 0$, whence for $|x| \le \delta\sqrt{n}$,

$$n\psi\left(\frac{x}{\sqrt{n}}\right) = \sum_{k=3}^{6} \frac{1}{n^{\frac{k}{2}-1}} \sum_{1 \le i_1,\dots,i_k \le d} a(i_1,\dots,i_k) \prod_{\nu=1}^{k} x_{i_\nu} + O\left(\frac{|x|^7}{n^{\frac{5}{2}}} \right),$$

and

$$e^{n\psi(\frac{x}{\sqrt{n}})} - 1 = \sum_{k=1}^{4} \frac{1}{n^{\frac{k}{2}}} \sum_{1 \le i_1,\ldots,i_{k+2} \le d} b(i_1,\ldots,i_{k+2}) \prod_{\nu=1}^{k+2} x_{i_\nu} + O\left(\frac{|x|^7}{n^{\frac{5}{2}}}\right)$$

as $x \to 0$. Thus

$$A_n - A = \int_{\mathbb{R}^d} e^{-\frac{\sigma(x)}{2}} \left(\sum_{k=1}^{4} \frac{1}{n^{\frac{k}{2}}} \sum_{1 \le i_1,\ldots,i_{k+2} \le d} b(i_1,\ldots,i_{k+2}) \prod_{\nu=1}^{k+2} x_{i_\nu} \right) dx + O\left(\frac{1}{n^{\frac{5}{2}}}\right)$$

$$= \sum_{k=1}^{4} \frac{c_k}{n^{\frac{k}{2}}} + O\left(\frac{1}{n^{\frac{5}{2}}}\right).$$

It follows that

$$u_n = u_n(1/2)\left(A + \sum_{k=1}^{4} \frac{c_k}{n^{\frac{k}{2}}} + O\left(\frac{1}{n^{5/2}}\right)\right)$$

as $n \to \infty$. Theorem 5.5.2 is now established by lemma 5.5.3, and corollary 5.4.5.
$\square$

5.5.4 PROPOSITION [A-Kea2].
Suppose that $d = 1, 2$, and that $f^{(1)}, f^{(2)}$ are aperiodic probabilities on $\mathbb{Z}^d$ satisfying (7), then
$T_{f^{(1)}}$ *and* $T_{f^{(2)}}$ *are similar.*

PROOF.
By theorem 5.5.2, the renewal sequences $u(f^{(1)})$ and $u(f^{(2)})$ are equivalent, whence by theorem 5.4.1, $T_{f^{(1)}}$ and $T_{f^{(2)}}$ are similar. $\square$

REMARK: ISOMORPHISM OF RANDOM WALKS (SEE [A-Kea2]).
By theorem 3.6 of [A-Kea2], irreducible, recurrent random walks $T_{f^{(1)}}$ and $T_{f^{(2)}}$ on (possibly different) discrete Abelian groups giving rise to equivalent renewal sequences $u(f^{(1)})$ and $u(f^{(2)})$ are isomorphic.

5.5.5 COROLLARY.
If f is a centred, aperiodic probability on $\mathbb{Z}$, satisfying (7),
then $T_f \times T_f \times T_f$ is isomorphic to $S : \mathbb{R}^2 \to \mathbb{R}^2$ defined by $Sx = x + 1$ and considered with respect to Lebesgue measure.

PROOF.
Write $u_n(f) = p_{0,0}^{(n)}$, then by theorem 5.5.2 $u(f) \sim u^{(\frac{1}{2})}$, whence by theorems 5.2.2 and 5.4.1,

$$T_f \to T_{u(f)} \sim T_{u^{(\frac{1}{2})}}.$$

Since $T_f \times T_f$ is conservative, it follows that

$$T_f \times T_f \sim T_{u^{(\frac{1}{2})}} \times T_{u^{(\frac{1}{2})}} \to T_{u^{(1)}} \leftarrow T_{u^{(\frac{1}{4})}} \times T_{u^{(\frac{3}{4})}},$$

whence

$$T_f \times T_f \times T_f \sim S := T_{u^{(\frac{1}{4})}} \times T_{u^{(\frac{3}{4})}} \times T_{u^{(\frac{1}{2})}}.$$

The transformation $T_{u^{(\frac{3}{4})}} \times T_{u^{(\frac{1}{2})}}$ is dissipative. If W is a generating wandering set for it, then $X_{T_{u^{(\frac{1}{4})}}} \times W$ is an infinite generating wandering set for S.

It follows that $T_f \times T_f \times T_f$ (being similar to S) also has an infinite generating wandering set, and is therefore isomorphic to $z \mapsto z + 1$ on $\mathbb{R}^2$ considered with respect to Lebesgue measure. $\qquad\qquad\qquad\qquad\qquad\qquad\qquad\qquad\qquad$ $\square$

REMARK.

The moment assumption (7) on f in theorem 5.5.2 has been reduced in [**A-Lig-P**] to

$$(2\sqrt{\log}) \qquad\qquad\qquad \sum_{x \in \mathbb{Z}^d} \|x\|^2 \sqrt{\log^+ \|x\|} f_x < \infty.$$

CHAPTER 6

Inner functions

§6.1 Inner functions on the unit disc

Let U be the unit disc in $\mathbb{C}$ and let $\mathbb{T} = \mathbb{R}/\mathbb{Z} \cong [0,1)$, then $\partial U = \{e^{2\pi i t} : t \in \mathbb{T}\}$.

DEFINITION: POISSON MEASURE.
For $z \in \mathbb{C}$, let $\pi_z \in \mathcal{P}(\mathbb{T})$ be the *Poisson measure* (at z) defined by

$$\int_{\mathbb{T}} f d\pi_z = \tilde{f}(z)$$

for $f \in C(\mathbb{T})$ where $\tilde{f} : \overline{U} \to \mathbb{C}$ is the unique function which is harmonic in U, satisfying $\tilde{f}|_{\mathbb{T}} = f$ (see [**Rudi**]).

A system of measures with this property is called a *harmonic measure*.
It follows that, for $n \geq 0$,

$$\hat{\pi}_z(-n) = \int_0^1 e^{2\pi i n t} d\pi_z(t) = z^n$$

whence

$$\frac{d\pi_z}{dm}(x) = \sum_{n=0}^{\infty} z^n e^{-2\pi i n x} + \sum_{n=1}^{\infty} \overline{z}^n e^{2\pi i n x} = \mathrm{Re}\left(\frac{e^{2\pi i x} + z}{e^{2\pi i x} - z}\right) := p_z(x),$$

where $m \equiv \pi_0$ is Lebesgue measure on $\mathbb{T}$.

DEFINITION: INNER FUNCTION OF U. An analytic function $f : U \to U$ is called an *inner function* (of U), if, for m-a.e. $x \in \mathbb{T}$,

$$f(re^{2\pi i x}) \to e^{2\pi i T(x)} \in \mathbb{T} \text{ as } r \to 1.$$

The measurable map $T : \mathbb{T} \to \mathbb{T}$ is called the $\mathbb{T}$-*restriction* of f.

For example, *Möbius transformations* of form

$$z \mapsto e^{i\theta} \frac{z - \alpha}{1 - \overline{\alpha}z}$$

are inner functions.

It turns out that the restrictions of inner functions are non-singular transformations of $(\mathbb{T}, \mathcal{B}, m)$.

6.1.1 PROPOSITION (SEE [**No**] [**Let**]). *Suppose that $f : U \to U$ is inner, with $\mathbb{T}$-restriction T, then*

$$\pi_z \circ T^{-1} = \pi_{f(z)} \quad \forall \ z \in U.$$

In particular, $(\mathbb{T}, \mathcal{B}, m, T)$ is a non-singular transformation and

$$\widehat{T} p_z = p_{f(z)} \quad \forall \ z \in U.$$

PROOF. Fix $z \in U$. For each $n \geq 0$, $z \mapsto f(z)^n$ is a bounded analytic function on U and $f(re^{2\pi x})^n \to e^{2\pi n i T(x)}$ as $r \to 1$ for a.e. $x \in \mathbb{T}$, whence

$$f(z)^n = \int_0^1 e^{2\pi n i T(x)} d\pi_z = \hat{\pi}_{f(z)}(n).$$

It follows that $\pi_z \circ T^{-1} = \pi_{f(z)}$, whence $m \circ T^{-1} \sim m$ and $\widehat{T} p_z = p_{f(z)}$. □

REMARK. A converse to this result is given in [**Let**]:

If $T : \mathbb{T} \to \mathbb{T}$ satisfies $\pi_z \circ T^{-1} = \pi_{f(z)}$ $(z \in U)$ for some $f : U \to U$, then either f is inner or $\overline{f}$ is inner; and T is the $\mathbb{T}$-restriction of f.

Using proposition 6.1.1, we'll be able to study the ergodic theory of T on $\mathbb{T}$, by iterating f on U. For example

6.1.2 COROLLARY. *If f fixes some point $z \in U$, then π_z is a T-invariant probability.*

We'll see in the sequel that the converse of this result is also true: If there is a m-absolutely continuous, T-invariant probability, then f fixes some point in U.

We'll need

SCHWARZ'S LEMMA [**Rudi**]. *Let $g : U \to U$ be analytic, and suppose that $g(0) = 0$, then $|g(x)| \leq |x| \ \forall \ x \in U$ and $|g'(0)| \leq 1$ with equality in either inequality for some $z \in U$ if and only if $\exists \ \lambda \in \partial U$ such that $g(z) = \lambda z$.*

6.1.3 PROPOSITION [**Nor**]. *Suppose that f is a non-Möbius inner function with a fixed point in U, then its $\mathbb{T}$-restriction is exact.*

PROOF.

Assume that $f(0) = 0$ (otherwise conjugate to this situation with a Möbius transformation). By Schwarz's lemma $\exists \ \epsilon > 0$ and $\rho < 1$ such that $|f(z)| \leq \rho|z| \ \forall \ |z| \leq \epsilon$ whence $f^n(z) \to 0$ as $n \to \infty \ \forall \ |z| \leq \epsilon$ where $f^1(z) = f(z)$, and $f^{n+1}(z) = f(f^n(z))$.

By the normal families theorem (see [**Rudi**])

$$f^n(z) \to 0 \text{ uniformly on compact subsets of } U, \text{ as } n \to \infty.$$

This implies that, for $z \in U$,

$$\widehat{T}^n p_z = p_{f^n(z)} \to p_0 = 1 = \int_{\mathbb{T}} p_z dm \text{ uniformly on } \mathbb{T} \text{ as } n \to \infty.$$

Let $\varphi \in C(\mathbb{T})$, and set

$$\tilde{\varphi}(z) = \int_{\mathbb{T}} \varphi p_z dm, \ \varphi_r(x) := \tilde{\varphi}(re^{2\pi i x}),$$

then

$$\varphi_r(x) = \tilde{\varphi}(re^{2\pi ix})$$

$$= \int_0^1 \varphi(y) p_{re^{2\pi ix}}(y) dm$$

$$= \int_0^1 \varphi(y) p_{re^{2\pi iy}}(x) dm,$$

and it follows from Fubini's theorem that

$$\widehat{T}^n \varphi_r(x) = \int_0^1 \varphi(y) p_{f^n(re^{2\pi iy})}(x) dm \to \int_{\mathbb{T}} \varphi \, dm \text{ uniformly on } \mathbb{T} \text{ as } n \to \infty.$$

Since

$$\varphi_r(x) = \tilde{\varphi}(re^{2\pi ix}) \to \varphi(x) \text{ uniformly on } \mathbb{T} \text{ as } r \to 1,$$

we obtain that

$$\widehat{T}^n \varphi \xrightarrow{L^1(m)} \int_{\mathbb{T}} \varphi \, dm \text{ as } n \to \infty.$$

This is true by approximation for every $\varphi \in L^1(m)$, and so T is exact by theorem 1.3.3. $\square$

REMARK.

M. Craizer has shown in [**Cra**] that the natural extension of a non-Möbius inner function with a fixed point in U is isomorphic to a Bernoulli shift.

The above proposition was actually a "warm up" for our characterisation of the exactness of $\mathbb{T}$-restrictions of inner functions.

The next lemma will be instrumental in translating iteration theory for f to ergodic theory for T. It shows that the geometry of the Poisson kernels in $L^1(m)$ is hyperbolic (see chapter 7).

Set

$$[z, \omega] := \left| \frac{z - \omega}{1 - \overline{\omega}z} \right| \quad (z, \omega \in U, \ z \neq \omega).$$

6.1.4 LEMMA.

$$\int_{\mathbb{T}} |p_z - p_\omega| dm = \frac{4}{\pi} \sin^{-1}[z, \omega].$$

PROOF.

Set $d(z, \omega) := \int_{\mathbb{T}} |p_z - p_\omega| dm$. Suppose that $\gamma : U \to U$ is Möbius. Its $\mathbb{T}$-restriction g is an invertible non-singular transformation of $(\mathbb{T}, \mathcal{B}, m)$, and using proposition 6.1.1,

$$d(z, \omega) = \sup_{\varphi \in L^\infty(m)} \int_{\mathbb{T}} (p_z - p_\omega) \varphi \, dm$$

$$= \sup_{\varphi \in L^\infty(m)} \int_{\mathbb{T}} (p_z - p_\omega) \varphi \circ g \, dm$$

$$= \sup_{\varphi \in L^\infty(m)} \int_{\mathbb{T}} (p_{\gamma(z)} - p_{\gamma(\omega)}) \varphi \, dm$$

$$= d(\gamma(z), \gamma(\omega)).$$

It follows from this that

$$d(z,\omega) = d(0, |\frac{z-\omega}{1-\overline{\omega}z}|),$$

and it's not hard to show directly that for $r \in [0,1)$

$$d(0,r) = \frac{4}{\pi}\sin^{-1} r,$$

whence the lemma.   □

The following is a consequence of Schwarz's lemma.

SCHWARZ-PICK LEMMA.   *Let $g : U \to U$ be analytic, then*

$$[gx, gy] \leq [x,y] \quad \text{for } x \neq y \in U$$

with equality for some $x \neq y$ if and only if g is a Möbius transformation.

6.1.5 THEOREM [**A6**].

$$T \text{ is exact } \Leftrightarrow [f^n(z), f^n(\omega)] \to 0 \text{ as } n \to \infty \ \forall \, z, \omega \in U.$$

PROOF.   Suppose that T is exact, then by theorem 1.3.3

$$\|\widehat{T}^n g\|_1 \to 0 \ \forall \, g \in L^1, \ \int g = 0,$$

in particular, for $z, \omega \in U$,

$$[f^n(z), f^n(\omega)] = \sin\left(\frac{\pi}{4}\|\widehat{T}^n(p_z - p_\omega)\|_1\right) \to 0.$$

On the other hand, for $g \in L_0^1$, and $r \in [0,1)$, as before

$$\widehat{T}^n g_r(x) = \int_0^1 g(y) p_{f^n(re^{2\pi i y})}(x) dy$$

$$= \int_0^1 g(y)(p_{f^n(re^{2\pi i y})}(x) - p_{f^n(0)}(x)) dy$$

whence

$$\|\widehat{T}^n g_r\|_1 \leq \int_0^1 |g(y)| \|p_{f^n(re^{2\pi i y})} - p_{f^n(0)}\|_1 dy$$

$$= \frac{4}{\pi} \int_0^1 |g(y)| \sin^{-1}[f^n(re^{2\pi i y}), f^n(0)] dy$$

$$\to 0 \text{ as } n \to \infty,$$

by lemma 6.1.4 and $\|\widehat{T}^n g\|_1 \to 0$ follows since

$$g_r \xrightarrow{L^1} g \text{ as } r \to 1.$$

Thus again by theorem 1.3.3, T is exact.   □

In order proceed further, we need more information concerning the iteration of analytic functions $f : U \to U$.

DENJOY-WOLFF THEOREM   [**De**] [**Wo**].

Suppose that $f : U \to U$ is analytic, and not Möbius, then there is a point $\beta \in \overline{U}$ such that

(1)
$$\frac{1 - |f(z)|^2}{|1 - \overline{\beta} f(z)|^2} \geq \frac{1 - |z|^2}{|1 - \overline{\beta} z|^2} \ \forall \ z \in U,$$

and

(2)
$$f^n(z) \to \beta \ as \ n \to \infty \ \forall \ z \in U.$$

PROOF [**Hei**].

Suppose first that $f(\beta) = \beta \in U$. If $\varphi_\beta(z) := \frac{z - \beta}{1 - \overline{\beta} z}$, then, by the Schwarz-Pick lemma

$$[f(z), \beta] = [f(z), f(\beta)] \leq [z, \beta] \ \forall \ z \in U$$

implying (1) since

$$1 - [x, y]^2 = \frac{(1 - |x|^2)(1 - |y|^2)}{|1 - \overline{x} y|^2} \ \forall \ x, y \in U.$$

For these reasons, for $\beta \in U$, the conditions (1) and $f(\beta) = \beta$ are equivalent. If one of them holds, then as in the proof of theorem 6.1.4,

$$f^n(z) \to \beta \ \text{as} \ n \to \infty \ \forall \ z \in U.$$

Next, we show that, in case f fixes no point in U, there is a point $\beta \in \partial U$ satisfying (1). Let $r \in (0, 1)$ and define $f_r : U \to U$ by $f_r(z) = f(rz)$. It follows that f_r is analytic in U and continuous on $\overline{U}$, and also

$$\sup_{z \in U} |f_r(z)| < 1,$$

since otherwise f would be a constant in ∂U. By Rouché's theorem (see [**Rudi**]), the functions $f_r(z) - z$, and z have the same number of zeroes in U, and there is a unique fixed point $f_r(\beta_r) = \beta_r \in U$.

It now follows, as above, that

$$\frac{1 - |f(rz)|^2}{|1 - \overline{\beta_r} f(rz)|^2} \geq \frac{1 - |z|^2}{|1 - \overline{\beta_r} z|^2} \ \forall \ z \in U, \ r \in (0, 1).$$

There is a sequence $r_k \to 1$ such that β_{r_k} converges to some point β, which by continuity, satisfies (1), and hence is on ∂U.

Let $\beta \in \partial U$ satisfy (1), then

$$\frac{1 - |f^n(z)|^2}{|\beta - f^n(z)|^2} \geq \frac{1 - |z|^2}{|\beta - z|^2} > 0 \ \forall \ n \geq 0, \ z \in U.$$

In order to prove (2), it suffices to show that

$$|f^n(z)| \to 1 \ \text{as} \ n \to \infty.$$

Suppose otherwise, then

$$\exists \ z \in U, \ n_k \to \infty \ \ni \ f^{n_k}(z) \to p \in U.$$

By assumption, $f(p) \neq p$ and so by the Schwarz-Pick lemma

$$[f^{n+1}(z), f^n(z)] \downarrow [f(p), p] \text{ as } n \to \infty,$$

whence $[f^2(p), f(p)] = [f(p), p]$, which contradicts that f is not Möbius. $\square$

DEFINITION: DENJOY-WOLFF POINT. The point β produced in the Denjoy-Wolff theorem is clearly unique. It is called the *Denjoy-Wolff point* of f.

REMARK: DENJOY-WOLFF POINTS FOR MÖBIUS TRANSFORMATIONS.

A Möbius transformation f satisfying $f^n \neq$ Id $\forall\, n \geq 1$ also has a unique point $\beta \in \overline{U}$ satisfying (1) This point is also called the Denjoy-Wolff point of f and it can be shown that $f^n(z) \to \beta$ for some (and hence all) $z \in U$ if and only if $\beta \in \partial U$.

The following proposition deals with the ergodic properties of $\mathbb{T}$-restrictions of inner functions whose Denjoy-Wolff points are on the boundary. The advertised converse to our first corollary follows from it.

6.1.6 PROPOSITION. *Let f be an inner function with restriction T, and Denjoy-Wolff point $\beta = e^{2\pi i b} \in \partial U$, then for every $\epsilon > 0$, and $h \in L^1(m)_+$,*

$$\int_{[|x-b| \geq \epsilon]} \widehat{T}^n h(x) dx \to 0 \text{ as } n \to \infty.$$

In particular, there is no m-absolutely continuous, T-invariant probability on $\mathbb{T}$.

PROOF.   For $z \in U$ by theorem 6.1.2,

$$\widehat{T}^n p_z(x) = \frac{1 - |f^n(z)|^2}{|e^{2\pi x} - f^n(z)|^2} \to 0 \text{ as } n \to \infty$$

uniformly in x on compact subsets of $\mathbb{T} \setminus \{b\}$. Given $\varphi \in L^1(m)_+$, set

$$\varphi_r(x) = \int_\mathbb{T} \varphi p_{re^{2\pi ix}} dm = \int_\mathbb{T} \varphi(y) p_{re^{2\pi iy}}(x) dy,$$

then $\varphi_r \xrightarrow{L^1(m)} \varphi$ as $r \to 1$, and

$$\widehat{T}^n \varphi_r(x) = \int_\mathbb{T} \varphi(y) p_{f^n(re^{2\pi iy})}(x) dy.$$

This implies that

$$\int_{[|x-b| \geq \epsilon]} \widehat{T}^n \varphi_r dm = \int_\mathbb{T} \varphi(y) \left(\int_{[|x-b| \geq \epsilon]} p_{f^n(re^{2\pi iy})}(t) dt \right) dy \to 0 \text{ as } n \to \infty,$$

whence, by approximation as $r \to 1$,

$$\int_{[|x-b| \geq \epsilon]} \widehat{T}^n \varphi dm \to 0 \text{ as } n \to \infty.$$

Now suppose that μ is a m-absolutely continuous, T-invariant measure with $\frac{d\mu}{dm} = h \in L^1(m)_+$. It follows that

$$\mu([|x - b| \geq \epsilon]) = \mu(T^{-n}[|x - b| \geq \epsilon]) = \int_{[|x-b|\geq\epsilon]} \widehat{T}^n h \, dm \to 0 \text{ as } n \to \infty$$

for every $\epsilon > 0$, whence $\mu \equiv 0$. $\square$

DEFINITION: BLASCHKE SERIES.
The *Blaschke series* of the inner function $f : U \to U$ at $z \in U$ is

$$\mathfrak{b}_f(z) := \sum_{n=1}^{\infty} 1 - |f^n(z)| = \infty.$$

An inner function with Denjoy-Wolff point in U has divergent Blaschke series at each $z \in U$.

6.1.7 PROPOSITION   [**A5**].
The $\mathbb{T}$-restriction T of an inner function $f : U \to U$ is conservative, or totally dissipative according to whether f's Blaschke series diverges or converges (respectively) at some and hence all $z \in U$.

PROOF.
There is no loss of generality in supposing that the Denjoy-Wolff point is $\beta = e^{2\pi i b} \in \partial U$.

For $z \in U$, and $x \in \mathbb{T} \setminus \{b\}$,

$$\widehat{T}^n p_z(x) = p_{f^n(z)}(x) = \frac{1 - |f^n(z)|^2}{|e^{2\pi i x} - f^n(z)|^2} \sim \frac{1 - |f^n(z)|^2}{|e^{2\pi i x} - \beta|^2}$$

(since $f^n(z) \to \beta$) as $n \to \infty$. Therefore,

$$\sum_{n=1}^{\infty} \widehat{T}^n p_z(x) = \infty \quad \Leftrightarrow \quad \sum_{n=1}^{\infty} 1 - |f^n(z)| = \infty.$$

Thus, if the Blaschke series diverges for some $z \in U$, then by proposition 1.3.1, T is conservative,

$$\sum_{n=1}^{\infty} \widehat{T}^n g = \infty \text{ a.e. } \forall \, g \in L^1(m), \ g > 0,$$

and f's Blaschke series diverges for all $z \in U$.

If the Blaschke series converges for some $z \in U$, then by proposition 1.3.1, T is totally dissipative,

$$\sum_{n=1}^{\infty} \widehat{T}^n g < \infty \text{ a.e. } \forall \, g \in L^1(m), \ g > 0,$$

and f's Blaschke series converges for all $z \in U$. $\square$

6.1.8 THEOREM   [**Neu**].
If the $\mathbb{T}$-restriction T of an inner function $f : U \to U$ is conservative, then it is ergodic.

PROOF.

By proposition 6.1.8, f's Blaschke series diverges $\forall\, z \in U$.

Let $g : \mathbb{T} \to \mathbb{R}$ be a measurable, bounded, T-invariant function. If

$$\tilde{g}(z) := \int_{\mathbb{T}} \left(\frac{e^{2\pi i x} + z}{e^{2\pi i x} - z} \right) g(x)dx,$$

then $\tilde{g} \in H^1(U)$, and $\operatorname{Im} \tilde{g} \circ f \equiv \operatorname{Im} \tilde{g}$. It follows that there is a constant $c \in \mathbb{R}$ such that

$$\tilde{g} \circ f \equiv \tilde{g} + c.$$

There is a function $g^* \in L^1(m)$ such that

$$\tilde{g}(z) = \int_{\mathbb{T}} g^* p_z dm,$$

and it follows that

$$g^* \circ T = g^* + c \text{ a.e. on } \mathbb{T}.$$

The conservativity of T now forces $c = 0$, and the Blaschke series divergence implies, by Blaschke's theorem, that $\tilde{g}$ is constant, whence g is constant.   □

6.1.9 PROPOSITION [**Mañ-Do**]. *Suppose that f is an inner function, with restriction T and Denjoy-Wolff point $\beta = e^{2\pi i b} \in \partial U$.*

If f is not conservative then

$$T^n(x) \to b \text{ in } \mathbb{T}, \text{ for a.e. } x \in \mathbb{T}.$$

PROOF. Since T is dissipative,

$$\sum_{n=1}^{\infty} \widehat{T}^n p_z = \sum_{n=1}^{\infty} p_{f^n(z)}$$

converges uniformly on compact subsets of $\mathbb{T} \setminus \{b\}$, whence $\forall\, \epsilon > 0$,

$$\sum_{n=1}^{\infty} m(T^{-n}[|x - b| \geq \epsilon]) = \sum_{n=1}^{\infty} \int_{[|x-b|\geq\epsilon|]} \widehat{T}^n 1 dm = \int_{[|x-b|\geq\epsilon|]} \sum_{n=1}^{\infty} p_{f^n(0)} dm < \infty$$

and $T^n(x) \to b$ for a.e. $x \in \mathbb{T}$.   □

§6.2 Inner functions on the upper half plane

Given an inner function f on U, it is usually hard to get a formula for its restriction $T : \mathbb{T} \to \mathbb{T}$. The situation is better on

$$\mathbb{R}^{2+} = \{z \in \mathbb{C} : \operatorname{Im} z > 0\}$$

which is conformally equivalent with U, and where it is also more convenient to analyse inner functions with Denjoy- Wolff points on the boundary.

The harmonic measures on $\mathbb{R}^{2+}$ are the *Cauchy measures* on $\mathbb{R}$ defined by the densities

$$\frac{dP_\omega}{dt}(t) = \varphi_\omega(t) := \frac{1}{\pi} \operatorname{Im} \frac{1}{t - \omega}.$$

Note that $\varphi a + ib(t) = \frac{1}{\pi} \frac{b}{(x-a)^2 + b^2}$.

DEFINITION: INNER FUNCTION OF $\mathbb{R}^{2+}$. An inner function of $\mathbb{R}^{2+}$ is an analytic map $f : \mathbb{R}^{2+} \to \mathbb{R}^{2+}$ such that for a.e. $x \in \mathbb{R}$

$$f(x + iy) \to T(x) \in \mathbb{R} \text{ as } y \downarrow 0.$$

The measurable map $T : \mathbb{R} \to \mathbb{R}$ is called the $\mathbb{R}$-*restriction* of f to $\mathbb{R}$.

Since it is now given in the same parametrisation as f, we'll write T both for the inner function, and its restriction.

Now it is easy to write down

EXAMPLES.

$$Tx = \alpha x + \beta + \sum_{k=1}^{\infty} \frac{p_k}{t_k - x}$$

is an inner function when

$$\alpha, p_1, \ldots \geq 0, \quad 0 < \alpha + \sum_{n=1}^{\infty} p_n < \infty, \quad \beta, t_1, \ldots, t_n \in \mathbb{R};$$

and so is

$$Tx = \alpha x + \beta \tan x,$$

when $\alpha, \beta > 0$.

Let $\lambda \in \partial U$, and set

$$\phi_\lambda(z) = i\left(\frac{\lambda + z}{\lambda - z}\right),$$

then $\phi_\lambda : U \to \mathbb{R}^{2+}$ is a an analytic bijection.

6.2.1 THEOREM   [**Ts**].
If $T : \mathbb{R}^{2+} \to \mathbb{R}^{2+}$ is an analytic function, then,

$$T(z) = \alpha_T z + \beta_T + \int_{\mathbb{R}} \left(\frac{1 + tz}{t - z}\right) d\mu_T(t)$$

where $\alpha, \beta \in \mathbb{R}$, $\alpha \geq 0$, and μ is a positive measure on $\mathbb{R}$, which is singular iff T is inner;
and

T is inner if, and only if $\phi_\lambda^{-1} \circ T \circ \phi_\lambda$ is an inner function of U (for some and hence all $\lambda \in \partial U$).

PROOF.
Let $v := \operatorname{Im} T$, then $v : \mathbb{R}^{2+} \to \mathbb{R}_+$ is harmonic.

Set $\phi(z) = \phi_{-1}(z) = i\left(\frac{1-z}{1+z}\right)$, then $\phi(e^{2\pi i t}) = \tan \pi t$ and $v \circ \phi : U \to \mathbb{R}_+$ is harmonic.

By Herglotz's theorem (see [**Rudi**]) $\exists$ a positive measure μ on $\mathbb{T}$ such that

$$v(z) = \int_0^1 p_{\phi^{-1}(z)}(t) d\mu(t).$$

Now

$$p_{\phi^{-1}(z)}(x) = \operatorname{Im} \frac{1 + \tan \pi x z}{\tan \pi x - z},$$

whence

$$v(z) = \alpha_T z + \int_0^1 \operatorname{Im} \frac{1 + \tan \pi x z}{\tan \pi x - z} = \int_{\mathbb{R}} \operatorname{Im} \frac{1 + tz}{t - z} d\mu_T(t)$$

where $\alpha_T := \mu(\{\frac{1}{2}\})$ and $\int_{\mathbb{R}} h(t) d\mu_T(t) := \int_{\mathbb{T} \setminus \{\frac{1}{2}\}} h(\tan \pi x) d\mu(x)$.

The function $z \mapsto \alpha_T z + \int_{\mathbb{R}} \frac{1+tz}{t-z} d\mu_T(t)$ is evidently analytic, and has the same real part as T, whence

$$T(z) = \alpha_T z + \beta_T + \int_{\mathbb{R}} \left(\frac{1 + tz}{t - z} \right) d\mu_T(t)$$

for some $\beta_T \in \mathbb{R}$.

The measure μ is singular if and only if $v \circ \phi$ has (Lebesgue) a.e. zero boundary value, and this is when $\phi^{-1} \circ T \circ \phi$ is an inner function of U. Clearly, μ_T is singular whenever μ is singular and this is when $v(x + iy) \to 0$ a.e. as $y \to 0$, or in other words T is inner. □

6.2.2 PROPOSITION.
If T is a non-Möbius inner function of $\mathbb{R}^{2+}$, then its $\mathbb{R}$-restriction is a non-singular transformation of $(\mathbb{R}, \mathcal{B}, m)$;

$$P_\omega \circ T^{-1} = P_{T(\omega)} \quad \forall \ \omega \in \mathbb{R}^{2+};$$

there is a T-invariant m-absolutely continuous probability if and only if T fixes some point in $\mathbb{R}^{2+}$;

T is exact if, and only if $\langle T^n z, T^n \omega \rangle \to 0$ as $n \to \infty \ \forall \ z, \omega \in \mathbb{R}^{2+}$ where $\langle z, \omega \rangle := [\phi^{-1} z, \phi^{-1} \omega] = |\frac{z - \omega}{z - \bar{\omega}}|$;
and

$\exists \ x \in \overline{\mathbb{R}^{2+}} \cup \{\infty\}$ *such that $T^n z \to x$ as $n \to \infty \ \forall \ z \in \mathbb{R}^{2+}$.*

PROOF.
For $\phi(z) = \phi_{-1}(z) = i \left(\frac{1-z}{1+z} \right)$, define $\tilde{\phi} : \mathbb{T} \to \mathbb{R}$ by $\tilde{\phi}(t) := \phi(e^{2\pi i t})$, then

$$\frac{d\pi_z \circ \tilde{\phi}^{-1}}{dm}(x) = \operatorname{Im} \left(\frac{1}{x - \phi(z)} \right) := \frac{dP_{\phi(z)}}{dm}(x),$$

where m is Lebesgue measure on $\mathbb{R}$.

The rest follows from results in §6.1 applied to $\phi^{-1} \circ T \circ \phi$. □

The point $x \in \overline{\mathbb{R}^{2+}} \cup \{\infty\}$ such that $T^n z \to x$ as $n \to \infty \ \forall \ z \in \mathbb{R}^{2+}$ is called the *Denjoy-Wolff* point for T.

6.2.3 PROPOSITION **[A5]**.
The non-Möbius inner function T of $\mathbb{R}^{2+}$ has Denjoy-Wolff point ∞ if, and only if $\alpha_T \geq 1$, and in this case T is conservative if and only if

$$\sum_{n=1}^{\infty} \operatorname{Im} \frac{-1}{T^n(z)} = \infty$$

for some and hence all $z \in \mathbb{R}^{2+}$;
and totally dissipative if

$$\sum_{n=1}^{\infty} \operatorname{Im} \frac{-1}{T^n(z)} < \infty$$

for some and hence all $z \in \mathbb{R}^{2+}$.

PROOF. Choose $\beta \in \partial U$ and set $f := \phi_\beta^{-1} \circ T \circ \phi_\beta$, then f has Denjoy-Wolff point β, $\alpha_T \geq 1$ following from (1) in the Denjoy-Wolff theorem. The second part follows from proposition 6.1.7 as $\sum_{n=1}^\infty \operatorname{Im} \frac{-1}{T^n(z)}$ is the Blaschke series for f. $\square$

6.2.4 PROPOSITION [Let].
If $T : \mathbb{R}^{2+} \to \mathbb{R}^{2+}$ is an inner function, then, $m \circ T^{-1} = \frac{1}{\alpha_T} m$.

In particular, if $\alpha_T = 1$, then T is a measure preserving transformation of $(\mathbb{R}, \mathcal{B}(\mathbb{R}), m)$.

PROOF. We use that $\pi b P_{a+ib}(A) \to m(A)$ as $b \to \infty$, $\frac{|a|}{b} \to 0$. If T is inner, then $\frac{\operatorname{Im} T(ib)}{b} \to \alpha_T$ as $b \to \infty$ whence

$$m(T^{-1}A) \;\leftarrow\; \pi b P_{ib}(T^{-1}A) = \pi b P_{T(ib)}(A) \;\to\; \frac{m(A)}{\alpha_T}.$$

$\square$

6.2.5 THEOREM [A5].
Suppose that $T : \mathbb{R}^{2+} \to \mathbb{R}^{2+}$ is an inner function with Denjoy-Wolff point ∞. If $\alpha_T > 1$, then T is totally dissipative.

If $\alpha_T = 1$ and T is conservative, then T is pointwise dual ergodic with return sequence given by

$$a_n(T) \;\sim\; \frac{1}{\pi} \sum_{k=1}^n \operatorname{Im} \frac{-1}{T^k(z)} \quad \forall\, z \in \mathbb{R}^{2+}.$$

All bounded sets in $\mathbb{R}$ are uniform sets.

PROOF.
Noting that $\alpha_{T^n} = \alpha_T^n$ we see that

$$\operatorname{Im} \frac{-1}{T^k(z)} \leq \frac{1}{\operatorname{Im} T^k(z)} \leq \frac{1}{\alpha_T^k \operatorname{Im} z}$$

whence for $\alpha_T > 1$, $\sum_{n=1}^\infty \operatorname{Im} \frac{-1}{T^n(z)} = \infty$ and T is totally dissipative.

Now assume that $\alpha_T = 1$ and T is conservative, then fixing $z \in \mathbb{R}^{2+}$, we have that

$$\sum_{k=1}^n \widehat{T}^k \varphi_z(t) = \sum_{k=1}^n \varphi_{T^k(z)}(t) = \frac{1}{\pi} \sum_{k=1}^n \operatorname{Im} \frac{1}{t - T^k(z)} \sim \frac{1}{\pi} \sum_{k=1}^n \operatorname{Im} \frac{-1}{T^k(z)}$$

uniformly as t ranges in compact subsets of $\mathbb{R}$.

By theorem 6.1.8, T is ergodic, whence by Hurewicz's ergodic theorem, T is pointwise dual ergodic with return sequence given by

$$a_n(T) \;\sim\; \frac{1}{\pi} \sum_{k=1}^n \operatorname{Im} \frac{-1}{T^k(z)} \quad (z \in \mathbb{R}^{2+}).$$

$\square$

§6.3 The dichotomy

The main result of this section is that any ergodic non-Möbius inner function is exact.

We shall also see that there are exact, totally dissipative inner functions preserving Lebesgue measure.

The main tool is the following theorem of Pommerenke.

6.3.1 POMMERENKE'S THEOREM [**Pom**]. *Suppose that* $T : \mathbb{R}^{2+} \to \mathbb{R}^{2+}$ *is analytic with Denjoy-Wolff point* ∞, *and let* $T^n(i) = a_n + ib_n$, *then*

$$\exists \lim_{n \to \infty} \frac{T^n(z) - a_n}{b_n} := F(z) \ \forall \ z \in \mathbb{R}^{2+}.$$

Moreover,

$$F \circ T \equiv aF + b$$

where $b \in \mathbb{R}$, $a \geq \alpha_T$ *and*

$$F \circ T \equiv F \ \Leftrightarrow \ F \equiv i.$$

PROOF.   Without loss of generality, T is not Möbius. Set

$$F_n(z) = \frac{T^n(z) - a_n}{b_n},$$

then $F_n(i) = i$, and

$$\langle F_n(z), F_n(z') \rangle = \langle T^n(z), T^n(z') \rangle \downarrow \ \text{as } n \uparrow .$$

Clearly,

$$\langle T^n(z), T^n(z') \rangle \downarrow 0 \ \text{as } n \uparrow \infty \ \forall \ z, z' \in \mathbb{R}^{2+}$$

if and only if

$$F_n(z) \to i \ \forall \ z \in \mathbb{R}^{2+}.$$

Suppose that

$$\langle T^n(z), T^n(z') \rangle \downarrow L(z, z') = L(Tz, Tz').$$

If $F_{n_k} \to G$, then $G(i) = i$, and

$$L(z, z') = \langle G(z), G(z') \rangle.$$

Since

$$\langle G(Tz), G(Tz') \rangle = L(Tz, Tz') = L(z, z') = \langle G(z), G(z') \rangle,$$

we have by the Schwarz-Pick lemma that

$$G \circ T = A \circ G$$

where A is Möbius. We claim first that if A has a fixed point $p \in \mathbb{R}^{2+}$, then $G \equiv i$. To see this, note first that

$$\text{Im} \, T^n(z) \uparrow \ \text{as Im} \, z \uparrow$$

for all $n \geq 1$ because $\alpha_{T^n} = \alpha_T^n \geq 1$, whence

$$\text{Im} \, G(z) \uparrow \ \text{as Im} \, z \uparrow .$$

Now $\text{Im} \, T^n(i) \uparrow$ as $n \uparrow$, whence

$$\text{Im} \, A^n(i) = \text{Im} \, A^n \circ G(i) = \text{Im} \, G \circ T^n(i) \uparrow \ \text{as } n \uparrow .$$

Since $\langle A^n(i), p \rangle = \langle i, p \rangle$ for $n \geq 1$, we have that $A^2(i) = i$. Since $A^2(p) = p$, we must have $A(i) = i$. Now let $\varphi : U \to \mathbb{R}^{2+}$ be an analytic bijection with $\varphi(0) = i$ and set

$$f = \varphi^{-1} \circ T \circ \varphi, \ g = \varphi^{-1} \circ G \circ \varphi, \ a = \varphi^{-1} \circ A \circ \varphi,$$

then

$$g(0) = 0, \ g(f(z)) = \lambda g(z) \ (\text{ for some } \lambda \in \partial U),$$

and

$$g(f^n(0)) = 0 \ \forall \ n \geq 0.$$

It follows from Blaschke's theorem (see [**Rudi**]) that

$$|g(z)| \leq \prod_{n=0}^{\infty} [z, f^n(0)],$$

whence, for every $N \in \mathbb{N}$,

$$|g(z)| = |g(f^n(z))| \leq \prod_{k=0}^{N} [f^n(z), f^{k+n}(0)] \underset{n \to \infty}{\longrightarrow} |g(z)|^N.$$

This implies that $g \equiv 0$, or $G \equiv i$, whence $L \equiv 0$, and $F_n(z) \to i$.

In case A has no fixed point in $\mathbb{R}^{2+}$, we have that $A^n(i) \to \infty$, whence $A(z) = az + b$ where $a \geq 1$ and $b \in \mathbb{R}$. We now claim, in case $A \neq \text{Id}$, that $F_n \to G$. Suppose that $F_{m_\ell} \to H$. We show that $H = G$. We have that $H \circ T = B \circ H$ with $B^n(z) \to \infty$. Also, $\langle H(z), i \rangle = L(z, i) = \langle G(z), i \rangle$, whence $H = C \circ G$ where C is Möbius and $C(i) = i$. But since $A^n(z), B^n(z) \to \infty$, we have in addition that $C(\infty) = \infty$, whence $C = \text{Id}$, and $G = H$.

Lastly, we must show that $a \geq \alpha_T$. Clearly,

$$\frac{b_{n+1}}{b_n} \to a \text{ as } n \to \infty.$$

Using theorem 6.2.1, we have that

$$\frac{b_{n+1}}{b_n} = \alpha_T + \int_{\mathbb{R}} \left(\frac{1 + t^2}{(t - a_n)^2 + b_n^2} \right) d\mu(t) \geq \alpha_T.$$

$\square$

6.3.2 DICHOTOMY [**A6**].

Let $T : \mathbb{R}^{2+} \to \mathbb{R}^{2+}$ be a non-Möbius inner function.

Either T is exact, or there is a non-constant, T-invariant, bounded analytic function $h : \mathbb{R}^{2+} \to U$ (and T is not ergodic).

PROOF.

If T's Denjoy-Wolff point is in $\mathbb{R}^{2+}$ then T is exact. By possibly conjugating by a Möbius transformation we may assume that T's Denjoy-Wolff point is ∞. By theorem 6.3.1,

either $\langle T^n(x), T^n(y) \rangle \to 0$ as $n \to \infty \ \forall \ x, y \in \mathbb{R}^{2+}$ and T is exact by proposition 6.2.2;

or $\exists \ F : \mathbb{R}^{2+} \to \mathbb{R}^{2+}$ analytic, satisfying $F \circ T \equiv aF + b$ for some $b \in \mathbb{R}$, $a \geq \alpha_T$, $(a, b) \neq (1, 0)$.

Under these conditions, $\exists\, H : \mathbb{R}^{2+} \to U$ analytic and non-constant such that $H(az+b) = H(z)$. The required non-constant, T-invariant function is $h := H \circ F : \mathbb{R}^{2+} \to U$. $\qquad\square$

6.3.3 COROLLARY.
If $\alpha_T > 1$, then T is totally dissipative and non-ergodic.

PROOF.
Total dissipativity was shown in theorem 6.2.5. Non-ergodicity follows from Pommerenke's theorem as $\exists\, F : \mathbb{R}^{2+} \to \mathbb{R}^{2+}$ analytic such that

$$F \circ T \equiv aF + b$$

where $b \in \mathbb{R}$, $a \geq \alpha_T > 1$ and it follows that $\exists\, H : \mathbb{R}^{2+} \to U$ analytic and non-constant such that $H(az+b) = H(z)$, $H \circ F$ being bounded, analytic, non-constant and T-invariant. $\qquad\square$

§6.4 Examples

In this section, we consider inner functions $T : \mathbb{R}^{2+} \to \mathbb{R}^{2+}$ with $\alpha_T = 1$ having special forms.

6.4.1 THEOREM [A5] [A6].
Suppose that $T : \mathbb{R}^{2+} \to \mathbb{R}^{2+}$ is inner and

$$T(z) = z + \beta + \int_{\mathbb{R}} \frac{d\nu(t)}{t - z}$$

where $\beta \in \mathbb{R}$ and ν is singular and compactly supported on $\mathbb{R}$.

If $\beta = 0$, then T is exact and pointwise dual ergodic with return sequence $a_n(T) \sim \frac{1}{\pi}\sqrt{\frac{2n}{\nu(\mathbb{R})}}$.

If $\beta \neq 0$, then T is totally dissipative and non-ergodic.

PROOF.
Write $T^n z = u_n(z) + i v_n(z)$, then

$$v_{n+1} = v_n + v_n \int_{\mathbb{R}} \frac{d\nu(t)}{(t - u_n)^2 + v_n^2}, \quad u_{n+1} = u_n + \beta + \int_{\mathbb{R}} \frac{(t - u_n)d\nu(t)}{(t - u_n)^2 + v_n^2}.$$

Suppose first that $\beta = 0$, and note that

$$|u_{n+1} - u_n| \leq \int_{\mathbb{R}} \frac{|t - u_n|d\nu(t)}{(t - u_n)^2 + v_n^2} \leq \frac{\nu(\mathbb{R})}{2v_n} \leq \frac{\nu(\mathbb{R})}{2v_0} := K.$$

Choosing $L > K$ such that $\nu([-L, L]^c) = 0$ we have

$$u_n > L \implies 0 < u_{n+1} < u_n, \quad u_n < -L \implies 0 > u_{n+1} > u_n.$$

It follows that $\sup_{n \geq 1} |u_n| < \infty$ whence $v_n \uparrow \infty$ (since $T^n z \to \infty$),

$$v_{n+1}^2 - v_n^2 \approx 2v_n^2 \int_{\mathbb{R}} \frac{d\nu(t)}{(t - u_n)^2 + v_n^2} \to 2\nu(\mathbb{R})$$

and $v_n \sim \sqrt{2\nu(\mathbb{R})n}$ as $n \to \infty$.

It follows from theorem 6.2.3 T is conservative hence pointwise dual ergodic by theorem 6.2.5 with return sequence

$$a_n(T) \sim \frac{1}{\pi} \sum_{k=1}^{n} \operatorname{Im} \frac{-1}{T^k(z)} \sim \sum_{k=1}^{n} \frac{1}{v_k} \sim \frac{1}{\pi} \sqrt{\frac{2n}{\nu(\mathbb{R})}}.$$

Exactness follows from the dichotomy.

Now suppose that $\beta \neq 0$, then $v_{n+1} \leq v_n + \frac{\nu(\mathbb{R})}{v_n}$ whence $v_{n+1}^2 - v_n^2 = O(1)$ and $v_n = O(\sqrt{n})$ as $n \to \infty$.

Next, since $T^n(z) \to \infty$, $u_{n+1} - u_n \to \beta$ and $|u_n| \sim |\beta|n$ as $n \to \infty$, whence

$$v_{n+1} - v_n = v_n \int_{\mathbb{R}} \frac{d\nu(t)}{(t - u_n)^2 + v_n^2} \sim \frac{v_n \nu(\mathbb{R})}{n^2 \beta^2} = O\left(\frac{1}{n^{\frac{3}{2}}}\right)$$

and $v_n \uparrow v < \infty$.

In view of this, as $n \to \infty$:

$$u_{n+1} - u_n = \beta - \frac{\nu(\mathbb{R})}{u_n} + O\left(\frac{1}{n^2}\right)$$

whence

$$u_{n+1} - u_n = \beta - \frac{\nu(\mathbb{R})}{\beta n} + O\left(\frac{1}{n}\right)$$

and

$$u_n = \beta n - \frac{\nu(\mathbb{R})}{\beta} \log n + o\left(\log n\right).$$

Substituting the third back in the first, as $n \to \infty$:

$$u_{n+1} - u_n = \beta - \frac{\nu(\mathbb{R})}{\beta n - \frac{\nu(\mathbb{R})}{\beta} \log n + o\left(\log n\right)} + +O\left(\frac{1}{n^2}\right) = \beta - \frac{\nu(\mathbb{R})}{\beta n} + O\left(\frac{\log n}{n^2}\right)$$

and

$$\exists \lim_{n \to \infty} u_n - \beta n + \frac{\nu(\mathbb{R})}{\beta} \log n.$$

We have shown that

$$T^n(z) - \beta n + \frac{\nu(\mathbb{R})}{\beta} \log n \to F(z)$$

(say) as $n \to \infty$.

Evidently $F : \mathbb{R}^{2+} \to \mathbb{R}^{2+}$ is analytic, and $F(Tz) = F(z) + \beta$, whence T is not ergodic. $\square$

EXAMPLE: GENERALISED BOOLE TRANSFORMATIONS.
The *generalised Boole transformation*

$$Tx = x + \beta + \sum_{k=1}^{n} \frac{p_k}{t_k - x}$$

where $n \geq 1$, $p_1, \ldots, p_n \geq 0$, $\beta, t_1, \ldots, t_n \in \mathbb{R}$ is as in theorem 6.4.1 where the measure ν is supported on a finite set. Boole's (original) transformation $x \mapsto x - \frac{1}{x}$

was shown to preserve Lebesgue measure in [**Bo**], and to be ergodic in [**Ad-W**]. The generalised Boole transformation with $\beta = 0$ was shown to be ergodic in [**Li-Schw**].

6.4.2 COROLLARY.

Suppose that $T : \mathbb{R}^{2+} \to \mathbb{R}^{2+}$ is inner, analytic around $x_0 \in \mathbb{R}$ and $T(x_0) = x_0$, $T'(x_0) = 1$ then $\nu_{x_0} \circ T^{-1} = \nu_{x_0}$ where $d\nu_{x_0}(t) = \frac{dt}{(t-x_0)^2}$.

Moreover, T is conservative if and only if $T''(x_0) = 0$ and in this case T is exact and pointwise dual ergodic with return sequence $a_n(T) \propto \sqrt{n}$.

If $T''(x_0) \neq 0$, then T is not ergodic.

PROOF.

Assume that $x_0 = 0$ and let $T''(x_0) = a$. We deduce the corollary from theorem 6.4.1 by showing that

$$R(z) := \frac{-1}{T(\frac{-1}{z})} = z + a + \int_{\mathbb{R}} \frac{d\nu(t)}{t - z}$$

for some singular measure ν compactly supported on $\mathbb{R}$.

Firstly,

$$\frac{1}{T(z)} - \frac{1}{z} \to -a \text{ as } z \to 0.$$

It follows that $\exists\, K > 1$ such that

$$(1) \qquad R(z) = z + a + \sum_{n=1}^{\infty} \frac{b_n}{z^n} \ \forall\, |z| \geq K$$

whence $\alpha_R = 1$ and

$$(2) \qquad R(z) = z + a + \beta + \int_{\mathbb{R}} \frac{1 + tz}{t - z} d\mu(t)$$

for some singular measure μ on $\mathbb{R}$ and $\beta \in \mathbb{R}$.

Set $g(z) := R(z) - z - a$, then by (1),

$$-iyg(iy) \to b_1 \text{ as } y \to \infty.$$

But by (2),

$$-iyg(iy) = -iy\left(\beta - y^2 \int_{\mathbb{R}} \frac{t d\mu(t)}{t^2 + y^2}\right) + iy \int_{\mathbb{R}} \frac{t d\mu(t)}{t^2 + y^2} + y^2 \int_{\mathbb{R}} \frac{1 + t^2}{t^2 + y^2} d\mu(t)$$

and so

$$\int_{\mathbb{R}} (1 + t^2) d\mu(t) < \infty$$

by the convergence of the real part, and

$$\int_{\mathbb{R}} t d\mu(t) = \beta$$

by the convergence of the imaginary part.

It follows that

$$R(z) = z + a + \int_{\mathbb{R}} \frac{d\nu(t)}{t - z}$$

where $d\nu(t) := (1 + t^2) d\mu(t)$.

To see that ν is compactly supported, recall that g is uniformly continuous on compact subsets of $[|z| \geq K]$ whence

$$h_y(x) := \operatorname{Im} g(x + iy) \to 0 \text{ as } y \to 0$$

uniformly as x ranges in compact subsets of $(-K, K)^c$. If $dq_y := h_y dm$, then $q_y = P_{iy} * \nu$, whence if f is a continuous function, compactly supported outside $(-K, K)$,

$$\int_{\mathbb{R}} f d\nu \; \leftarrow \; \int_{\mathbb{R}} f h_y dm \; \to \; 0$$

as $y \to 0$. $\qquad\qquad\qquad\qquad\qquad\qquad\qquad\qquad\qquad\qquad\qquad\qquad\qquad$ $\square$

EXAMPLE: SOME TANGENT FUNCTIONS.
The functions $T_\alpha(x) := \alpha x + (1 - \alpha) \tan x$ for $\alpha \in [0, 1)$ satisfy the conditions of corollary 6.4.2 as

$$T_\alpha(0) = 0, \quad T_\alpha'(0) = 1 \; \& \; T_\alpha''(0) = 0.$$

Ergodicity of T_0 was established in [**Schw2**]. Note that since T_α is conservative and ergodic with the infinite invariant measure ν_0, there can be no finite, absolutely continuous, T_α-invariant probability.

DEFINITION: ODD FUNCTION.
The analytic function $T : \mathbb{R}^{2+} \to \mathbb{R}^{2+}$ is called *odd* if $\beta_T = 0$ and μ_T is symmetric (i.e. $\int_{\mathbb{R}} h(t) d\mu(t) = \int_{\mathbb{R}} h(-t) d\mu(t)$).

6.4.3 PROPOSITION [**A5**].
Let $T : \mathbb{R}^{2+} \to \mathbb{R}^{2+}$ *be analytic. The following are equivalent:*
a) T *is odd,*
b) $\operatorname{Re} T|_{i\mathbb{R}} \equiv 0$,
c) $T(-\overline{\omega}) = -\overline{T(\omega)}$ $\forall \omega \in \mathbb{R}^{2+}$.
An inner function $T : \mathbb{R}^{2+} \to \mathbb{R}^{2+}$ *is odd if and only if*
d) $T(-x) = -T(x)$ *for a.e.* $x \in \mathbb{R}$.

PROOF.
The implications a) $\Longrightarrow$ c) $\Longrightarrow$ b), c) $\Longrightarrow$ d) (for inner T) are elementary. It follows from Schwarz's reflection principle (see [**Rudi**]) that b) $\Longrightarrow$ c), and d) $\Longrightarrow$ c) (for inner T) because $e^{itf(\omega)} = \int_{\mathbb{R}} e^{itT(x)} dP_\omega(x)$.
It remains to show c) $\Longrightarrow$ a). Assume that T satisfies c), then $\operatorname{Re} T|_{i\mathbb{R}} \equiv 0$ and $\beta_T = 0$. We must show that μ_T is symmetric. The equations $\operatorname{Im} T(-a + ib) = \operatorname{Im} T(a + ib)$ are

$$\int_{\mathbb{R}} \varphi_{ib}(t - a)(1 + t^2) d\mu_T(t) = \int_{\mathbb{R}} \varphi_{ib}(t + a)(1 + t^2) d\mu_T(t).$$

Let $g : \mathbb{R} \to \mathbb{R}$ be continuous with compact support and set $g_b := g * \varphi_{ib}$ $(b > 0)$. Using the above, and Fubini's theorem,

$$\int_{\mathbb{R}} g_b(-t) d\mu_T(t) = \int_{\mathbb{R}} \int_{\mathbb{R}} \varphi_{ib}(t + a) g(a) d\mu_T(t) da$$

$$= \int_{\mathbb{R}} \int_{\mathbb{R}} \varphi_{ib}(t - a) g(a) d\mu_T(t) da$$

$$= \int_{\mathbb{R}} g_b(t) d\mu_T(t).$$

Since $g_b \to g$ as $b \to 0$ and $\sup_{t \in \mathbb{R},\ b>0} |g_b(t)| < \infty$, we have by bounded convergence that

$$\int_{\mathbb{R}} g(-t) d\mu_T(t) = \int_{\mathbb{R}} g(t) d\mu_T(t)$$

and that μ_T is symmetric.   □

Let $T : \mathbb{R}^{2+} \to \mathbb{R}^{2+}$ be a non-Möbius odd inner function and suppose that

$$T(z) = \alpha z + \int_{\mathbb{R}} \left(\frac{1 + tz}{t - z} \right) d\mu(t).$$

Evidently

$$\frac{\operatorname{Im} T(ib)}{b} = \alpha_T + \int_{\mathbb{R}} \left(\frac{1 + t^2}{t^2 + b^2} \right) d\mu(t) \to \begin{cases} \alpha_T \in [0, \infty) & \text{as } b \to \infty, \\ \gamma_T := \alpha_T + \int_{\mathbb{R}} \left(\frac{1+t^2}{t^2} \right) d\mu(t) \in (0, \infty]. \end{cases}$$

The ergodic theory of odd inner functions is described by the next two results.

6.4.4 PROPOSITION.
Let $T : \mathbb{R}^{2+} \to \mathbb{R}^{2+}$ be a non-Möbius odd inner function.

a) If $\alpha_T < 1 < \gamma_T$ then $\exists$ a m-absolutely continuous, T-invariant probability and T is exact.

b) If either $\alpha_T > 1$ or $\gamma_T < 1$ then T is totally dissipative and non-ergodic.

PROOF.
a) If $\alpha_T < 1 < \gamma_T$ then by continuity $\exists\ b > 0$ such that $\frac{\operatorname{Im} T(ib)}{b} = 1$ whence $T(ib) = ib$ (by oddness). By proposition 6.2.2, P_{ib} is an m-absolutely continuous, T-invariant probability and T is exact.

b) If $\gamma_T < 1$, and $R(z) := \frac{-1}{T(\frac{-1}{z})}$, then R is also odd and $\alpha_R = \frac{1}{\gamma_T} > 1$. Thus, there is no loss of generality in assuming that $\alpha_T > 1$ and the result follows from corollary 6.3.3.   □

The next result shows that a non-Möbius, odd inner function with $\alpha = 1$ is exact, independently of conservativity.

6.4.5 THEOREM [A6].
If $T : \mathbb{R}^{2+} \to \mathbb{R}^{2+}$ is odd, inner and $\alpha_T = 1$ then T is exact.

PROOF.
Write $T^n(i) = ib_n$, then

$$\frac{b_{n+1}}{b_n} = 1 + \int_{\mathbb{R}} \left(\frac{1 + t^2}{t^2 + b_n^2} \right) d\mu(t)$$

$$\leq 1 + \frac{\mu(\mathbb{R})}{b_n}$$

$$\to 1 \text{ as } n \to \infty.$$

It follows from Pommerenke's theorem that

$$\frac{T^n(z)}{b_n} \to F(z) \text{ as } n \to \infty\ \forall\ z \in \mathbb{R}^{2+},$$

where $F : \mathbb{R}^{2+} \to \mathbb{R}^{2+}$ is analytic and odd, and $F(Tz) = F(z)$, whence $F \equiv i$.

Consequently, by proposition 6.2.2, T is exact.   □

EXAMPLE: MORE TANGENT FUNCTIONS. Let

$$Tx = \alpha x + \beta \tan x,$$

where α, $\beta > 0$, then T is odd and

$$\alpha_T = \alpha, \quad \gamma_T = \alpha + \beta.$$

By theorem 6.4.5:

if $\alpha < 1 < \alpha + \beta$, $\exists$ a T-invariant, m-absolutely continuous probability and T is exact;

if either $\alpha > 1$ or $\alpha + \beta < 1$, T is totally dissipative and non-ergodic;

if $\alpha = 1$, then T is exact.

Note that the case $\alpha + \beta = 1$ is treated in the previous example "some tangent functions".

6.4.6 PROPOSITION   [A5].

$Tx = x + \beta \tan x$ is conservative, exact, and pointwise dual ergodic $\forall \beta > 0$ with return sequence $a_n(T) \sim \frac{\log n}{\beta}$.

PROOF.

Write $T^n(ib) = ib_n$, then

$$b_{n+1} = b_n + \beta \tanh b_n \ \sim \ \beta n \text{ as } n \to \infty.$$

Conservativity, pointwise dual ergodicity and the form of the return sequence follow from theorem 6.2.5, and exactness follows as above.   □

We show finally that $\exists$ a dissipative, exact inner function. The construction is of an odd inner function of form

$$Tz = z + \int_{\mathbb{R}} \left(\frac{1 + tz}{t - z} \right) d\mu(t)$$

where μ is a positive, singular, symmetric measure on $\mathbb{R}$ and the main task is to control the growth of $\operatorname{Im} T^n(i)$ as $n \to \infty$.

The techniques also yield pointwise dual ergodic inner functions with arbitrary regularly varying return sequence of index in $(0, 1/2)$.

6.4.7 LEMMA.

Let μ be a positive, symmetric measure on $\mathbb{R}$.

Suppose that $c(x) := \mu((-x, x)^c)$ is regularly varying with index $-\alpha$ where $0 < \alpha < 2$,

then

a) $F(b) := \int_{-\infty}^{\infty} \frac{t^2 + 1}{t^2 + b^2} \mu(dt) \sim d_\alpha c(b)$ as $b \to \infty$ where $d_\alpha := 2 \int_0^\infty \frac{z^{1-\alpha}}{(z^2+1)^2} dz < \infty$
and

b) if $b_n > 0$, $b_{n+1} = b_n(1 + F(b_n))$, then $b_n \sim \alpha^{\frac{1}{\alpha}} C^{-1}(n)$ as $n \to \infty$ $\forall b > 0$ where $C(y) := \frac{1}{F(y)}$.

PROOF.

a) Write

$$F(b) = \int_{-\infty}^{\infty} \frac{t^2+1}{t^2+b^2}\, \mu(dt).$$

Next, changing variables

$$F(b) = b^{-2} + 2(1-b^{-2})\int_0^{\infty} c(bz)\frac{z}{(z^2+1)^2}\, dz.$$

By the representation theorem for slowly varying functions, $\exists\; M > 1$ and $k :
[M,\infty) \to \mathbb{R}_+$, $\epsilon : [M,\infty) \to \mathbb{R}$ measurable, such that $|\epsilon(t)| < \frac{\alpha}{2}\; \forall\; t \geq M$,
$\epsilon(t) \to 0$, $k(t) \to k \in \mathbb{R}_+$ as $t \to \infty$ and such that

$$c(y) = \frac{k(y)}{y^{\alpha}}L(y) \;\;\forall\; y \geq M$$

where

$$L(y) := e^{\int_M^y \frac{\epsilon(t)}{t}\, dt} \quad (y \geq M).$$

Consequently

$$\frac{c(by)}{c(b)} \leq \frac{1}{y^{\frac{\alpha}{2}}} \;\;\forall\; b > 0,\; y \geq \frac{M}{b},$$

whence by dominated convergence,

$$\int_{\frac{M}{b}}^{\infty} \frac{c(bz)}{c(b)}\frac{z}{(z^2+1)^2}\, dz \to d_{\alpha}$$

as $b \to \infty$.

To complete the proof of a), note that $\frac{1}{b^2 c(b)} \to 0$ as $b \to \infty$ and

$$\int_0^{\frac{M}{b}} \frac{zc(bz)}{(z^2+1)^2}\, dz = O\!\left(\frac{1}{b^2}\right) = o(c(b))$$

as $b \to \infty$.

b) Now set $C(b) := \frac{1}{F(b)}$, then $\frac{C(xy)}{C(x)} \to y^{\alpha}$ as $x \to \infty\;\; \forall\; y > 0$, and

$$C(b_{n+1}) = \frac{b_n^{\alpha}}{L(b_{n+1})}\left(1 + \frac{L(b_n)}{b_n}\right)^{\alpha}$$

$$= C(b_n)\frac{L(b_n)}{L(b_{n+1})}\left(1 + \frac{\alpha}{C(b_n)} + o\!\left(\frac{1}{C(b_n)}\right)\right).$$

Now

$$\left|\log\frac{L(b_n)}{L(b_{n+1})}\right| \leq \int_{b_n}^{b_n(1+F(b_n))} \frac{|\epsilon(t)|dt}{t} = o(\log(1+F(b_n))) = o\!\left(\frac{1}{C(b_n)}\right)$$

whence

$$\frac{L(b_n)}{L(b_{n+1})} = 1 + o\!\left(\frac{1}{C(b_n)}\right)$$

and

$$C(b_{n+1}) = C(b_n)\left(1 + o\!\left(\frac{1}{C(b_n)}\right)\right)\left(1 + \frac{\alpha}{C(b_n)} + o\!\left(\frac{1}{C(b_n)}\right)\right) = C(b_n) + \alpha + o(1).$$

It follows that $C(b_{n+1}) - C(b_n) \to \alpha$ as $n \to \infty$ and $C(b_n) \sim n\alpha$.

Finally, by the inverse theorem, C^{-1} is regularly varying at ∞ with index $\frac{1}{\alpha}$ and

$$b_n \sim C^{-1}(\alpha n) \sim \alpha^{\frac{1}{\alpha}} C^{-1}(n).$$

$\square$

6.4.8 THEOREM   [A6].
$\exists$ *a dissipative, exact inner function.*

PROOF. If

$$Tz = z + \int_{\mathbb{R}} \left(\frac{1+tz}{t-z} \right) d\mu(t)$$

where μ is a positive, singular, symmetric measure on $\mathbb{R}$ with $c(x) \sim \frac{1}{x^{\frac{1}{2}}}$ as $x \to \infty$, then $b_n \propto n^2$ and T is dissipative. $\square$

6.4.9 THEOREM   [A-De-Fi].
Suppose $0 < \gamma < 1/2$ and that $b(n)$ is regularly varying with index γ as $n \to \infty$, then there is an odd inner function

$$T(x) = x + \int_{\mathbb{R}} \frac{1+tx}{t-x}\, d\mu(t)$$

such that $a_n(T) \sim b(n)$ as $n \to \infty$.

PROOF. Let

$$T(x) = x + \int_{\mathbb{R}} \frac{1+tx}{t-x}\, d\mu(t)$$

be an odd inner function where the symmetric measure μ has a tail distribution

$$c_\mu(b) = \mu(\{t : \ |t| \geq b\}),$$

which is regularly varying at infinity with index $-\alpha$, where $\alpha \in (1,2)$.

Set $\operatorname{Im} T^n(i) := b_n$ and $C(y) := \frac{1}{F(y)}$. By lemma 6.4.7,

$$b_n \sim \alpha^{\frac{1}{\alpha}} C^{-1}(n)$$

as $n \to \infty$, and is $\frac{1}{a}$-regularly varying.

Since $1 < \alpha < 2$, $\sum_{n \geq 1} \frac{1}{b_n} = \infty$, whence T is conservative by proposition 6.2.3 and pointwise dual ergodic by theorem 6.2.5 with return sequence

$$a_n(T) \sim \frac{1}{\pi} \sum_{k=1}^{n} \frac{1}{b_k} \sim \frac{n}{b_n \pi (1 - \frac{1}{\alpha})} \sim \frac{1}{\pi \alpha^{\frac{1}{\alpha}} (1 - \frac{1}{\alpha})} \frac{n}{C^{-1}(n)}.$$

In order to obtain the proposition from the last statement, let $0 < \gamma < 1/2$ and a γ–regularly varying sequence $b(n)$ be given. Define $\alpha = \frac{1}{1-\gamma} \in (1,2)$ and set

$$\psi(x) = \frac{1}{\pi(1 - \alpha^{-1})\alpha^{1/\alpha}} \frac{x}{b(x)}.$$

Let μ be a symmetric measure on $\mathbb{R}$, singular with respect to Lebesgue measure, such that

$$c_\mu(b) \sim \frac{1}{d_\alpha \psi^{-1}(b)}$$

as $b \to \infty$. From the above, it follows that if

$$T(x) = x + \int_{\mathbb{R}} \frac{1 + tx}{t - x}\, d\mu(t)$$

$a_n(T) \sim b(n)$.   □

EXERCISE 6.4.1. Suppose that T is an odd inner function of form

$$T(x) = x + \int_{\mathbb{R}} \frac{1 + tx}{t - x}\, d\mu(t).$$

Show that

1) $a_n(T) = O\left(\sqrt{n}\right)$ as $n \to \infty$,

   and

2) $T \times T \times T$ is totally dissipative.

CHAPTER 7

Hyperbolic geodesic flows

In order to introduce the subject concisely, we define first a version of the geodesic flow on

$$SL(2, \mathbb{R}) := \{ g = \begin{pmatrix} a & b \\ c & d \end{pmatrix} : a, b, c, d \in \mathbb{R}, \ \det g = 1 \}$$

- a closed subset of $\mathbb{R}^4$, and a locally compact, Polish, topological group under matrix multiplication, when equipped with the inherited topology.

Let m be a left Haar measure on $SL(2, \mathbb{R})$. As shown in [**La**], p.2, $SL(2, \mathbb{R})$ is *unimodular* in the sense that right translations leave m invariant.

DEFINITION: G-FLOW.
The *G-flow* transformation $\Phi^t : SL(2, \mathbb{R}) \to SL(2, \mathbb{R})$ is defined by

$$\Phi^t(x) = x g_t \text{ where } g_t := \begin{pmatrix} e^{\frac{t}{2}} & 0 \\ 0 & e^{-\frac{t}{2}} \end{pmatrix}.$$

We'll see in the sequel that this G-flow is isomorphic to the geodesic flow on the space of line elements in hyperbolic space. By unimodularity, $m \circ \Phi^t = m$.

EXERCISE 7.0.1. Show that $\Phi^t : SL(2, \mathbb{R}) \to SL(2, \mathbb{R})$ is totally dissipative.

A countable, closed subgroup of $SL(2, \mathbb{R})$ is called *discrete* or a *Fuchsian* group. Given a discrete subgroup $\Gamma \subset SL(2, \mathbb{R})$ we consider the the G-flow on

$$X_\Gamma := \Gamma \backslash SL(2, \mathbb{R}) := \{ \Gamma g : \ g \in SL(2, \mathbb{R}) \}.$$

This is defined by

$$\Phi^t_\Gamma(\Gamma g) := \Gamma g g_t.$$

EXERCISE 7.0.2.
1) Show that $\exists \ V \subset SL(2, \mathbb{R})$ open so that $\{ \gamma V : \ \gamma \in \Gamma \}$ are disjoint, and $SL(2, \mathbb{R}) = \bigcup_{\gamma \in \Gamma} \gamma V \bmod m$.
2) Show that $m_\Gamma(A\Gamma) := m(A\Gamma \cap V)$ $(A \in \mathcal{B}(SL(2, \mathbb{R})))$ defines a measure on X_Γ and that $m_\Gamma \circ \Phi^t_\Gamma = m_\Gamma$.

The theorems of Hopf and Tsuji ([**Hop3**], [**Ts**]) established in §§7.3-4 show that φ_Γ is either totally dissipative, or conservative and ergodic, according to whether the Poincaré series

$$\sum_{\begin{pmatrix} a & b \\ c & d \end{pmatrix} \in \Gamma} \frac{1}{a^2 + d^2 + b^2 + c^2 + 2}$$

converges or diverges respectively.

Indeed, in case φ_Γ is conservative, it is rationally ergodic, with return sequence given by the asymptotic Poincaré series ([**A-Su**], see §7.5).

§7.1 Hyperbolic space models

Hyperbolic space can be defined as $H := U = \{z \in \mathbb{C} : |z| < 1\}$ equipped with the arclength element

$$ds(u, v) := \frac{2\sqrt{du^2 + dv^2}}{1 - u^2 - v^2},$$

and the area element

$$dA(u, v) := \frac{4dudv}{(1 - u^2 - v^2))^2}.$$

DEFINITION: HYPERBOLIC DISTANCE, GEODESIC. The *hyperbolic distance* between $x, y \in H$ can be defined by

$$\rho(x, y) = \inf\left\{\int_\gamma ds : \gamma \text{ is a piecewise smooth arc joining } x \text{ and } y\right\},$$

where, for $\gamma : [a, b] \to H$ piecewise smooth,

$$\int_\gamma ds := \int_a^b \frac{2|\gamma'(t)|}{1 - |\gamma(t)|^2} dt.$$

The *geodesics* in H are arcs in H with the property that the ds-length of any of their segments is the hyperbolic distance between the endpoints of the segment.

This section is a review of the formula for hyperbolic distance, and the characterisation of the geodesics.

Möbius transformations. The *Möbius transformations* of Ω, a sub-domain of the Riemann sphere S^2, are the analytic bijections of Ω whose inverses are also analytic. We denote the collection of Möbius transformations of Ω by Möb (Ω). It is a topological group under composition and the compact-open topology.

We'll see that the Möbius transformations of H preserve hyperbolic arclength and area.

The collection of Möbius transformations of the Riemann sphere S^2 is given by

$$\text{Möb}\,(S^2) = PSL(2, \mathbb{C}) := SL(2, \mathbb{C})/\{\pm I\}$$

where

$$SL(2, \mathbb{C}) := \left\{g = \begin{pmatrix} a & b \\ c & d \end{pmatrix} : a, b, c, d \in \mathbb{C}, \ \det g = 1\right\}$$

and for each $g = \begin{pmatrix} a & b \\ c & d \end{pmatrix} \in SL(2, \mathbb{C})$ the associated Möbius transformation of S^2 is defined by $g(z) = \frac{az+b}{cz+d}$.

It can be checked that

$$\text{Möb}\,(H) = \left\{z \mapsto \lambda \frac{z - \alpha}{1 - \bar{\alpha}z} : |\lambda| = 1, \ \alpha \in H\right\} \cong PSUU(1)$$

where

$$SUU(1) := \left\{g = \begin{pmatrix} \beta & \alpha \\ \bar{\alpha} & \bar{\beta} \end{pmatrix} : \alpha, \beta \in \mathbb{C}, \ |\beta|^2 - |\alpha|^2 = 1\right\}$$

and $PSUU(1) = SUU(1)/\{\pm I\}$.

Similarly, $PSL(2, \mathbb{R}) := SL(2, \mathbb{R})/\{\pm I\} \cong \text{Möb}\,(\mathbb{R}^{2+})$ where $\mathbb{R}^{2+} = \{x + iy : y > 0\}$ - the upper half plane.

If Ω is a simply connected proper sub-domain of $\mathbb{C}$, then by the Riemann mapping theorem, Ω is conformally isomorphic to $\mathbb{R}^{2+}$. It follows that $\text{Möb}\,(\Omega) \cong PSL(2, \mathbb{R})$, the isomorphism being conjugation by a conformal isomorphism.

A conformal isomorphism $\vartheta : H \to \mathbb{R}^{2+}$ is given by

$$\vartheta = \begin{pmatrix} \frac{1+i}{2} & \frac{1+i}{2} \\ -\frac{1-i}{2} & \frac{1-i}{2} \end{pmatrix} \in SL(2, \mathbb{C})$$

and indeed

$$SUU(1) = \vartheta^{-1} SL(2, \mathbb{R}) \vartheta.$$

Note that

$$\vartheta^{-1} g^t \vartheta = \begin{pmatrix} \cosh \frac{t}{2} & \sinh \frac{t}{2} \\ \sinh \frac{t}{2} & \cosh \frac{t}{2} \end{pmatrix} := \gamma^t.$$

We use the same notation to denote the G-flow $\Phi^t : SUU(1) \to SUU(1)$ which is defined by $\Phi^t(g) := g\gamma^t$.

Let $\Omega \subsetneq \mathbb{C}$ be a simply connected domain, and let $\vartheta_\Omega : H \to \Omega$ be a conformal isomorphism. Let $[x, y]_\Omega := [\vartheta_\Omega^{-1} x, \vartheta_\Omega^{-1} y]$ for $x, y \in \Omega$.

The Schwarz-Pick Lemma on Ω becomes:

$$[gx, gy]_\Omega \leq [x, y]_\Omega \ \forall \ x, y \in \Omega$$

and $g : \Omega \to \Omega$ analytic with equality if and only if $g \in \text{Möb}\,(\Omega)$.

In particular, $[\cdot, \cdot]_\Omega$ is independent of the conformal isomorphism $\vartheta_\Omega : H \to \Omega$. By a calculation using $\vartheta_{\mathbb{R}^{2+}}(z) = i\frac{1+z}{1-z}$,

$$[x, y]_{\mathbb{R}^{2+}} = |\frac{x - y}{x - \overline{y}}| \quad (x, y \in \mathbb{R}^{2+}).$$

7.1.1 PROPOSITION. Let $\Omega \subsetneq \mathbb{C}$ be a simply connected domain.
1. Given a, b, z, $\omega \in \Omega$ such that $[a, b]_\Omega = [z, \omega]_\Omega$, $\exists \ g \in \text{Möb}\,(\Omega)$ such that $g(a) = z$ and $g(b) = \omega$.
2. $\forall \ a, z \in \Omega$ and $\theta \in [0, 1)$, there is a unique $g \in \text{Möb}\,(\Omega)$ such that $g(a) = z$ and $\arg g'(a) = \theta$ (i.e. $g'(a) = |g'(a)|e^{2\pi i\theta}$).

PROOF. We establish (1) on H where, given $z, \omega \in H$ with $[z, \omega] = r$, one finds $g \in \text{Möb}\,(H)$ with $g(0) = z$, $g(r) = \omega$.

(2) is proved on $\mathbb{R}^{2+}$ where a calculation shows that given $z = x + iy \in \mathbb{R}^{2+}$, $g(i) = z$ if and only if

$$g = \begin{pmatrix} \frac{x\cos\theta}{\sqrt{y}} + \sqrt{y}\sin\theta & \frac{x\sin\theta}{\sqrt{y}} - \sqrt{y}\cos\theta \\ \frac{\cos\theta}{\sqrt{y}} & \frac{\sin\theta}{\sqrt{y}} \end{pmatrix} \quad (\theta \in [0, 2\pi)).$$

$\square$

7.1.2 PROPOSITION.
Möbius transformations of H preserve hyperbolic area, arclength and distance.

PROOF. The proof is standard using that for $g \in \text{Möb}\,(H)$,

$$\frac{|g'(z)|}{1 - |g(z)|^2} = \frac{1}{1 - |z|^2}.$$

$\square$

7.1.3 PROPOSITION. *The hyperbolic distance is given by*

$$\rho(x, y) = 2\tanh^{-1}[x, y],$$

and the geodesics in H are diameters of H, and circles orthogonal to ∂H.

PROOF. We prove that for $\gamma : [0, 1] \to H$ be a simple, piecewise smooth arc joining $x, y \in H$,

$$\int_\gamma ds \geq 2\tanh^{-1}[x, y]$$

with equality if and only if $\gamma([0, 1])$ is a segment of a diameter, or circle orthogonal to ∂H.

Let first $\gamma = u + iv : [0, 1] \to H$ be a smooth arc joining 0 to $x \in (0, 1) \subset H$. The point here is that u is also a smooth arc joining 0 to x, and since

$$\frac{|\gamma'(t)|}{1 - |\gamma(t)|^2} \geq \frac{u'(t)}{1 - u(t)^2} \ \forall \ t \in [0, 1],$$

we have that

$$\int_\gamma ds = \int_0^1 \frac{2|\gamma'(t)|}{1 - |\gamma(t)|^2} dt \geq \int_0^1 \frac{2u'(t)}{1 - u(t)^2} dt = \int_0^x \frac{2}{1 - s^2} ds = 2\tanh^{-1} x.$$

Since $u' \geq 0$, there is equality if and only if $v \equiv 0$ and $u' \geq 0$, we have

$$\int_\gamma ds \geq 2\tanh^{-1} x$$

with equality if and only if $\gamma([0, 1]) = [0, x]$. This establishes the proposition when $x \in (0, 1)$, and $y = 0$.

Now suppose that $x, y \in H$ and $\gamma : [0, 1] \to H$ is a smooth arc joining x to y. Let $z := [x, y] \in (0, 1)$ and $g \in \text{Möb}\,(H)$ be such that $x = g(0)$, $y = g(z)$, then $g^{-1} \circ \gamma : [0, 1] \to H$ is a smooth arc joining 0 to z, and by the above

$$\int_{g^{-1} \circ \gamma} ds \geq 2\tanh^{-1} z$$

with equality if and only if $g^{-1} \circ \gamma([0, 1]) = [0, z]$. It follows from proposition 7.1.2 that

$$\int_\gamma ds = \int_{g^{-1} \circ \gamma} ds \geq 2\tanh^{-1} z = 2\tanh^{-1}[x, y]$$

with equality if and only if $\gamma([0, 1]) = g([0, z])$.

It can be checked that the collection

$$\{g([0, r]) : \ r \in (0, 1), \ g \in \text{Möb}\,(H)\}$$

is precisely the collection of segments of diameters, and segments of circles orthogonal to ∂H.

$\square$

REMARKS.

1) Note that in particular $\rho(0, x) = 2 \tanh^{-1} |x|$, whence

$$1 - |x| = 1 - \tanh \frac{\rho(0, x)}{2} = \frac{2}{e^{\rho(0, x)} + 1} \sim 2e^{-\rho(0, x)} \text{ as } |x| \to 1.$$

2) Set for $x \in H$ $N_\rho(x, \epsilon) := \{y \in H : \rho(x, y) < \epsilon\}$, then $N_\rho(x, \epsilon) = \varphi_x N_\rho(0, \epsilon)$ where $\varphi_x \in \text{Möb}(H)$ is defined by $\varphi_x(z) := \frac{z + x}{1 + \overline{x}z}$, and hence

$$A(N_\rho(x, \epsilon)) = A(N_\rho(0, \epsilon)) := A(\epsilon) = 4\pi \sinh^2 \frac{\epsilon}{2} \sim 4\pi\epsilon^2 \text{ as } \epsilon \to 0.$$

As mentioned above, any simply connected proper sub domain Ω of $\mathbb{C}$ is conformally isomorphic to H and so can be used as a model for hyperbolic space. Hyperbolic arclength, area and geodesics are as transported from H by a conformal isomorphism.

For example, on $\mathbb{R}^{2+} = \{x + iy : y > 0\}$,

$$ds(x, y) = \frac{\sqrt{dx^2 + dy^2}}{y}, \quad \& \quad dA(x, y) = \frac{dxdy}{y^2}$$

and the geodesics are vertical lines and circles orthogonal to $\mathbb{R}$.

§7.2 The geodesic flow of H

The geodesic flow of H is not defined on H. It will be defined on $H \times \mathbb{T}$ ($\cong$ $H \times [0, 1)$), the space of *line elements* of H. The space $H \times \mathbb{T}$ is equipped with the metric

$$\overline{\rho}(\omega, \omega') = \rho(x(\omega), x(\omega')) + \|\theta(\omega) - \theta(\omega')\|$$

where $\omega = (x(\omega), \theta(\omega))$ and $\|\theta\| = \theta \wedge (1 - \theta)$; and the measure m defined by

$$dm(x, \theta) = dA(x)d\theta.$$

The isometries (or Möbius transformations) act on $H \times \mathbb{T}$ (as differentiable maps) by

$$g(\omega) = (g(x(\omega)), \theta(\omega) + \arg g'(x(\omega)).$$

Here $\arg : \mathbb{C}\backslash\{0\} \to \mathbb{T} \cong [0, 1)$ is defined by $z = |z|e^{2\pi i \arg(z)}$. Clearly, by proposition 7.1.2, Möbius transformations preserve the measure m.

A geodesic meets ∂H at two points, and is characterised by them, and can be *directed* (or oriented) by ordering these points. We'll denote the directed geodesic from $e^{2\pi i a}$ to $e^{2\pi i b}$ by

$$\overrightarrow{[e^{2\pi i a}, e^{2\pi i b}]}.$$

To each line element ω there corresponds a unique directed geodesic passing through $x(\omega)$ whose directed tangent at $x(\omega)$ makes an angle $\theta(\omega)$ with the radius $(0, 1)$.

DEFINITION: GEODESIC FLOW ON $H \times \mathbb{T}$.

The *geodesic flow* transformation $\varphi^t : H \times \mathbb{T} \to H \times \mathbb{T}$ $(t \in \mathbb{R})$ is defined as follows at $\omega \in H \times \mathbb{T}$:

If $t > 0$, the point $x(\varphi^t \omega)$ is the unique point on the geodesic at distance t from $x(\omega)$ in the direction of the geodesic, and if $t < 0$, the point $x(\varphi^t \omega)$ is the unique point on the geodesic at distance $-t$ against the direction of the geodesic. The angle $\theta(\varphi^t \omega)$ is the angle made by the directed tangent to the geodesic at the point $x(\varphi^t \omega)$ with the radius $(0, 1)$.

7.2.1 PROPOSITION. *The geodesic flow commutes with the action of Möbius transformations on $H \times \mathbb{T}$:*

$$\varphi^t \circ g = g \circ \varphi^t \quad (t \in \mathbb{R}, \ g \in \mathrm{M\ddot{o}b}\,(H)).$$

PROOF.

Choose $g \in \mathrm{M\ddot{o}b}\,(H)$ and let $\omega \in H \times \mathbb{T}$. We'll show that $\varphi^t(g(\omega)) = g(\varphi^t \omega)$ for $t \in \mathbb{R}$.

Let the directed geodesic passing through $x(\omega)$ with the tangent direction $\theta(\omega)$ be $\overrightarrow{[e^{2\pi i a}, e^{2\pi i b}]}$. As Möbius transformations preserve arclength,

$$g\big(\overrightarrow{[e^{2\pi i a}, e^{2\pi i b}]}\big) = \overrightarrow{[g(e^{2\pi i a}), g(e^{2\pi i b})]},$$

and $g(\varphi^t \omega) \in \overrightarrow{[g(e^{2\pi i a}), g(e^{2\pi i b})]}$, whence

$$\forall\, t \in \mathbb{R} \ \exists\, \tau(\omega, t) \in \mathbb{R} \text{ such that } g(\varphi^t \omega) = \varphi^{\tau(\omega, t)} g(\omega).$$

Clearly,

$$\tau(\omega, s+t) = \tau(\omega, s) + \tau(\varphi^s \omega, t) \quad (s, t \in \mathbb{R}, \ \omega \in H \times \mathbb{T}),$$

moreover, since g is an isometry of (H, ρ),

$$|\tau(\omega, t)| = \rho(x(g(\omega)), x(\varphi^{\tau(\omega, t)} g(\omega))) = \rho(x(g(\omega)), x(g(\varphi^t \omega))) = |t|.$$

It follows that either $\tau(\omega, t) = -t$, or $\tau(\omega, t) = t$.

The case $\tau(\omega, t) = -t$ is impossible because if it were so,

$$g(e^{2\pi i b}) \leftarrow g(x(\varphi^t \omega)) = x(\varphi^{-t} g(\omega)) \to g(e^{2\pi i a})$$

as $t \to +\infty$. $\qquad\qquad\qquad\qquad\qquad\qquad\qquad\qquad\qquad\qquad\qquad\square$

Define $\eta : \mathrm{M\ddot{o}b}\,(H) \to H \times \mathbb{T}$ by $\eta(\gamma) = \gamma(0,0) = (\gamma(0), \arg \gamma'(0))$. By proposition 7.1.1 (2), η is a bijection.

PROPOSITION 7.2.2.
1. $m \circ \eta^{-1}$ *is a left Haar measure on* $\mathrm{M\ddot{o}b}\,(H)$.
2. *The geodesic flow on* $H \times \mathbb{T}$ *is conjugate to the G-flow on* $\mathrm{M\ddot{o}b}\,(H)$. *Specifically* $\eta \circ \Phi^t = \varphi^t \circ \eta$.
3. $m \circ \varphi^t = m \quad (t \in \mathbb{R})$.

PROOF.

Clearly $\eta \circ L_\gamma = \gamma \circ \eta$ where $L_\gamma : \text{Möb}(H) \to \text{Möb}(H)$ is defined by $L_\gamma g = \gamma g$. Since $m \circ \gamma = m \ \forall \ \gamma \in \text{Möb}(H)$, $m \circ \eta^{-1} \circ L_\gamma = m \circ \gamma \circ \eta^{-1} = m \circ \eta^{-1}$ proving (1).

Note that

$$\eta(\Phi^t(I)) = \eta(\gamma^t) = (\gamma^t(0), \arg \gamma^{t\prime}(0)) = (\tanh \frac{t}{2}, 0) = \varphi^t(0,0) = \varphi^t \eta(I).$$

The rest of the proof of (2) is based on the facts that the G-flow commutes with left translation: $g\Phi^t(h) = gh\gamma^t = \Phi^t(gh)$, the geodesic flow commutes with Möbius transformations, and $\eta(gh) = g(\eta(h))$:

$$\eta(\Phi^t(g)) = \eta(g\Phi^t(I)) = g(\eta(\Phi^t(I))) = g(\varphi^t \eta(I)) = \varphi^t(\eta(g)).$$

(3) follows from (1), (2) and the fact that Φ^t preserves left Haar measure on $\text{Möb}(H)$. □

§7.3 Asymptotic geodesics

DEFINITION: STABLE AND UNSTABLE MANIFOLDS. The *stable-*, and *unstable* manifolds of φ^t at ω are defined by

$$W^+(\omega) := \{\omega' \in H \times \mathbb{T} : \ \bar{\rho}(\varphi^t \omega, \varphi^t \omega') \to 0 \text{ as } t \to \infty\}$$

and

$$W^-(\omega) := \{\omega' \in H \times \mathbb{T} : \ \bar{\rho}(\varphi^t \omega, \varphi^t \omega') \to 0 \text{ as } t \to -\infty\}$$

respectively.

In this section, we prove

7.3.1 HOPF'S LEMMA   [**Hop2**].

Suppose that $f_\pm : H \times \mathbb{T} \to \mathbb{R}$ are measurable functions such that

(1) $f_- = f_+ \ m - a.e.$

(2) $\omega' \in W^-(\omega) \ \Rightarrow \ f_-(\omega) = f_-(\omega'), \quad \omega' \in W^+(\omega) \ \Rightarrow \ f_+(\omega) = f_+(\omega'),$

(3) $f_\pm \circ \varphi^t \equiv f_\pm \ \forall \ t \in \mathbb{R},$

then $\exists \ a \in \mathbb{R}$ such that $f_+ = f_- = a$ a.e..

Define $\pi_\pm : H \times \mathbb{T} \to \mathbb{T}$ by

$$\pi_\pm(\omega) = \lim_{t \to \pm\infty} x(\varphi^t \omega).$$

DEFINITION: HOROCYCLE.

A *horocycle* in H is a circle inside H, except for one point on ∂H. Given $x \in H$, and $a \in \partial H$, there is a unique horocycle $\mathcal{H}(x,a)$ passing through x and a.

7.3.2 PROPOSITION. *The stable manifold of φ^t at $\omega \in H \times \mathbb{T}$ consists of those line elements $(y, \theta(y))$ where $y \in \mathcal{H}(x(\omega), \pi_+(\omega))$ and $\theta(y)$ is the direction towards the centre of $\mathcal{H}(x(\omega), \pi_+(\omega))$.*

PROOF. First, note that if $\omega' \in W^+(\omega)$, then $\pi_+(\omega) = \pi_+(\omega') := c$ and $\exists \, a, b \in \partial H$ such that

$$x(\omega) \in \overrightarrow{[a, c]}, \quad \& \quad x(\omega') \in \overrightarrow{[b, c]}.$$

We pass to the upper half plane model, and assume that $\pi_+(\omega) = \pi_+(\omega') = \infty$ which forces $\theta(\omega) = \theta(\omega') = \frac{\pi}{2}$.

Here, $\varphi^t(z, \frac{\pi}{2}) = (z + ie^t \operatorname{Im} z, \frac{\pi}{2})$, whence writing $x(\omega) = u + iv$, $x(\omega') = u' + iv'$, we have

$$\overline{\rho}(\varphi^t \omega, \varphi^t \omega') = \rho(u + i(v + t), u' + i(v' + y))$$

$$= \tanh^{-1} \frac{\sqrt{(u - u')^2 + e^{2t}(v - v')^2}}{2\sqrt{(u - u')^2 + e^{2t}(v + v')^2}}$$

$$\to \tanh^{-1} \frac{|v - v'|}{2|v + v'|} \text{ as } t \to \infty.$$

This shows that if $\omega \in \mathbb{R}^{2+} \times \mathbb{T}$, and $\pi_+(\omega) = \infty$, then

$$W^+(\omega) = \{(y, \frac{\pi}{2}): \ y \in \mathbb{R}^{2+}, \ \operatorname{Im} y = \operatorname{Im} x(\omega)\}.$$

Transforming by the appropriate conformal maps yields the result. $\qquad\square$

Proposition 7.3.2 establishes the "Anosov property" and the "local product structure" $W_-(\omega) \times W_+(\omega) \times \mathbb{R}$ for $H \times \mathbb{T}$. See chapter III.2 in [**Mañ**].

7.3.3 COROLLARY [**Hop2**].

$$\pi_\pm(\omega) = \pi_\pm(\omega') \quad \Leftrightarrow \quad \exists \, a \in \mathbb{R} \ni \overline{\rho}(\varphi^{t+a} \omega, \varphi^t \omega') \to 0 \ as \ t \to \pm\infty.$$

Define $\chi : \mathbb{T} \times \mathbb{T} \times \mathbb{R} \to H \times \mathbb{T}$ as follows. Given $a, b \in \mathbb{T}$, let:
$x(\chi(a, b, 0))$ be the symmetric point of the directed geodesic $\overrightarrow{[e^{2\pi i a}, e^{2\pi i b}]}$,
$\theta(\chi(a, b, 0))$ be the direction of the directed tangent to $\overrightarrow{[e^{2\pi i a}, e^{2\pi i b}]}$ at $x(\chi(a, b, 0))$;
and let $\chi(a, b, t) := \varphi^t \chi(a, b, 0)$. Clearly

$$\chi(a, b, s + t) := \varphi^t \chi(a, b, s).$$

Möb (H) acts on $\mathbb{T} \times \mathbb{T} \times \mathbb{R}$ by

$$g(a, b, t) := \chi^{-1} g \chi(a, b, t) \quad ((a, b, t) \in \mathbb{T} \times \mathbb{T} \times \mathbb{R}, \ g \in \text{Möb}\,(H)).$$

Now since

$$g(\overrightarrow{[e^{2\pi i a}, e^{2\pi i b}]}) = \overrightarrow{[g(e^{2\pi i a}), g(e^{2\pi i b})]},$$

$$g(a, b, 0) = (ga, gb, t + \tau_g(a, b))$$

whence by proposition 7.2.1,

$$g(a, b, t) = \chi^{-1} g \varphi^t \chi(a, b, 0) = \chi^{-1} \varphi^t g \chi(a, b, 0) =$$

$$\chi^{-1} \varphi^t \chi(g(a), g(b), \tau_g(a, b)) = (ga, gb, t + \tau_g(a, b)).$$

The next result establishes absolute continuity of the local product structure $W_-(\omega) \times W_+(\omega) \times \mathbb{R}$ for $H \times \mathbb{T}$.

7.3.4 PROPOSITION   [**Hop2**].
There exists a constant $0 < c < \infty$ such that

$$dm \circ \chi^{-1}(a, b, t) = c \frac{dadbdt}{|e^{2\pi ia} - e^{2\pi ib}|^2}.$$

PROOF.
Define a measure μ on $\mathbb{T} \times \mathbb{T}$ by

$$d\mu(a, b) = \frac{dadb}{|e^{2\pi ia} - e^{2\pi ib}|^2}.$$

Considering the action of $g \in \text{Möb}(H)$ by $g(x, y) = (g(x), g(y))$, we have

$$d\mu \circ g(a, b) = \frac{|g'(e^{2\pi ia})||g'(e^{2\pi ib})|}{|g(e^{2\pi ia}) - g(e^{2\pi ib})|^2} dadb \equiv \frac{dadb}{|e^{2\pi ia} - e^{2\pi ib}|^2} = d\mu(a, b),$$

and it follows that

$$(\mu \times \lambda) \circ g = \mu \times \lambda \ \forall \ g \in \text{Möb}(H)$$

where λ denotes Lebesgue measure on $\mathbb{R}$. Thus $(\mu \times \lambda) \circ \chi \circ \eta^{-1}$ is a left Haar measure on $\text{Möb}(H)$. By proposition 7.2.2 (1), $m \circ \eta^{-1}$ is also a left Haar measure for $\text{Möb}(H)$, and the unicity of such implies that $(\mu \times \lambda) \circ \chi = cm$ for some $c > 0$. $\square$

PROOF OF HOPF'S LEMMA.
Let $g_\pm = f_\pm \circ \chi : \mathbb{T} \times \mathbb{T} \times \mathbb{R} \to \mathbb{R}$, then by (3), $g_\pm(a, b, t+s) = g_\pm(a, b, t) \ \forall \ t \in \mathbb{R}$, and $\exists \ h_\pm : \mathbb{T} \times \mathbb{T} \to \mathbb{R}$ such that $g_\pm(a, b, t) = h_\pm(a, b)$ a.e..
By (2),

$$h_-(a, b) = h_-(a', b), \quad h_+(a, b) = h_+(a, b') \ \forall \ a, a', b, b' \in \mathbb{T},$$

whence $\exists \ k_\pm : \mathbb{T} \to \mathbb{R}$ such that $h_-(a, b) = k_-(a)$ and $h_+(a, b) = k_+(b)$.
By (1) and proposition 7.3.4, $k_-(a) = k_+(b)$ for a.e. $(a, b) \in \mathbb{T} \times \mathbb{T}$ with respect to product measure and it follows that $k_- = k_+$ is constant a.e.. The lemma follows from this. $\square$

§7.4 Surfaces

Let Γ be a discrete group of isometries of H, and define the surface

$$\Gamma \backslash H \text{ by } \{\Gamma x : x \in H\}$$

equipped with the metric

$$\rho_\Gamma(\Gamma z, \Gamma z') = \inf_{\gamma \in \Gamma} \rho(z, \gamma z').$$

The space of line elements of $\Gamma \backslash H$ is

$$X_\Gamma := (\Gamma \backslash H) \times \mathbb{T} = \Gamma \backslash (H \times \mathbb{T}) \cong \Gamma \backslash \text{Möb}(H)$$

equipped with the metric

$$\bar\rho_\Gamma(\Gamma \omega, \Gamma \omega') = \inf_{\gamma \in \Gamma} \bar\rho(\omega, \gamma \omega').$$

Since $g \circ \varphi_t \equiv \varphi_t \circ g$ on H $\forall \ t \in \mathbb{R}$ and $g \in \mathrm{M\ddot{o}b}(H)$, the geodesic flow transformations on X_Γ can be defined by

$$\varphi_\Gamma^t \Gamma(\omega) = \Gamma \varphi^t(\omega).$$

7.4.1 PROPOSITION. *The G-flow on X_Γ is isomorphic to the geodesic flow on X_Γ.*

PROOF.
Define $\eta_\Gamma : \Gamma \backslash \mathrm{M\ddot{o}b}\,(H) \to \Gamma \backslash (H \times \mathbb{T})$ by $\eta_\Gamma(\Gamma g) := \Gamma \eta(g)$ where $\eta : \mathrm{M\ddot{o}b}\,(H) \to H \times \mathbb{T}$ is defined (as in §7.2) by $\eta(\gamma) = (\gamma(0), \arg \gamma'(0))$. It follows from proposition 7.2.2 that

$$\eta_\Gamma \circ \Phi_\Gamma^t(\Gamma g) = \eta_\Gamma(\Gamma g \gamma^t) = \Gamma \eta(g \gamma^t) = \Gamma \varphi^t \eta(g) = \varphi_\Gamma^t \Gamma \eta(g) = \varphi_\Gamma^t \circ \eta_\Gamma(\Gamma g).$$

$\square$

Let $\pi_\Gamma : H \to \Gamma \backslash H$, $\overline{\pi}_\Gamma : H \times \mathbb{T} \to X_\Gamma$ be the projections $\pi_\Gamma(z) = \Gamma z$, $\overline{\pi}_\Gamma(\omega) = \Gamma \omega$, and let F be a fundamental domain for Γ in H satisfying

$$F^o := \{x \in H : \rho(y, x) < \rho(\gamma(y), x) \ \forall \ \gamma \in \Gamma \setminus \{e\}\}$$

where $y \in H$ (such a fundamental domain is called a *Dirichlet fundamental domain*), then π_Γ and $\overline{\pi}_\Gamma$ are 1-1 on F and $F \times \mathbb{T}$, and so the measures $A_{|F}$ and $m_{|F}$ induce measures A_Γ and m_Γ on $\Gamma \backslash H$ and $X_\Gamma = \Gamma \backslash H \times \mathbb{T}$ respectively. Indeed, with boundary identifications, $(\Gamma \backslash H, \rho_\Gamma) \cong (F, \rho)$ and $(X_\Gamma, \overline{\rho}_\Gamma) \cong (F \times \mathbb{T}, \overline{\rho})$.

Since $\overline{\pi}_\Gamma \circ \gamma = \overline{\pi}_\Gamma \ \forall \ \gamma \in \Gamma$, we have that $(f \circ \overline{\pi}_\Gamma) \circ \gamma = (f \circ \overline{\pi}_\Gamma) \ \forall \ \gamma \in \Gamma$ whenever $f : X_\Gamma \to \mathbb{R}$. It follows that if $f : X_\Gamma \to \mathbb{R}_+$ is measurable, and $\tilde{f} : H \times \mathbb{T} \to \mathbb{R}_+$ is defined by

$$\tilde{f} := 1_{F \times \mathbb{T}} f \circ \overline{\pi}_\Gamma,$$

then

$$\int_{X_\Gamma} f dm_\Gamma = \int_{F \times \mathbb{T}} f \circ \overline{\pi}_\Gamma dm,$$

and

$$f \circ \overline{\pi}_\Gamma \equiv \sum_{\gamma \in \Gamma} \tilde{f} \circ \gamma.$$

For $f : X_\Gamma \to \mathbb{R}$ measurable and $0 \leq a < b$, set

$$S_{(a,b)}^{\varphi_\Gamma}(f) := \int_a^b f \circ \varphi_\Gamma^s ds, \quad S_t^{\varphi_\Gamma}(f) := S_{(0,t)}^{\varphi_\Gamma}(f).$$

Similarly, for $h : H \times \mathbb{T} \to \mathbb{R}$, measurable and $0 \leq a < b$, set

$$S_{(a,b)}^{\varphi}(h) := \int_a^b h \circ \varphi^s ds, \quad S_t^{\varphi}(h) := S_{(0,t)}^{\varphi}(h).$$

7.4.2 PROPOSITION. *Let $f : X_\Gamma \to \mathbb{R}$ be measurable, then for $0 \leq a < b$,*

$$S_{(a,b)}^{\varphi_\Gamma}(f) \circ \overline{\pi}_\Gamma = S_{(a,b)}^{\varphi}(f \circ \overline{\pi}_\Gamma) = \sum_{\gamma \in \Gamma} S_{(a,b)}^{\varphi}(\tilde{f} \circ \gamma).$$

PROOF.

$$
\begin{aligned}
S^{\varphi_\Gamma}_{(a,b)}(f) \circ \overline{\pi}_\Gamma &= \int_a^b f \circ \varphi^s_\Gamma \circ \overline{\pi}_\Gamma ds \\
&= \int_a^b f \circ \overline{\pi}_\Gamma \circ \varphi^s ds \\
&= S^{\varphi}_{(a,b)}(\tilde{f} \circ \overline{\pi}_\Gamma) \\
&= \int_a^b \left(\sum_{\gamma \in \Gamma} \tilde{f} \circ \gamma \right) \circ \varphi^s ds \\
&= \sum_{\gamma \in \Gamma} S^{\varphi}_{(a,b)}(\tilde{f} \circ \gamma).
\end{aligned}
$$

$\square$

7.4.3 HOPF'S THEOREM [Hop2].
The geodesic flow φ_Γ is either totally dissipative, or conservative and ergodic.

PROOF.
We claim that

$$
D := \{x \in X_\Gamma : \overline{\rho}_\Gamma(x, \varphi^t_\Gamma x) \to \infty \text{ as } t \to \pm\infty\} = \mathfrak{D}(\varphi_\Gamma) \quad \mod m.
$$

Here $\mathfrak{D}(\varphi_\Gamma)$ is the dissipative part of the flow φ_Γ.

If $W \subset D$ is compact, then clearly $\forall\, s \in \mathbb{R}$,

$$
\sum_{k=1}^{\infty} 1_W \circ \varphi^{ks}_\Gamma < \infty
$$

a.e. on W proving that $W \subset \mathfrak{D}(\varphi^s_\Gamma)$. This shows that $D \subset \mathfrak{D}(\varphi_\Gamma)$.

Now fix $f \in L^1(m_\Gamma)$ continuous and positive. We have by theorem 1.6.4 that

$$
\mathfrak{D}(\varphi_\Gamma) = D' := \left\{ x \in X_\Gamma : \int_0^\infty f(\varphi^s_\Gamma x) ds < \infty \right\} \quad \mod\ m.
$$

For $x \notin D$, $\exists\, t_k \to \infty$ and $M > 0$ such that $\overline{\rho}_\Gamma(x, \varphi^{t_k}_\Gamma x) \le M$. Since f is positive, $\exists\, \epsilon > 0$ such that $f(y) \ge \epsilon$ whenever $\overline{\rho}_\Gamma(x, y) \le M + 1$. It follows that

$$
\int_0^\infty f(\varphi^s_\Gamma x) ds \ge \sum_{k \ge 1} \int_{t_k-1}^{t_k+1} f(\varphi^s_\Gamma x) ds = \infty
$$

whence $x \notin D'$. This shows $D \supset \mathfrak{D}(\varphi_\Gamma)$ establishing the claim.

We'll first use Hopf's lemma to show that $\mathfrak{D}(\varphi_\Gamma)$ is trivial mod m.
Let

$$
B_\pm = \{\omega \in H \times \mathbb{T} : \overline{\rho}_\Gamma(\overline{\pi}_\Gamma(\omega), \overline{\pi}_\Gamma(\varphi^t_\Gamma \omega)) \to \infty \text{ as } t \to \pm\infty\}.
$$

Clearly $B_\pm = \overline{\pi}_\Gamma^{-1} D$ mod m, and $\varphi^t B_\pm = B_\pm$.

As shown above,

$$
\pi_\pm(\omega) = \pi_\pm(\omega') \iff \exists\, a \in \mathbb{R} \ni \overline{\rho}(\varphi^{t+a}\omega, \varphi^t\omega') \to 0 \text{ as } t \to \pm\infty,
$$

whence, if $\omega \in B_\pm$, $\pi_\pm(\omega') = \pi_\pm(\omega)$, then $\forall \gamma \in \Gamma$,

$$\overline{\rho}(\gamma\omega', \varphi^t\omega') \geq \overline{\rho}(\gamma\omega, \varphi^{t+a}\omega) - \overline{\rho}(\gamma\omega', \gamma\omega) - \overline{\rho}(\varphi^{t+a}\omega, \varphi^t\omega')$$
$$\geq \overline{\rho}_\Gamma(\overline{\pi}_\Gamma(\omega), \varphi_\Gamma^{t+a}\overline{\pi}_\Gamma(\omega)) - \overline{\rho}(\omega', \omega) - \overline{\rho}(\varphi^{t+a}\omega, \varphi^t\omega').$$

$$\therefore \ \overline{\rho}_\Gamma(\overline{\pi}_\Gamma(\omega'), \varphi_\Gamma^t\overline{\pi}_\Gamma(\omega')) \geq \overline{\rho}_\Gamma(\overline{\pi}_\Gamma(\omega), \varphi_\Gamma^{t+a}\overline{\pi}_\Gamma(\omega)) - \overline{\rho}(\omega', \omega) - \overline{\rho}(\varphi^{t+a}\omega, \varphi^t\omega')$$
$$\to \infty \text{ as } t \to \pm\infty,$$

and $\omega' \in B_\pm$.

It now follows from Hopf's lemma (with $f_\pm = 1_{B_\pm}$), that 1_D is constant and $\mathfrak{D}(\varphi_\Gamma)$ is trivial.

Thus, if the geodesic flow is not totally dissipative, then it is conservative.

Now suppose that the geodesic flow φ_Γ is conservative.

By Hopf's ratio ergodic theorem, for $f, p \in L^1(X_\Gamma)$, $p > 0$, $\int_{X_\Gamma} p\,dm = 1$,

$$\frac{S_t^{\varphi_\Gamma}(f)}{S_t^{\varphi_\Gamma}(p)} \underset{t \to \pm\infty}{\longrightarrow} E_p\left(\frac{f}{p}\Big|\mathfrak{I}\right) \text{ a.e.}$$

where $E_p(h|\mathfrak{I})$ denotes the conditional expectation of h with respect to $\mathfrak{I}$ (the φ_Γ-invariant sets) considered under the probability $p\,dm$.

Fix $p : X_\Gamma \to \mathbb{R}_+$ satisfying

$$\sup_{\overline{\rho}(\omega,\omega') \leq \epsilon} \frac{|p \circ \overline{\pi}_\Gamma(\omega) - p \circ \overline{\pi}_\Gamma(\omega')|}{p \circ \overline{\pi}_\Gamma(\omega)} \underset{\epsilon \to 0}{\longrightarrow} 0, \ \& \ \int_{X_\Gamma} p\,dm_\Gamma = 1;$$

for example p defined by

$$p \circ \overline{\pi}_\Gamma(\omega) = ce^{-4\min_{\gamma \in \Gamma} \rho(x(\omega), \gamma(0))}$$

where $c \in \mathbb{R}_+$ is chosen so that $\int_{X_\Gamma} p\,dm_\Gamma = 1$.

Consider the collection

$$\mathcal{H}_p = \{f \in L^1(X_\Gamma, m_\Gamma) : \sup_{\overline{\rho}(\omega,\omega') \leq \epsilon} \frac{|f \circ \overline{\pi}_\Gamma(\omega) - f \circ \overline{\pi}_\Gamma(\omega')|}{p \circ \overline{\pi}_\Gamma(\omega)} \underset{\epsilon \to 0}{\longrightarrow} 0\}.$$

Standard approximation techniques show that the $\overline{\mathcal{H}_p} = L^1(X_\Gamma, m_\Gamma)$.

We'll use Hopf's lemma to show that for $f \in \mathcal{H}_p$, $E_p\left(\frac{f}{p}|\mathfrak{I}\right)$ is constant, whence by approximation, $E_p\left(\frac{f}{p}|\mathfrak{I}\right)$ is constant for any $f \in L^1(X_\Gamma)$ and φ_Γ is ergodic.

Let f be as above, and set

$$X_\pm = \left\{\omega \in H \times \mathbb{T} : S_t^\varphi(p \circ \overline{\pi}_\Gamma)(\omega) : \underset{|t| \to \infty}{\longrightarrow} \infty, \ \& \ \exists \lim_{t \to \pm\infty} \frac{S_t^\varphi(f \circ \overline{\pi}_\Gamma)(\omega)}{S_t^\varphi(p \circ \overline{\pi}_\Gamma)(\omega)}\right\},$$

$$f_\pm = \begin{cases} \lim_{t \to \pm\infty} \frac{S_t^\varphi(f)(\omega)}{S_t(p)(\omega)} & \omega \in X_\pm \\ 0 & \text{else.} \end{cases}$$

Evidently, $f_\pm \circ \varphi^t = f_\pm$.

By proposition 7.4.2, and Hopf's ergodic theorem,

$$f_+ = f_- = E_p\left(\frac{f}{p}\Big|\mathfrak{I}\right) \text{ a.e.}$$

From the nature of f, it follows that

$$\pi_{\pm}(\omega) = \pi_{\pm}(\omega') \;\Rightarrow\; f_{\pm}(\omega) = f_{\pm}(\omega'),$$

and by Hopf's lemma, $E_p\left(\frac{f}{p}\big|\mathfrak{I}\right)$ is constant. $\qquad\qquad\qquad\qquad\qquad\square$

§7.5 The Poincaré series

Let $\Gamma \subset \mathrm{M\ddot{o}b}\,(H)$ be a discrete subgroup.

DEFINITION: ABELIAN POINCARÉ SERIES.
The *Abelian Poincaré series* of Γ with base points $x, y \in H$ is the function

$$s \mapsto \mathfrak{p}_{\Gamma}(x, y; s) := \sum_{\gamma \in \Gamma} e^{-s\rho(x, \gamma y)} \le \infty \quad (s > 0).$$

It follows from the triangle inequality for ρ that
$e^{-\rho(x, \gamma y)} = e^{\pm c} e^{-\rho(x', \gamma y')} \; \forall \; x, x', y, y' \in H$ and $\gamma \in \Gamma$ where $c := \rho(x, x') + \rho(y, y')$.

Thus, the convergence or divergence of the Abelian Poincaré series does not depend on the base points for any given $s > 0$.

DEFINITION: POINCARÉ SERIES.
The number $\mathfrak{p}_{\Gamma}(x, y; 1) \le \infty$ is known as the *Poincaré series* of Γ with base points $x, y \in H$.

Recalling that $1 - |\gamma(0)| \sim 2e^{-\rho(0, \gamma(0))}$ as $\gamma \to \infty$, we see that the Poincaré series $\mathfrak{p}_{\Gamma}(0, 0; s)$ converges or diverges together with the series

$$\sum_{\gamma \in \Gamma} (1 - |\gamma(0)|)^s.$$

DEFINITION: ASYMPTOTIC POINCARÉ SERIES.
The *asymptotic Poincaré series* of Γ with base points $x, y \in H$ is the distribution function

$$t \mapsto a_{\Gamma}(x, y; t) := \sum_{\gamma \in \Gamma, \; \rho(x, \gamma y) \le t} e^{-\rho(x, \gamma y)}.$$

We'll also consider the associated measures $a_{\Gamma}(x, y; \cdot)$ defined by

$$a_{\Gamma}(x, y; J) := \sum_{\gamma \in \Gamma, \; \rho(x, \gamma y) \in J} e^{-\rho(x, \gamma y)} \quad \text{intervals } J \subset [0, \infty).$$

Note that $a_{\Gamma}(x, y; t) = a_{\Gamma}(x, y; [0, t])$ and that $\mathfrak{p}_{\Gamma}(x, y; s) = \int_0^{\infty} e^{-st} a_{\Gamma}(x, y; dt)$.
For $x \in H$ and $\epsilon > 0$ define

$$\Delta(x, \epsilon) := \{\omega \in X_{\Gamma} : \; \overline{\pi}_{\Gamma}^{-1}\omega \in \Gamma N_{\rho}(x, \epsilon) \times \mathbb{T}\}.$$

If there is a Dirichlet fundamental domain F for Γ in H with $N_{\rho}(x, \epsilon) \subset F^o$ (equivalently if $\epsilon \le \epsilon_{\Gamma}(x) := \frac{1}{2} \min_{\gamma \in \Gamma \setminus \{\mathrm{Id}\}} \rho(x, \gamma(x)))$, then

$$m_{\Gamma}(\Delta(x, \epsilon)) = m(\overline{\pi}_{\Gamma}^{-1}\Delta(x, \epsilon) \cap (F \times \mathbb{T})) = m(N_{\rho}(x, \epsilon) \times \mathbb{T}) = A(\epsilon),$$

whence $\tilde{1}_{\Delta(x, \epsilon)} = 1_{N_{\rho}(x, \epsilon) \times \mathbb{T}}$.

NOTATION.

For functions $a, b : \mathbb{R}_+$ or $\mathbb{N} :\to \mathbb{R}_+$,

$a(t) \gtrsim b(t)$ means $\liminf_{t\to\infty} \frac{a(t)}{b(t)} \geq 1$, and

$a(t) \lesssim b(t)$ means $\limsup_{t\to\infty} \frac{a(t)}{b(t)} \leq 1$.

Clearly $a(t) \gtrsim b(t)$ and $a(t) \lesssim b(t)$ if and only if $a(t) \sim b(t)$.

7.5.1 PROPOSITION.   *Fix $x, y \in H$ and $\epsilon > 0$, then*

(i) $$\int_u^v m_\Gamma(\Delta(x,\epsilon) \cap \varphi_\Gamma^{-s}\Delta(y,\epsilon))ds \gtrsim \frac{e^{-2\epsilon}A(\epsilon)^2}{2\pi} a_\Gamma(x,y;(u+\epsilon,v-\epsilon))$$

as $u, v \to \infty$, $v - u > 2\epsilon$; and

(ii) $$\int_u^v m_\Gamma(\Delta(x,\epsilon) \cap \varphi_\Gamma^{-s}\Delta(y,\epsilon))ds \lesssim e^{2\epsilon}A(\epsilon)^2 \frac{a_\Gamma(x,y;(u-\epsilon,v-\epsilon)))}{2\pi}$$

as $u, v \to \infty$, $v - u > 2\epsilon$.

PROOF.

$$\int_u^v m_\Gamma(\Delta(x,\epsilon) \cap \varphi_\Gamma^{-s}\Delta(y,\epsilon))ds = \int_{\Delta(x,\epsilon)} S_{(u,v)}^{\varphi_\Gamma}(1_{\Delta(y,\epsilon)})dm_\Gamma$$

$$= \int_{N_\rho(x,\epsilon)\times\mathbb{T}} S_{(u,v)}^{\varphi_\Gamma}(1_{\Delta(y,\epsilon)}) \circ \overline{\pi}_\Gamma dm$$

$$\overset{7.4.2}{=} \int_{N_\rho(x,\epsilon)\times\mathbb{T}} \sum_{\gamma\in\Gamma} S_{(u,v)}^{\varphi}(1_{N_\rho(y,\epsilon)\times\mathbb{T}} \circ \gamma)dm$$

$$= \int_{N_\rho(x,\epsilon)\times\mathbb{T}} \sum_{\gamma\in\Gamma} S_{(u,v)}^{\varphi}(1_{\gamma N_\rho(y,\epsilon)\times\mathbb{T}})dm$$

$$= \int_{N_\rho(x,\epsilon)} \Phi(u,v;z)dA(z)$$

where

$$\Phi(u,v;z) := \sum_{\gamma\in\Gamma} \int_\mathbb{T} S_{(u,v)}^{\varphi}(1_{\gamma N_\rho(y,\epsilon)\times\mathbb{T}})(z,\theta)d\theta.$$

Set $\varphi_z(\omega) = \frac{z+\omega}{1+\bar{z}\omega}$. Using $\varphi_z\varphi^t = \varphi^t\varphi_z$, and $\varphi^t(0,\theta) = (\tanh\frac{t}{2}e^{2\pi i\theta}, \theta)$, we have

$$\Phi(u,v;z) = \sum_{\gamma\in\Gamma} \int_\mathbb{T} S_{(u,v)}^{\varphi}(1_{\gamma N_\rho(y,\epsilon)\times\mathbb{T}})(z,\theta)d\theta$$

$$= \sum_{\gamma\in\Gamma} \int_\mathbb{T} S_{(u,v)}^{\varphi}(1_{\varphi_z^{-1}\gamma N_\rho(y,\epsilon)\times\mathbb{T}})(0,\theta)d\theta$$

$$= \sum_{\gamma\in\Gamma} \int_\mathbb{T} \int_u^v 1_{\varphi_z^{-1}\gamma N_\rho(y,\epsilon)}(\tanh(\tfrac{s}{2})e^{2\pi i\theta})dsd\theta$$

$$= \sum_{\gamma\in\Gamma} \int_\mathbb{T} \int_0^{\tanh(\frac{t}{2})} 1_{\varphi_z^{-1}\gamma N_\rho(y,\epsilon)}(re^{2\pi i\theta})\frac{drd\theta}{1-r^2}$$

$$= \frac{1}{8\pi} \sum_{\gamma\in\Gamma} \int_{N_\rho(0,v)\backslash N_\rho(0,u)} \frac{1-|\omega|^2}{|\omega|} 1_{N_\rho(\varphi_z^{-1}\gamma(y),\epsilon)}(\omega)dA(\omega).$$

The last is because $\frac{dr\,d\theta}{1-r^2} = \frac{(1-|\omega|^2)dA(\omega)}{8\pi|\omega|}$ where $\omega := (r\cos 2\pi\theta, r\sin 2\pi\theta)$.

Now,

$$\omega \in N_\rho(\varphi_z^{-1}\gamma(y), \epsilon) \Rightarrow \rho(0, \omega) = \rho(\gamma(y), z) \pm \epsilon,$$

so

$$\Phi(u, v; z)$$

$$\geq \frac{1}{8\pi}\sum_{\gamma\in\Gamma}(1 - \tanh^2(\frac{\rho(\gamma(y), z) + \epsilon}{2}))A(N_\rho(0, v) \setminus N_\rho(0, u) \cap N_\rho(\varphi_z^{-1}\gamma(y), \epsilon))$$

$$\geq \frac{1}{8\pi}\sum_{\gamma\in\Gamma,\ u+\epsilon<\rho(\gamma(y), z)\leq v-\epsilon}(1 - \tanh^2(\frac{\rho(\gamma(y), z) + \epsilon}{2}))A(N_\rho(y, \epsilon)).$$

Since

$$1 - \tanh^2(\frac{\rho(\gamma(y), z) + \epsilon}{2}) \sim 4e^{-(\rho(\gamma(y), z)+\epsilon)} \geq 4e^{-2\epsilon}e^{-\rho(\gamma(y), x)} \text{ as } \gamma \to \infty$$

uniformly on $N_\rho(x, \epsilon)$, we have that

$$\Phi(u, v; z) \gtrsim e^{-2\epsilon}\frac{1}{2\pi}a_\Gamma(x, y; (u + \epsilon, v - \epsilon))A(\epsilon)$$

uniformly in $z \in N_\rho(x, \epsilon)$, whence

$$\int_u^v m_\Gamma(\Delta(x, \epsilon) \cap \varphi_\Gamma^{-s}\Delta(y, \epsilon))ds \gtrsim e^{-2\epsilon}\frac{1}{2\pi}A(\epsilon)^2 a_\Gamma(x, y; (u + \epsilon, v - \epsilon)) \to \infty$$

establishing (i).

On the other hand,

$$\Phi(u, v; z) \leq$$

$$\frac{1}{8\pi}\sum_{\gamma\in\Gamma}(1 - \tanh^2(\frac{\rho(\gamma(y), z) - \epsilon}{2}))\int_{N_\rho(0, v)\setminus N_\rho(0, u)}\frac{1}{|\omega|}1_{N_\rho(\varphi_z^{-1}\gamma(y), \epsilon)}(\omega)dA(\omega)$$

$$\leq \frac{1}{8\pi}\sum_{\gamma\in\Gamma,\ u-\epsilon\leq\rho(\gamma(y), z)\leq v+\epsilon}(1 - \tanh^2(\frac{\rho(\gamma(y), z) - \epsilon}{2}))\int_{N_\rho(\varphi_z^{-1}\gamma(y), \epsilon)}\frac{1}{|\omega|}dA(\omega)$$

$$\sim e^{2\epsilon}\frac{1}{2\pi}A(\epsilon)a_\Gamma(x, y; (u - \epsilon, v + \epsilon))$$

as $u, v \to \infty$, $u < v$ since

$$1 - \tanh^2(\frac{\rho(\gamma(y), z) - \epsilon}{2}) \sim 4e^{-(\rho(\gamma(y), z)-\epsilon)} \leq 4e^{2\epsilon}e^{-\rho(\gamma(y), x)} \text{ as } \gamma \to \infty$$

uniformly on $N_\rho(x, \epsilon)$ whence (ii). $\square$

7.5.2 COROLLARY. *For any Fuchsian group Γ, $x, y \in H$,*

(1) $$\mathfrak{p}_\Gamma(x, y; s) = O\left(\frac{1}{s - 1}\right) \text{ as } s \downarrow 1,$$

and

(2) $$a_\Gamma(x, y; t) \sim a_\Gamma(x, y; t + h) \text{ as } t \to \infty \ \forall \ h > 0.$$

for any Fuchsian group Γ.

PROOF.

$$\mathfrak{p}_\Gamma(x, y; 1 + u) = \sum_{\gamma \in \Gamma} e^{-(1+u)\rho(x,\gamma y)}$$

$$= \sum_{n=0}^{\infty} \sum_{\gamma \in \Gamma, \; n < \rho(x,\gamma y) \leq (n+1)} \sum_{\gamma \in \Gamma} e^{-(1+u)\rho(x,\gamma y)}$$

$$\leq \sum_{n=0}^{\infty} e^{-nu} a_\Gamma(x, y; (n, n+1])$$

$$\leq M \sum_{n=0}^{\infty} e^{-nu} \leq \frac{M'}{u}$$

establishing (1).

It follows from proposition 7.5.1 that $a_\Gamma(x, y; t + h) - a_\Gamma(x, y; t) = O(1)$ as $t \to \infty$ whence (2) in case $a_\Gamma(x, y; t) \to \infty$.

If $a_\Gamma(x, y; t) \uparrow M < \infty$, then $a_\Gamma(x, y; (u, v)) \to 0$ as $u, v \to \infty$, $v > u$ whence (2). $\qquad\square$

We prove the Hopf-Tsuji theorem, that the geodesic flow φ_Γ is conservative and ergodic if and only if Γ has divergent Poincaré series.

The proof (from [**A-Su**]) uses proposition 1.1.8. This involves establishing the Renyi inequality (lemma 7.5.4) which will also establish rational ergodicity.

7.5.3 LEMMA.
Let Γ be a Fuchsian group, and let $x, y \in H$, then for $\epsilon > 0$ sufficiently small,

$$\int_0^\infty m_\Gamma((\Delta(x, \epsilon) \cap \varphi_\Gamma^{-s} \Delta(y, \epsilon)) ds = \infty$$

if and only if the Poincaré series diverges.
In this case, for $0 < \epsilon < \epsilon_\Gamma(y)$ and t large,

$$\frac{1}{a_\Gamma(x, y; t)} \int_0^t m_\Gamma(\Delta(x, \epsilon) \cap \varphi_\Gamma^{-s} \Delta(y, \epsilon)) ds = e^{\pm 5\epsilon} \frac{A(\epsilon)^2}{2\pi}.$$

PROOF.
By proposition 7.5.1 and corollary 7.5.2 (2), $\exists N_\epsilon > 0$ such that $\forall t > N_\epsilon + 2\epsilon$,

$$\int_u^v m_\Gamma(\Delta(x, \epsilon) \cap \varphi_\Gamma^{-s} \Delta(y, \epsilon)) ds = \frac{e^{\pm 3\epsilon} A(\epsilon)^2}{2\pi} a_\Gamma(x, y; (N_\epsilon, t]).$$

The lemma follows from this. $\qquad\square$

7.5.4 LEMMA   [**A-Su**].

$\forall \; x, y \in H$, $\epsilon > 0$ *sufficiently small* , $\exists \; M \in \mathbb{R}_+$ *such that*

$$\int_{\Delta(y,\epsilon)} S_t^{\varphi_\Gamma}(1_{\Delta(x,\epsilon)})^2 dm_\Gamma \leq M \left(\int_{\Delta(y,\epsilon)} S_t^{\varphi_\Gamma}(1_{\Delta(x,\epsilon)}) dm_\Gamma \right)^2 \quad \forall \; t > 0.$$

PROOF.

Let $\epsilon > 0$ be so small that there is a Dirichlet fundamental domain F for Γ in H with $N_\rho(x, \epsilon) := \{y \in H : \rho(x,y) < \epsilon\} \subset F^o$.

$$\int_{\Delta(y,\epsilon)} S_t^{\varphi_\Gamma}(1_{\Delta(x,\epsilon)})^2 dm_\Gamma$$

$$= 2 \int_{\Delta(y,\epsilon)} \int_0^t \int_u^t 1_{\Delta(x,\epsilon)} \circ \varphi_\Gamma^u \ 1_{\Delta(x,\epsilon)} \circ \varphi_\Gamma^v du dv dm_\Gamma$$

$$= 2 \int_{N_\rho(x,\epsilon) \times \mathbb{T}} \int_0^t \int_u^t 1_{\Delta(x,\epsilon)} \circ \varphi_\Gamma^u \circ \overline{\pi}_\Gamma \ 1_{\Delta(x,\epsilon)} \circ \varphi_\Gamma^v du dv dm$$

$$= 2 \sum_{\beta,\gamma \in \Gamma} \int_{N_\rho(x,\epsilon) \times \mathbb{T}} \int_0^t \int_u^t 1_{\beta N_\rho(x,\epsilon) \times \mathbb{T}} \circ \varphi^u \ 1_{\gamma N_\rho(x,\epsilon) \times \mathbb{T}} \circ \varphi^v du dv dm$$

$$= 2 \int_{N_\rho(y,\epsilon)} \psi(t,z) dA(z)$$

where

$$\psi(t,z) := \sum_{\beta,\gamma \in \Gamma} \int_0^{2\pi} \int_0^t \int_u^t 1_{\beta N_\rho(x,\epsilon) \times \mathbb{T}} \circ \varphi^u(z,\theta) 1_{\gamma N_\rho(x,\epsilon) \times \mathbb{T}} \circ \varphi^v(z,\theta) du dv d\theta.$$

As in the proof of proposition 7.5.1, we set $\varphi_z(\omega) = \frac{z+\omega}{1+\bar{z}\omega}$ and use $\varphi_z \varphi^t = \varphi^t \varphi_z$ and $\varphi^t(0,\theta) = (\tanh \frac{t}{2} e^{2\pi i\theta}, \theta)$.

We have

$$\psi(t,z) = \sum_{\beta,\gamma \in \Gamma} \int_0^1 \int_0^t \int_u^t 1_{\beta N_\rho \times \mathbb{T}} \circ \varphi^u(z,\theta) 1_{\gamma N_\rho \times \mathbb{T}} \circ \varphi^v(z,\theta) du dv d\theta$$

$$= \sum_{\beta,\gamma \in \Gamma} \int_0^1 \int_0^t \int_u^t 1_{\varphi_z^{-1} \beta N_\rho \times \mathbb{T}}(\tanh(\frac{u}{2}) e^{2\pi i\theta}, \theta) 1_{\varphi_z^{-1} \gamma N_\rho \times \mathbb{T}}(\tanh(\frac{v}{2}) e^{2\pi i\theta}, \theta) du dv d\theta$$

$$= \int_0^t \sum_{\beta \in \Gamma_u} \sum_{\gamma \in \Gamma(\beta)} \int_0^1 1_{\varphi_z^{-1} \beta N_\rho}(\tanh(\frac{u}{2}) e^{2\pi i\theta}) \int_u^t 1_{\varphi_z^{-1} \gamma N_\rho}(\tanh(\frac{v}{2}) e^{2\pi i\theta}) dv d\theta du,$$

where

$$\Gamma_u = \Gamma_u(z) := \{\beta \in \Gamma : \rho(\beta(x), z) = u \pm \epsilon\},$$

and

$$G(\beta) =$$

$$\left\{ \gamma \in \Gamma : \int_0^t \int_0^1 \int_u^t 1_{\varphi_z^{-1} \beta N_\rho}(\tanh(\frac{u}{2}) e^{2\pi i\theta}) 1_{\varphi_z^{-1} \gamma N_\rho}(\tanh(\frac{u}{2}) e^{2\pi i\theta}) dv d\theta du > 0 \right\}.$$

We have

$$(3) \qquad \psi(t,z) \leq \int_0^t \sum_{\beta \in \Gamma_u} \sum_{\gamma \in \Gamma(\beta)} \int_0^1 \int_u^t 1_{\varphi_z^{-1} \gamma N_\rho}(\tanh(\frac{v}{2}) e^{2\pi i\theta}) dv d\theta du.$$

Fix u, $\beta \in \Gamma_u$, $\gamma \in \Gamma(\beta)$, then

$$
\int_0^1 \int_u^t 1_{\varphi_z^{-1}\gamma N_\rho(x,\epsilon)}\left(\tanh(\tfrac{v}{2})e^{2\pi i\theta}\right)dv\,d\theta
$$
$$
= \int_0^1 \int_{\tanh(\frac{u}{2})}^{\tanh(\frac{t}{2})} 1_{\varphi_z^{-1}\gamma N_\rho}\left(re^{2\pi i\theta}\right)\frac{dr\,d\theta}{1-r^2}
$$
$$
= \int_{N_\rho(0,t)\setminus N_\rho(0,u)\cap \varphi_z^{-1}\gamma N_\rho(x,\epsilon)} \frac{1-|\omega|^2}{|\omega|}dA(\omega)
$$
$$
\lesssim 2e^\epsilon e^{-\rho(z,\gamma(x))}A(N_\rho(\varphi_z^{-1}\gamma(x),\epsilon)\cap(N_\rho(0,t)\setminus N_\rho(0,u))).
$$

Hence

$$
\sum_{\gamma\in\Gamma(\beta)}\int_0^1\int_u^t 1_{\varphi_z^{-1}\gamma N_\rho}\left(\tanh(\tfrac{v}{2})e^{2\pi i\theta}\right)dv\,d\theta \lesssim
$$
$$
2e^\epsilon \sum_{\gamma\in\Gamma(\beta),\ \rho(z,\gamma(x))\leq t+\epsilon} e^{-\rho(z,\gamma(x))}A(N_\rho(\varphi_z^{-1}\gamma(x),\epsilon)\cap(N_\rho(0,t)\setminus N_\rho(0,u)))
$$

(4)
$$
\leq 2e^\epsilon \sum_{\gamma\in\Gamma(\beta),\ \rho(z,\gamma(x))\leq t+\epsilon} e^{-\rho(z,\gamma(x))}A(N_\rho(0,\epsilon)).
$$

To proceed, we need to analyse $\Gamma(\beta)$. To do this, we need some elementary facts from analytic hyperbolic geometry.

Consider the angle set subtended by $N_\rho(y,\epsilon)$ at $0 \notin N_\rho(y,\epsilon)$,
$$\Lambda(y,\epsilon) := \{\theta \in [0,2\pi] : \exists\, r \in (0,1)\ \text{such that}\ \rho(y,re^{2\pi i\theta}) < \epsilon\}.$$

Since $N_\rho(y,\epsilon)$ is the Euclidean ball with centre $\frac{(1-\delta^2)y}{1-\delta^2|y|^2}$ and radius $\frac{\delta(1-|y|^2)}{1-\delta^2|y|^2}$)
where $\delta = \tanh(\tfrac{\epsilon}{2})$, we have that

$$
\Lambda(y,\epsilon) = \left\{\theta \in [0,2\pi] : \|\theta - \arg y\| < \sin^{-1}\left(\frac{(1-|y|^2)\tanh(\frac{\epsilon}{2})}{|y|(1-\tanh^2(\frac{\epsilon}{2}))}\right)\right\}
$$

where $\|\theta\| := \theta \wedge (2\pi - \theta)$ $\theta \in [0,2\pi)$. Noting that

$$
\sin^{-1}\left(\frac{(1-|y|^2)\tanh(\frac{\epsilon}{2})}{|y|(1-\tanh^2(\frac{\epsilon}{2}))}\right) \leq \frac{2e^{2\epsilon}}{\pi}(1-|y|^2),
$$

we see that the length of $\Lambda(y,\epsilon) \subset \mathbb{T}$ satisfies

(i)
$$
|\Lambda(y,\epsilon)| \leq \frac{4e^{2\epsilon}}{\pi}(1-|y|^2).
$$

Let Λ_γ denote the angle set subtended by $\varphi_z^{-1}\gamma N_\rho$ at 0, then

$$
\Gamma(\beta) = \{\gamma \in \Gamma : \rho(z,\gamma(x)) \geq \rho(z,\beta(x)) - \epsilon,\ \Lambda_\beta \cap \Lambda_\gamma \neq \emptyset\}.
$$

Suppose $\gamma \in \Gamma(\beta)$, then by (i),

$$\| \arg \varphi_z^{-1}\beta(x) - \arg \varphi_z^{-1}\gamma(x) \| \leq \frac{|\Lambda_\beta| + |\Lambda_\gamma|}{2}$$

$$\leq \frac{e^{2\epsilon}}{\pi}(2 - |\varphi_z^{-1}\beta(x)|^2 - |\varphi_z^{-1}\gamma(x)|^2)$$

$$\leq \frac{2e^{2\epsilon}}{\pi}(1 - |\varphi_z^{-1}\beta(x)|^2).$$

$\therefore$ $\Gamma(\beta) \subset \Gamma_1(\beta) := \{\gamma \in \Gamma : \| \arg \varphi_z^{-1}\beta(x) - \arg \varphi_z^{-1}\gamma(x) \| \leq \frac{2e^{2\epsilon}}{\pi}(1 - |\varphi_z^{-1}\beta(x)|^2)\}$,
and, by (3) and (4),

$$(5) \qquad \psi(t,z) \leq 4e^{2\epsilon} \int_0^t \sum_{\beta \in \Gamma_u} \sum_{\gamma \in \Gamma_1(\beta),\ \rho(x,\gamma(x)) \leq t+\epsilon} e^{-\rho(x,\gamma(x))}.$$

Set $\Gamma_2(\beta) = \beta^{-1}\Gamma_1(\beta)$.

We claim that

$$(6) \qquad \rho(z, \beta\gamma(x)) \geq \rho(z, \beta(x)) + \rho(x, \gamma(x)) - \frac{8e^{2\epsilon}}{\pi} \ \forall \ \gamma \in \Gamma_2(\beta).$$

To see this, we show first that for any $x, y \in H$, $|y| > |x|$

$$(7) \qquad \rho(0, y) \geq \rho(0, x) + \rho(x, y) - \frac{4\| \arg y - \arg x\|}{1 - |x|^2}.$$

To establish (7), let $x' \in H$, $|x'| = |x|$, and $\arg x' = \arg y$, then

$$\rho(0, y) = \rho(0, x') + \rho(x', y) \geq \rho(0, x) + \rho(x, y) - 2\rho(x', x)$$

and evidently $\rho(x', x) \leq \frac{2\| \arg y - \arg x\|}{1 - |x|^2}$ demonstrating (7).
    Returning to the proof of (6), we have by the definition of $\Gamma_1(\beta)$ that

$$\| \arg \varphi_z^{-1}\beta(x) - \arg \varphi_z^{-1}\beta\gamma(x) \| \leq \frac{4e^{2\epsilon}}{\pi}(1 - |\varphi_z^{-1}\beta(x)|^2),$$

and by (7),

$$\rho(z, \beta\gamma(x)) = \rho(0, \varphi_z^{-1}\beta\gamma(x))$$

$$\geq \rho(0, \varphi_z^{-1}\beta(x)) + \rho(\varphi_z^{-1}\beta(x), \varphi_z^{-1}\beta\gamma(x)) - \frac{2\| \arg \varphi_z^{-1}\beta(x) - \arg \varphi_z^{-1}\beta\gamma(x) \|}{1 - |\varphi_z^{-1}\beta(x)|^2}$$

$$\geq \rho(z, \beta(x)) + \rho(x, \gamma(x)) - \frac{8e^{2\epsilon}}{\pi}.$$

It follows from (6) that

$$\gamma \in \Gamma_2(\beta), \ \rho(z, \beta\gamma(x)) \leq t + \epsilon$$

$$\Rightarrow \ \rho(x, \gamma(x)) \leq \rho(z, \beta\gamma(x)) - \rho(y, \beta(x)) + \frac{8e^{2\epsilon}}{\pi} + \epsilon \leq t + \frac{8e^{2\epsilon}}{\pi} + \epsilon,$$

whence

$$\sum_{\gamma\in\Gamma_1(\beta),\ \rho(z,\gamma(x))\leq t+\epsilon} e^{-\rho(z,\gamma(x))} \leq e^{\frac{8e^{2\epsilon}}{\pi}} e^{-\rho(z,\beta(x))} \sum_{\gamma\in\Gamma_2(\beta),\ \rho(z,\beta\gamma(x))\leq t+\epsilon} e^{-\rho(z,\gamma(x))}$$

$$\leq e^{\frac{8e^{2\epsilon}}{\pi}} e^{-\rho(z,\beta(x))} a_\Gamma(x,x;t+\frac{8e^{2\epsilon}}{\pi}+\epsilon),$$

$$\therefore\ \psi(t,z) \leq e^{\frac{8e^{2\epsilon}}{\pi}} a_\Gamma(x,x;t+\frac{8e^{2\epsilon}}{\pi}+\epsilon) \int_0^t \sum_{\beta\in\Gamma_u} e^{-\rho(y,\beta(x))} du$$

$$\leq a_\Gamma(x,x;t+\frac{8e^{2\epsilon}}{\pi}+\epsilon) \sum_{\beta\in\Gamma} e^{-\rho(y,\beta(x))} \int_{0\vee(\rho(z,\beta(x))-\epsilon)}^{t\wedge(\rho(z,\beta(x))+\epsilon)} du$$

$$\leq a_\Gamma(x,x;t+\frac{8e^{2\epsilon}}{\pi}+\epsilon)2\epsilon \sum_{\beta\in\Gamma,\ \rho(z,\beta(x))<t+\epsilon} e^{-\rho(y,\beta(x))}$$

$$\leq 2\epsilon a_\Gamma(x,x;t+\frac{8e^{2\epsilon}}{\pi}+\epsilon)a_\Gamma(y,x,t+2\epsilon)$$

$$= O\Big(a_\Gamma(y,x,t)^2\Big)$$

as $t\to\infty$.   □

7.5.5 HOPF-TSUJI THEOREM   (C.F. [**Su**] [**Nic**]). *Let* $\Gamma \subset \text{Möb}(H)$ *be a Fuchsian group, then the geodesic flow on* X_Γ *is conservative if and only if the Poincaré series diverges.*

PROOF.
If the geodesic flow is conservative, then $\forall\ x,y \in H\ \epsilon > 0$,

$$\int_0^\infty m_\Gamma((\Delta(x,\epsilon)\cap\varphi_\Gamma^{-s}\Delta(y,\epsilon)))ds = \infty$$

and by lemma 7.5.2, the Poincaré series diverges.

Conversely, if the Poincaré series diverges, then by lemmas 7.5.3 and 7.5.4, there the sets $\Delta := \Delta(x,\epsilon)$ for $x \in H$ and $\epsilon > 0$ sufficiently small satisfy

$$\int_0^\infty m_\Gamma((\Delta\cap\varphi_\Gamma^{-s}\Delta))ds = \infty,\ \&\ \int_\Delta S_t^{\varphi_\Gamma}(1_\Delta)^2 dm_\Gamma = O\left(\left(\int_\Delta S_t^{\varphi_\Gamma}(1_\Delta)dm_\Gamma\right)^2\right)$$

as $t \to \infty$ and by proposition 1.1.8, φ_Γ is not totally dissipative. It follows from theorem 7.4.3 that is conservative and ergodic.   □

7.5.6 THEOREM [**A-Su**]. *Let* $\Gamma \subset \text{Möb}(H)$ *be a Fuchsian group with divergent Poincaré series, then the geodesic flow on* X_Γ *is rationally ergodic, with return sequence given by*

$$a_n(\varphi_\Gamma^1) \sim \frac{a_\Gamma(x,y;n)}{2\pi}\ \forall\ x,y \in H.$$

PROOF.

Rational ergodicity follows from ergodicity and the Renyi inequality. Let the return sequence be $a(n) := a_n(\varphi_\Gamma^1)$, then by lemma 7.5.3, $\forall\ x, y \in H,\ \epsilon > 0$ small enough

$$m_\Gamma((\Delta(x, \epsilon))^2 a(t)\ sim \int_0^t m_\Gamma(\Delta(x, \epsilon) \cap \varphi_\Gamma^{-s}\Delta(y, \epsilon))ds$$

$$= e^{\pm 5\epsilon}\left(\frac{m_\Gamma((\Delta(x, \epsilon))^2}{2\pi}\right)a_\Gamma(x, y; t)$$

for $t > 0$ large enough. It follows (as $\epsilon \to 0$) that

$$a(t) \sim \frac{a_\Gamma(x, y; t)}{2\pi}\ \forall\ x, y \in H.$$

$\square$

In view of theorem 7.5.6, for each Fuchsian group with divergent Poincaré series, $\exists\ a_\Gamma(t)$ such that

$$a_\Gamma(x, y; t) \sim a_\Gamma(t)\ \text{as}\ t \to \infty\ \forall\ x, y \in H.$$

7.5.7 EQUIDISTRIBUTION COROLLARY.

Let $\Gamma \subset \text{Möb}\,(H)$ be a Fuchsian group with divergent Poincaré series, then $\forall\ x, y \in H$, and intervals $I \subset \mathbb{T}$

$$\sum_{\gamma \in \Gamma,\ \rho(x, \gamma y) \leq t,\ \arg \gamma y \in I} e^{-\rho(x, \gamma y)} \sim \pi_x(I)a_\Gamma(t)$$

where $d\pi_x(t) := \Re\left(\frac{e^{2\pi i t} + x}{e^{2\pi i t} - x}\right)dt.$

PROOF.

Suppose first that $x = 0$. It follows from rational ergodicity that $\forall\ y \in H$, and $\epsilon > 0$ small enough,

$$\frac{1}{a(t)}\int_0^t 1_{\Delta(y, \epsilon)} \circ \varphi_\Gamma^s ds \longrightarrow m(\Delta(y, \epsilon))\ \text{weakly in}\ L^2(\Delta(0, \epsilon)).$$

In particular

$$\int_{N_\rho(0, \epsilon) \times I}\int_0^t 1_{\Delta(y, \epsilon)} \circ \varphi_\Gamma^s ds dm \sim A(\epsilon)^2|I|a(t)\ \text{as}\ t \to \infty.$$

On the other hand, as in the proof of proposition 7.5.1, setting $\varphi_z(\omega) = \frac{z + \omega}{1 + \bar{z}\omega}$:

$$\int_{N_\rho(0, \epsilon) \times I} S_t^{\varphi_\Gamma}(1_{\Delta(y, \epsilon)})dm_\Gamma = \int_{N_\rho(0, \epsilon)} \Phi(t, z)dA(z)$$

where

$$\Phi(t,z) := \sum_{\gamma \in \Gamma} \int_I S_t^\varphi (1_{\gamma N_\rho(y,\epsilon) \times \mathbb{T}})(z,\theta) d\theta$$

$$= \sum_{\gamma \in \Gamma} \int_I S_t^\varphi (1_{\varphi_z^{-1} \gamma N_\rho(y,\epsilon) \times \mathbb{T}})(0,\theta) d\theta$$

$$= \sum_{\gamma \in \Gamma} \int_I \int_0^t 1_{\varphi_z^{-1} \gamma N_\rho(y,\epsilon)} (\tanh(\frac{s}{2}) e^{2\pi i \theta}) ds \, d\theta$$

$$= \sum_{\gamma \in \Gamma} \int_I \int_0^{\tanh(\frac{t}{2})} 1_{\varphi_z^{-1} \gamma N_\rho(y,\epsilon)} (r e^{2\pi i \theta}) \frac{dr \, d\theta}{1-r^2}$$

$$\sim 2 \sum_{\gamma \in \Gamma} \int_{N_\rho(0,t) \cap H_I} e^{-\rho(0,\omega)} 1_{N_\rho(\varphi_z^{-1} \gamma(y),\epsilon)}(\omega) dA(\omega)$$

where $H_I := \{\omega \in H : \arg \omega \in I\}$.

Setting $I = (a,b)$, it follows that

$$\Phi(t,z) \lesssim 2 e^{2\epsilon} A(\epsilon) \sum_{\gamma \in \Gamma, \ \rho(0,\gamma y) \leq t, \ a - 2\epsilon < \arg \gamma(y) < b + 2\epsilon} e^{-\rho(0,\gamma y)},$$

and

$$\Phi(t,z) \gtrsim 2 e^{-2\epsilon} A(\epsilon) \sum_{\gamma \in \Gamma, \ \rho(0,\gamma y) \leq t, \ a + 2\epsilon < \arg \gamma(y) < b - 2\epsilon} e^{-\rho(0,\gamma y)}.$$

Thus, putting things together

$$\sum_{\gamma \in \Gamma, \ \rho(0,\gamma y) \leq t, \ \arg \gamma(y) \in I} e^{-\rho(0,\gamma y)} \sim e^{\pm 2\epsilon} (|I| \pm 2\epsilon) a_\Gamma(t) \text{ as } t \to \infty$$

whence the result when $x = 0$.

For general $x, y \in H$ let $\varphi_x(z) := \frac{z+x}{1+\bar{x}z}$, let $\Gamma' := \varphi_x^{-1} \circ \Gamma \circ \varphi_x$ and set $y' = \varphi_x(y)$, then

$$\sum_{\gamma \in \Gamma, \ \rho(x,\gamma y) \leq t, \ \arg \gamma(y) \in I} e^{-\rho(x,\gamma y)} = \sum_{\gamma \in \Gamma', \ \rho(0,\gamma y') \leq t, \ \arg \gamma(y') \in \varphi_x^{-1} I} e^{-\rho(0,\gamma y')}$$

$$\sim |\varphi_x^{-1} I| a_{\Gamma'}(t)$$

$$\sim \pi_x(I) a_\Gamma(t)$$

because

$$a_{\Gamma'}(t) \sim a_{\Gamma'}(0,y';t) = a_\Gamma(x,y;t) \sim a_\Gamma(t).$$

$\square$

7.5.8 THEOREM [Hed].

Let $\Gamma \subset \text{Möb}(H)$ be a Fuchsian group with $A_\Gamma(H/\Gamma) < \infty$, then

(i) φ_G is mixing, and

(ii) $a_\Gamma(x,y;t+I) \to \frac{2\pi}{A_\Gamma(H/\Gamma)}$ as $t \to \infty \ \forall \ x,y \in H$.

PROOF OF (I).

This proof is taken from [**Tho**], theorem 1. By theorem 7.4.3, φ_Γ is ergodic, and we have an ergodic action of $SL(2,\mathbb{R})$, $T: SL(2,\mathbb{R}) \to \mathfrak{A}_0(X,\mathcal{B},m)$ as probability preserving transformations on the probability space $(X,\mathcal{B},m)$ for which

$$G_t := T_{\begin{pmatrix} e^{\frac{t}{2}} & 0 \\ 0 & e^{-\frac{t}{2}} \end{pmatrix}} = \varphi_\Gamma^t.$$

Write

$$H_s := T_{\begin{pmatrix} 1 & s \\ 0 & 1 \end{pmatrix}}, \quad \tilde{H}_s := T_{\begin{pmatrix} 1 & 0 \\ s & 1 \end{pmatrix}},$$

then $H_s \circ G_t = G_t \circ H_{se^{-2t}}$.

We claim first that H is ergodic. To prove this we show that H-invariant functions are G-invariant.

Let $f \in L^2(X,\mathcal{B},m)_0$ be such that $H_s f = f \ \forall \ s \in \mathbb{R}$ and fix $\tau > 0$. We'll show that $G_\tau f = f$ by showing that

$$|\langle G_\tau f, f \rangle - \|f\|_2^2| < \epsilon \ \forall \ \epsilon > 0.$$

Now fix $\epsilon > 0$ and choose $\delta > 0$ such that

$$|\langle \tilde{H}_a f, f \rangle - \|f\|_2^2| < \epsilon, \quad |\langle G_\tau \tilde{H}_b f, f \rangle - \langle G_\tau f, f \rangle| < \epsilon \quad \forall \ |a|, \ |b| < \delta,$$

and fix $0 < a < \delta e^{-\frac{\tau}{2}}$.

Since $H_s f = f \ \forall \ s \in \mathbb{R}$, we have

$$\langle \tilde{H}_a f, f \rangle = \langle H_s \tilde{H}_a H_t f, f \rangle \ \forall \ s, t \in \mathbb{R}.$$

Now

$$\begin{pmatrix} 1 & s \\ 0 & 1 \end{pmatrix} \begin{pmatrix} 1 & 0 \\ a & 1 \end{pmatrix} \begin{pmatrix} 1 & t \\ 0 & 1 \end{pmatrix} = \begin{pmatrix} 1+as & t+a(at+1) \\ a & at+1 \end{pmatrix},$$

so choosing $s, t \in \mathbb{R}$ such that $\frac{1}{1+at} = e^{\frac{\tau}{2}}$, and $s = \frac{-t}{at+1}$ we have

$$\begin{pmatrix} 1+as & t+a(at+1) \\ a & at+1 \end{pmatrix} = \begin{pmatrix} e^{\frac{\tau}{2}} & 0 \\ a & e^{-\frac{\tau}{2}} \end{pmatrix}.$$

Next, note that

$$\begin{pmatrix} e^{\frac{\tau}{2}} & 0 \\ b & e^{-\frac{\tau}{2}} \end{pmatrix} = \begin{pmatrix} e^{\frac{\tau}{2}} & 0 \\ 0 & e^{-\frac{\tau}{2}} \end{pmatrix} \begin{pmatrix} 1 & 0 \\ a & 1 \end{pmatrix}$$

where $b := e^{\frac{\tau}{2}} a < \delta$, whence

$$H_s \circ \tilde{H}_a \circ H_t = G_\tau \circ \tilde{H}_b$$

and

$$|\langle G_\tau \tilde{H}_b f, f \rangle - \|f\|_2^2| = |\langle H_s \tilde{H}_a H_t \rangle - \|f\|_2^2| = |\langle \tilde{H}_a f, f \rangle - \|f\|_2^2| < \epsilon.$$

It follows that

$$|\langle G_\tau f, f \rangle - \|f\|_2^2| \leq |\langle G_\tau \tilde{H}_b f, f \rangle - \langle G_\tau, f \rangle| + |\langle G_\tau \tilde{H}_b f, f \rangle - \|f\|_2^2|$$
$$< 2\epsilon.$$

This shows that f is G-invariant and hence constant.

To finish, we show that G is mixing. For $f, g \in L^2(X,\mathcal{B},m)_0$,

$$|\langle G_t(H_s f - f), g \rangle| = |\langle G_t f, G_t H_s G_{-t} g \rangle - \langle G_t f, g \rangle| = |\langle G_t f, H_{se^{2t}} g - g \rangle|$$
$$\leq \|f\|_2 \|H_{se^{2t}} g - g\|_2 \to 0$$

as $t \to -\infty$. Since H is ergodic,

$$\overline{\{H_s f - f : \ f \in L^2(X, \mathcal{B}, m)_0\}} = L^2(X, \mathcal{B}, m)_0$$

(see chapter 2), and G is mixing.

(ii) Follows from proposition 7.5.1. $\square$

Examples.

To see that there are indeed Fuchsian groups Γ with divergent Poincaré series and $A_\Gamma(H/\Gamma) = \infty$, we note the following results:

The Fuchsian group Γ has divergent Poincaré series if and only if its Riemann surface H/Γ has no Green's function ([**Myr**] see also [**Ts**]).

If $\Omega \subsetneq S^2$ is a domain and $|S^2 \setminus \Omega| \geq 3$, then Ω is hyperbolic and has a Green's function if and only if $\log$-$\operatorname{cap}(S^2 \setminus \Omega) > 0$ ([**Ahl**]); the hyperbolic area of Ω being finite if and only if $|S^2 \setminus \Omega| < \infty$ ([**Sie**], see also [**Ts**]).

Other examples of Fuchsian groups Γ with divergent Poincaré series and $A_\Gamma(H/\Gamma) = \infty$, include 1- and 2-dimensional Abelian covers of compact surfaces ([**Ree**]).

§7.6 Further results

Conservative geodesic flows are now known to have moment sets (see [**A-De4**]), whence those with regularly varying asymptotic Poincaré series exhibit distributional convergence phenomena.

There are cases where the asymptotic Poincaré series is known to be regularly varying. See [**Gui**], [**Pol-Sh**], [**A-De3**] and references therein.

CHAPTER 8

Cocycles and Skew Products

§8.1 Skew Products

DEFINITION: COCYCLES, COBOUNDARY, TRANSFER FUNCTION, SKEW PRODUCT. Let Γ be a countable semigroup, let $T : \Gamma \to \mathfrak{A}(X, \mathcal{B}, m)$ be an action of Γ as non-singular transformations of the standard probability space $(X, \mathcal{B}, m)$ and let G be a locally compact topological group.

A T-*cocycle* is a measurable function $\phi : X \times \Gamma \to G$ satisfying

$$\phi(x, ab) = \phi(T_b x, a)\phi(x, b).$$

The T-cocycle $\phi : X \times \Gamma \to G$ is said to be a T-*coboundary* if $\exists\, h : X \to G$ measurable (called the *transfer function*), such that $\phi(x, a) = h(T_a x)^{-1} h(x)$.

The *skew product* action T_ϕ is defined on $X \times G$ by $T_{\phi,a}(x, g) := (T_a x, \phi(x, a)g)$, the cocycle equation ensuring that $T_{\phi,ab} = T_{\phi,b} \circ T_{\phi,a}$.

It is known that for T an ergodic $\mathbb{Z}$ action, $\exists$ a cocycle $\phi : X \times \mathbb{Z} \to G$ with T_ϕ ergodic if and only if G is amenable (see [**Go-Sin**] for details).

In this chapter, we restrict attention to the situation in which
T is a probability preserving transformation of $(X, \mathcal{B}, m)$;
$\Gamma = \mathbb{N}$ (or $\mathbb{Z}$ if T is invertible), $T_n = T^n$;
    and
G is a locally compact, separable, Polish Abelian group with Haar measure m_G and equipped with the norm $\| \cdot \| = \| \cdot \|_G$ (see corollary 1.6.10).

Here, T-cocycles ϕ are generated by measurable functions $\varphi(\cdot) := \phi(\cdot, 1) : X \to G$ as follows:

$$\phi(x, n) = \varphi_n(x) := \begin{cases} \sum_{k=0}^{n-1} \varphi(T^k x) & n \geq 1, \\ 0 & n = 0, \\ -\sum_{k=1}^{|n|} \varphi(T^{-k} x) & T \text{ invertible and } n < -1, \end{cases}$$

and a coboundary is a function of form $\phi(x) = h(x) - h(Tx)$ where $h : X \to G$ is measurable.

DEFINITION: COHOMOLOGOUS.
Measurable functions $\alpha, \beta : X \to G$ are said to be *cohomologous* if $\exists\, h : X \to G$ measurable, such that $\alpha(x) = \beta(x) + h(x) - h(Tx)$ (i.e. $\alpha - \beta$ is a coboundary).

The skew product or G-*extension* $T_\phi : X \times G \to X \times G$ is now defined by $T_\phi(x, y) := (Tx, \phi(x) + y)$. It is a measure preserving transformation of the measure space $(X \times G, \mathcal{B} \otimes \mathcal{B}(G), m \times m_G)$. Note that $T_\phi^n(x, y) = (T^n x, y + \phi_n(x))$.

If $\varphi, \psi : X \to G$ measurable are cohomologous, then T_φ and T_ψ are isomorphic. Indeed suppose $h : X \to G$ is measurable and $\psi(x) = \varphi(x) + h(x) - h(Tx)$, then $\pi_h^{-1} \circ T_\varphi \circ \pi_h = T_\psi$ where $\pi_h(x,y) = (x, h(x) + y)$.

8.1.1 PROPOSITION (CONSERVATIVE, AND POSITIVE PARTS OF SKEW PRODUCTS).

Suppose that T is ergodic and let $\varphi : X \to G$ be measurable, then

$$\mathfrak{C}(T_\varphi), \ \mathfrak{P}(T_\varphi) \in \{\emptyset, X \times G\} \qquad \mathrm{mod} \ \ m \times m_G.$$

PROOF. Let $\Gamma \subset G$ be a countable dense subgroup of G. It follows from §1.6 that the action of Γ on G by translation is ergodic.

As in the proof of proposition 1.2.4 (ii), the $\mathbb{N} \times \Gamma$ action S on $(X \times G, \mathcal{B}(X \times G), m \times m_G)$ given by $S_{(n,a)}(x,y) := (T^n x, y + a)$ is ergodic.

Both $\mathfrak{C}(T_\varphi)$ and $\mathfrak{P}(T_\varphi)$ are T_φ- and Q_a-invariant $\forall \ a \in G$ where $Q_a(x,y) := (x, y + a)$, hence S-invariant.

Thus $\mathfrak{C}(T_\varphi), \ \mathfrak{P}(T_\varphi) \in \{\emptyset, X \times G\} \ \mathrm{mod} \ m$. $\qquad \square$

8.1.2 PROPOSITION. *T_ϕ is conservative iff*

$$\liminf_{n \to \infty} \|\phi_n(x)\| = 0 \ \text{for a.e.} \ x \in X.$$

PROOF.
Assume first that T_ϕ is conservative and let $\epsilon > 0$. By Halmos' recurrence theorem

$$\sum_{n=1}^{\infty} 1_{X \times B_G(0,\epsilon/2)} \circ T_\phi^n = \infty \ \text{a.e. on} \ X \times B_G(0,\epsilon/2).$$

So for a.e. $x \in X$, $y \in B_G(0,\epsilon/2)$,

$$\sum_{n=1}^{\infty} 1_{B_G(0,\epsilon/2)}(y + \phi_n(x)) = \infty,$$

whence for a.e. $x \in X$, $\liminf_{n \to \infty} \|\phi_n(x)\| \le \epsilon$.

Now assume that

$$\liminf_{n \to \infty} \|\phi_n(x)\| = 0 \ \text{for a.e.} \ x \in X.$$

Fix $f : G \to \mathbb{R}_+$ be continuous, positive and integrable and let $0 < \epsilon < \kappa_G$. For $y \in G$, let $\delta(y, \epsilon) := \inf_{B_G(y,\epsilon)} f$. By compactness of $B_G(y, \epsilon)$, $\delta(y, \epsilon) > 0$.

We have that $\forall \ y \in G$, for a.e. $(x,z) \in X \times B_G(y, \frac{\epsilon}{2})$,

$$\sum_{n=1}^{\infty} (1 \otimes f) \circ T_\phi^n(x,z) = \sum_{n=1}^{\infty} f(z + \phi_n(x)) \ge \delta(y,\epsilon) \sum_{n=1}^{\infty} 1_{B_G(0,\frac{\epsilon}{2})}(\phi_n(x)) = \infty$$

and T_ϕ is conservative by proposition 1.1.6. $\qquad \square$

REMARK. The "if" part of proposition 8.1.2 can be generalised to the case where T is a conservative, ergodic non-singular transformation (see theorem 5.5 in [Schm1]).

The "only if" part may fail in case T is a conservative, ergodic measure preserving transformation of an infinite measure space.

To see this, let τ be an ergodic probability preserving transformation of the probability space Ω, and let $\phi : X \to \mathbb{Z}$ be such that τ_ϕ is ergodic (for example, τ_ϕ as in example 1.2.8 when $p = \frac{1}{2}$ is ergodic by theorem 3.4.2). By proposition 8.1.2, $\liminf_{n\to\infty} |\phi_n| = 0$ a.e..

By corollary 5.3.4, $\exists$ a conservative, exact measure preserving transformation S of Y (say) such that $S \times \tau_\phi$ is dissipative.

By theorem 2.7.1, $T := S \times \tau$ is a conservative, ergodic measure preserving transformation (of $Y \times \Omega$) and $S \times \tau_\phi = T_{\tilde{\phi}}$ where $\tilde{\phi} : Y \times \Omega \to \mathbb{Z}$ is defined by $\tilde{\phi}(y,\omega) = \phi(\omega)$.

Evidently $T_{\tilde{\phi}} = S \times \tau_\phi$ is dissipative, but

$$\liminf_{n\to\infty} \left| \sum_{k=0}^{n-1} \tilde{\phi} \circ T_{\tilde{\phi}}^k(y,\omega) \right| = \liminf_{n\to\infty} |\phi_n(\omega)| = 0$$

a.e. on $Y \times \Omega$.

8.1.3 PROPOSITION. *If ϕ is a coboundary, then T_ϕ is conservative.*

PROOF. Evidently T_0 is conservative, and if ϕ is a coboundary, then T_ϕ is isomorphic to T_0. □

8.1.4 THEOREM [Atk], [Schm3]. *If T is an ergodic, probability preserving transformation and $\phi : X \to \mathbb{R}$ is measurable, $\frac{\phi_n}{n} \xrightarrow{m} 0$, then T_ϕ is conservative.*

PROOF. If T_ϕ is not conservative, then by proposition 8.1.2, $\liminf_{n\to\infty} |\phi_n| > 0$ on some $A \in \mathcal{B}_+$. By Egorov's theorem, $\exists B \in \mathcal{B}_+ \cap A$, $\epsilon > 0$, $n_0 \in \mathbb{N}$ such that

$$|\phi_n(x)| \geq \epsilon \ \forall \ x \in B, \ n \geq n_0.$$

Using Rokhlin towers, we get $C \in \mathcal{B}_+ \cap B$ such that $\varphi_C \geq n_0$ on C. Thus, fixing $x \in C$ and writing $a_k := \phi_{(\varphi_C)_k}(x)$, we have that the set $\{a_k : k \geq 1\}$ is ϵ-*separated* in the sense that for $m > n$,

$$|a_m - a_n| = |\phi_{(\varphi_C)_m}(x) - \phi_{(\varphi_C)_n}(x)| = |\phi_{(\varphi_C)_{m-n}}(T_C^n x)| \geq \epsilon.$$

Let $F \subset \mathbb{R}$ be finite and ϵ-separated. Evidently $\exists a \in F$ such that $|a| \geq \frac{|F|\epsilon}{3}$. We claim that $\forall \ 0 < p < 1$

$$|\{a \in F : \ |a| \geq \frac{p|F|\epsilon}{3}\}| \geq (1-p)|F|.$$

If this is not the case, then $\exists F' \subset F$, $|F'| > p|F|$ such that $|a| < \frac{p|F|\epsilon}{3} \ \forall \ a \in F'$. However since F' is also ϵ-separated $\exists a \in F'$ such that $|a| \geq \frac{|F'|\epsilon}{3} > \frac{p|F|\epsilon}{3}$ contradicting this.

Choosing $p = \frac{1}{2}$, we have that

$$\sum_{k=1}^{n} 1_{[|\phi_{(\varphi_C)_k}| \geq \frac{n\epsilon}{6}]} \geq \frac{n}{2} \; \forall \; n \geq 1 \text{ on } C.$$

It follows from Kac's formula and Birkhoff's ergodic theorem that $(\varphi_C)_n \sim \frac{n}{m(C)}$ a.e. on C with the consequence that a.e. on C, for large n:

$$\sum_{k=1}^{n} 1_{[|\phi_k| > \frac{k\epsilon m(C)}{6}]} \geq \sum_{k=1}^{n} 1_{[|\phi_k| > \frac{n\epsilon m(C)}{6}]}$$

$$\geq \sum_{k=1}^{\frac{nm(C)}{2}} 1_{[|\phi_{(\varphi_C)_k}| \geq \frac{n\epsilon m(C)}{6}]}$$

$$\geq \frac{nm(C)}{4}.$$

By Fatou's lemma,

$$\liminf_{n \to \infty} \frac{1}{n} \sum_{k=1}^{n} m\left(\left[|\phi_k| > \frac{k\epsilon m(C)}{6}\right]\right) \geq \frac{m(C)}{4}$$

and in particular $\frac{\phi_n}{n} \overset{m}{\nrightarrow} 0$. $\qquad\qquad \Box$

8.1.5 COROLLARY. *If T is an ergodic, probability preserving transformation and $\phi : X \to \mathbb{R}$ is integrable, then T_ϕ is conservative iff $\int_X \phi dm = 0$.*

PROOF.

Suppose that $\phi : X \to \mathbb{R}$ is integrable. then by Birkhoff's theorem,
$\exists \; \lim_{n \to \infty} \frac{\phi_n}{n} = \int_X \phi dm$. a.e. .

If $\int_X \phi dm \neq 0$ then $\phi_n \to \infty$ a.e. and T_ϕ is dissipative by proposition 8.1.2.

If $\int_X \phi dm = 0$, then by theorem 8.1.4 T_ϕ is conservative. $\qquad \Box$

§8.2 Persistencies and Essential values

DEFINITION: PERSISTENCIES, ESSENTIAL VALUES. Let $(X, \mathcal{B}, m)$ be a standard probability space, and let $T : X \to X$ be an ergodic, probability preserving transformation. Suppose that $\phi : X \to G$ is measurable. The collection of *persistencies* of ϕ is

$$\Pi(\phi) = \{a \in G : \forall \; A \in \mathcal{B}_+, \; \epsilon > 0, \; \exists \; n \geq 1, \; m(A \cap T^{-n} A \cap [\|\phi_n - a\| < \epsilon]) > 0\}.$$

Now let $T : X \to X$ be invertible.

The collection of *essential values* $E(\phi)$ of ϕ is

$$E(\phi) = \{a \in G : \forall \; A \in \mathcal{B}_+, \; \epsilon > 0, \; \exists \; n \in \mathbb{Z}, \; m(A \cap T^{-n} A \cap [\|\phi_n - a\| < \epsilon]) > 0\}.$$

REMARKS.

1) The concept of *essential value* was was introduced in [**Schm1**].

2) Evidently $0 \in E(\phi)$ and $\Pi(\phi) \subset E(\phi)$.

3) We'll see below that the only case where $E(\phi) \neq \Pi(\phi)$ is when T_ϕ is dissipative, in which case $\Pi(\phi) = \emptyset$, $E(\phi) = \{0\}$.

8.2.1 PROPOSITION [**Schm1**].
1) *Either $\Pi(\phi) = \emptyset$, or $\Pi(\phi)$ is a closed subgroup of G.*
2) *If T is invertible, then $E(\phi) = \Pi(\phi) \cup \{0\}$.*

PROOF.
To see that $\Pi(\phi)$ is closed let $a \in \overline{\Pi(\phi)}$ and let $\epsilon > 0$, $A \in \mathcal{B}_+$.
$\exists\, a' \in \Pi(\phi)$ such that $\|a - a'\| < \epsilon/2$, and $\exists\, n \geq 1$ such that
$m(A \cap T^{-n}A \cap [\|\varphi_n - a'\| < \epsilon/2]) > 0$.
It follows that
$m(A \cap T^{-n}A \cap [\|\varphi_n - a\| < \epsilon]) \geq m(A \cap T^{-n}A \cap [\|\varphi_n - a'\| < \epsilon/2]) > 0$. Thus,
$a \in \Pi(\phi)$ and $\Pi(\phi)$ is closed.

To show that $\Pi(\phi)$ is a group, we show that $a, b \in \Pi(\phi) \implies a - b \in \Pi(\phi)$.
Let $a, b \in \Pi(\phi)$, $\epsilon > 0$, $A \in \mathcal{B}_+$ and let $n \geq 1$ be such that $m(A \cap T^{-n}A \cap [\|\phi_n - a\| < \epsilon/2]) > 0$.
By Rokhlin's lemma, $\exists\, B \in \mathcal{B}_+$, $B \subset A \cap T^{-n}A \cap [\|\phi_n - a\| < \epsilon/2]$ such that $B \cap T^{-k}B = \emptyset$ for $1 \leq k \leq n$.
Since $b \in \Pi(\phi)$, $\exists\, N \geq 1$ such that $m(B \cap T^{-N}B \cap [\|\phi_N - b\| < \epsilon/2]) > 0$. The construction of B implies that $N > n$ whence

$$B \cap T^{-N}B \cap [\|\phi_N - b\| < \epsilon/2]$$
$$= B \cap T^{-N}B \cap [\|\phi_n - a\| < \epsilon/2] \cap [\|\phi_N - b\| < \epsilon/2]$$
$$\subset B \cap T^{-N}B \cap [\|\phi_{N-n} \circ T^n - (b - a)\| < \epsilon],$$

$$0 < m(B \cap T^{-N}B \cap [\|\phi_{N-n} \circ T^n - (b - a)\| < \epsilon])$$
$$\leq m(A \cap T^{-n}A \cap T^{-N}A \cap [\|\phi_{N-n} \circ T^n - (b - a)\| < \epsilon])$$
$$\leq m(A \cap T^{-(N-n)}A \cap [\|\phi_{N-n} - (b - a)\| < \epsilon])$$

and $b - a \in \Pi(\phi)$.
Finally to demonstrate 2) we show that if T is invertible, then $E(\phi) \setminus \{0\} \subset \Pi(\phi)$. This is sufficient as $\Pi(\phi) \subset E(\phi)$.
Let $0 \neq a \in E(\phi)$, $\epsilon > 0$ and $A \in \mathcal{B}_+$, and suppose (to obtain a contradiction) that $\nexists\, n' \geq 1$ such that $m(A \cap T^{-n'}A \cap [\|\phi_{n'} - a\| < \epsilon]) > 0$.
Since $0 \neq a \in E(\phi)$, $\forall\, B \in \mathcal{B}_+$ $B \subset A$ $\exists\, n \neq 0$ such that $m(B \cap T^{-n}B \cap [\|\phi_n - a\| < \epsilon/4]) > 0$. By our assumption, $n = -\ell$ for some $\ell \geq 1$ in which case

$$m(B \cap T^{-\ell}B \cap [\|\phi_\ell + a\| < \epsilon/4]) = m(B \cap T^{\ell}B \cap [\|\phi_\ell \circ T^\ell + a\| < \epsilon/4])$$
$$= m(B \cap T^{\ell}B \cap [\|\phi_{-\ell} - a\| < \epsilon/4]) > 0.$$

As above, by Rokhlin's lemma, $\exists\, A_1, A_2, A_3 \in \mathcal{B}(A)_+$ and $n_2 > n_1 \geq 1$, $n_3 > n_1 + n_2$ such that $m(A_i \cap T^{-n_i}A_i \cap [\|\phi_{n_i} + a\| < \epsilon/4]) > 0$ $(i = 1, 2, 3)$ and such that $A_{i+1} \subset A_i \cap T^{-n_i}A_i \cap [\|\phi_{n_i} + a\| < \epsilon/4]$ $(i = 1, 2)$.
It follows that

$$0 < m(A_3 \cap T^{-n_3}A_3 \cap [\|\phi_{n_3} + a\| < \epsilon/4])$$
$$\leq m(A \cap T^{-n_3-(n_1+n_2)}A \cap \|\phi_{n_3-(n_1+n_2)} - a\| < \epsilon)$$

contradicting our assumption. $\qquad\square$

8.2.2 PROPOSITION.
T_ϕ is conservative iff $0 \in \Pi(\phi)$.

PROOF. By proposition 8.1.2, $0 \in \Pi(\phi) \implies T_\phi$ is conservative. Conversely suppose that T_ϕ is conservative and let $A \in \mathcal{B}_+$, $\epsilon > 0$. $\exists\, n \geq 1$ such that $m \times m_G(A \times B_G(0, \epsilon/2) \cap T_\phi^{-n} A \times B_G(0, \epsilon/2)) > 0$. Since $A \times B_G(0, \epsilon/2) \cap T_\phi^{-n} A \times B_G(0, \epsilon/2) \subset (A \cap T^{-n} A \cap [\|\phi_n\| < \epsilon]) \times B_G(0, \epsilon/2)$, we have $m(A \cap T^{-n} A \cap [\|\phi_n\| < \epsilon]) > 0$ and $0 \in \Pi(\phi)$. $\square$

8.2.3 PROPOSITION.
Suppose that $\phi, \varphi : X \to G$ are cohomologous, then $\Pi(\phi) = \Pi(\varphi)$.

PROOF. By symmetry, it is sufficient to show that $\Pi(\phi) \subseteq \Pi(\varphi)$.
Suppose that $\varphi = \phi + h \circ T - h$ where $h : X \to G$ is measurable. Let $a \in \Pi(\phi)$ and let $A \in \mathcal{B}_+$, $\epsilon > 0$. Since X is a standard space, by Lusin's theorem $\exists\, B \subset A$, $B \in \mathcal{B}_+$ such that $\|h(x) - h(y)\| < \frac{\epsilon}{2} \,\forall\, x, y \in B$. Since $a \in \Pi(\phi)$, $\exists\, n \geq 1$ such that $m(B \cap T^{-n} B \cap [\|\phi_n - a\| < \frac{\epsilon}{2}]) > 0$. By construction of B, if $x \in B \cap T^{-n} B$, then $\|\varphi_n(x) - \phi_n(x)\| = \|h(T^n x) - h(x)\| < \frac{\epsilon}{2}$ whence $m(B \cap T^{-n} B \cap [\|\varphi_n - a\| < \epsilon]) \geq m(B \cap T^{-n} B \cap [\|\phi_n - a\| < \frac{\epsilon}{2}]) > 0$, and $a \in \Pi(\varphi)$. $\square$

DEFINITION: PERIODS.
Define the collection of *periods* for T_ϕ-invariant functions:

$$\mathrm{Per}\,(\phi) = \{a \in G : Q_a A = A \quad \mathrm{mod}\ \ m\ \forall\ A \in \mathfrak{I}(T_\phi)\}$$

where $Q_a(x, y) = (x, y + a)$.

8.2.4 THEOREM [**Schm1**].

$$E(\phi) = \mathrm{Per}\,(\phi).$$

PROOF.
$\mathrm{Per}\,(\phi) \subset E(\phi)$
Suppose $a \notin E(\phi)$, then $\exists\, \epsilon > 0$, and $A \in \mathcal{B}_+$ such that $m(A \cap T^{-n} A \cap [\|\phi_n - a\| < \epsilon]) = 0 \,\forall\, n \in \mathbb{Z}$. Set

$$B_1 = \bigcup_{n \in \mathbb{Z}} T_\phi^n \left(A \times B_G(0, \frac{\epsilon}{2}) \right), \quad B_2 = \bigcup_{n \in \mathbb{Z}} T_\phi^n \left(A \times B_G(a, \frac{\epsilon}{2}) \right).$$

We have that
$$m(B_1 \cap B_2) = 0$$
because $\forall\, n \in \mathbb{Z}$,

$$m \times m_G \left((A \times B_G(0, \frac{\epsilon}{2})) \cap T_\phi^{-n} A \times B_G(a, \frac{\epsilon}{2})) \right)$$
$$\leq m_G(B_G(0, \frac{\epsilon}{2})) m(A \cap T^{-n} A \cap [\|\phi_n - a\| < \epsilon])$$
$$= 0.$$

Also, $1_{B_i} \circ T_\phi = 1_{B_i}$ $(i = 1, 2)$, but $1_{B_1} \circ Q_a = 1_{B_2}$, whence $a \notin \mathrm{Per}\,(\phi)$.

$E(\phi) \subset \mathrm{Per}\,(\phi)$

Now assume that $a \notin \text{Per}\,(\phi)$, then $\exists\, A, B \in \mathfrak{I}(T_\phi)_+$ disjoint such that $B = Q_a A$. Set for $x \in X$,

$$A_x = \{y \in G : (x, y) \in A\}$$

Note that

$$A_{Tx} = \{y \in G : (Tx, y) = T_\phi(x, y - \phi(x)) \in A\} = A_x + \phi(x),$$

whence $m_G(A_x) = m_G(A_{Tx})$, and by ergodicity, $m_G(A_x) = m \times m_G(A) > 0$ for m-a.e. $x \in X$.

Using the disintegration theorem, one shows that $\exists\, \theta \in \mathcal{B}(A)$ such that

$$0 < m_G(\theta_x) < \infty.$$

Next, choose $P \in \mathcal{P}(G)$, $P \sim m_G$ and consider the Polish metric space $(\mathcal{S}, D)$ where $\mathcal{S} := \mathcal{S}(X, \mathcal{B}(G), P)$ is the measure algebra of $(X, \mathcal{B}(G), P)$ (as in §1.0) and $D(A, B) := P(A \triangle B)$.

By Fubini's theorem, $\forall\, C \in \mathcal{B}(X \times G)$, $x \mapsto [C_x]$ is measurable $(X \to \mathcal{S})$. Define $\omega : X \times X \times G \to \mathbb{R}_+$ by

$$\omega(x, y, t) = m_G(\theta_x \cap (\theta_y + t)).$$

This is measurable, and by Lusin's theorem, $\exists\, \epsilon > 0, \delta > 0$ and $X_0 \in \mathcal{B}(X)_+$ such that

$$m_G(\theta_x) \geq 2\epsilon\ \forall\, x \in X_0, \ \&\ \omega(x, y, t) \geq \epsilon\ \forall\, x, y \in X_0,\ \|t\| < \delta.$$

Lastly, we show that $a \notin E(\phi)$. This will follow from

$$X_0 \cap T^{-n} X_0 \cap [\|\phi_n(x) - a\| < \delta/2] = \emptyset\ \forall\, n \geq 1.$$

Indeed, supposing that $x, T^n x \in X_0$, we note that

$$\left(a + \theta_{T^n x}\right) \cap \left(\theta_x + \phi_n(x)\right) \subset B_{T^n x} \cap A_{T^n x} = \emptyset,$$

whence, for $x \in X_0$,

$$T^n x \in X_0 \ \Rightarrow\ \|\phi_n(x) - a\| \geq \delta.$$

$\square$

8.2.5 COROLLARY.
Suppose that T is an ergodic probability preserving transformation and that $\phi : X \to G$ is measurable, then T_ϕ is ergodic iff $E(\phi) = G$.

§8.3 Coboundaries

In this section, we investigate which functions $\varphi : X \to G$ are T-coboundaries. It turns out that this is connected to the existence of absolutely continuous T_φ-invariant probabilities.

Recall from proposition 8.1.1 that if T is ergodic and $\varphi : X \to G$ is measurable, then the positive part $\mathfrak{P}(T_\varphi) \in \{\emptyset,\ X \times G\}$ mod $m \times m_G$.

8.3.1 THEOREM.
Suppose that T is ergodic and let $\varphi : X \to G$ be measurable, then $\mathfrak{P}(T_\varphi) = X \times G$ mod $m \times m_G$ iff the distributions of $\{\varphi_n : n \geq 1\}$ are uniformly tight in the sense that $\forall\, \epsilon > 0 \ \exists\ $ a compact set $M \subset G$ such that $m([\varphi_n \notin M]) < \epsilon\ \forall\, n \geq 1$.

PROOF.

Suppose first that $\mathfrak{P}(T_\varphi) = X \times G$ mod $m \times m_G$, then $\exists$ a T_φ-invariant probability $P \sim m \times m_G$. Since T is ergodic, $P(A \times G) = m(A) \ \forall \ A \in \mathcal{B}$ whence $\int_G p(x,y) dm_G(y) = 1$ for a.e. $x \in X$ where $dP(x,y) = p(x,y) dm(x) dm_G(y)$.

Let $\epsilon > 0$, then $\exists \ M \subset G$ compact and $F \in \mathcal{B}$ such that $m(F) > 1 - \epsilon/4$ and $\int_M p(x,y) dm_G(y) > 1 - \epsilon/4 \ \forall \ x \in F$. It follows that $P(A \times M) > m(A) - \epsilon/2 \ \forall \ A \in \mathcal{B}$.

Let $M' := M - M$ which is also compact. Evidently $[\varphi_n \notin M'] \times M \subset X \times M \cap T_\varphi^{-n}(X \times M^c)$ whence

$$m([\varphi_n \notin M']) \leq P([\varphi_n \notin M'] \times M) + \epsilon/2$$
$$\leq P(X \times M \cap T_\varphi^{-n}(X \times M^c)) + \epsilon/2 < \epsilon$$

Conversely, suppose that the distributions of $\{\varphi_n : n \geq 1\}$ are uniformly tight. In particular, suppose that $a > 0$ and $M \subset G$ compact satisfy $m([\varphi_n \in M]) \geq a \ \forall \ n \geq 1$.

Evidently $[\varphi_n \in M] \times M \subset X \times M \cap T_\varphi^{-n}[X \times M']$ where $M' := M + M$ (also compact), whence

$$m \times m_G(X \times M \cap T_\varphi^{-n}[X \times M']) \geq a m_G(M) > 0 \ \forall \ n \geq 1.$$

If there is no $m \times m_G$-absolutely continuous, T_φ-invariant probability, then by theorem 1.4.4

$$\frac{1}{N} \sum_{n=1}^{N} \widehat{T}_\varphi^n 1_{X \times M} \to 0$$

in measure on $X \times M'$, whence by the bounded convergence theorem,

$$\frac{1}{N} \sum_{n=1}^{N} m \times m_G(X \times M \cap T_\varphi^{-n}[X \times M']) \to 0$$

as $N \to \infty$ contradicting the assumed uniform tightness. Thus, $\mathfrak{P}(T_\varphi) \neq \emptyset$ mod $m \times m_G$. $\qquad\square$

REMARK. It follows from the proof of theorem 8.3.1 that the distributions of $\{\varphi_n : n \geq 1\}$ are uniformly tight if and only if $\exists \ M \subset G$ compact such that

$$\limsup_{N \to \infty} \frac{1}{N} \sum_{n=1}^{N} m([\varphi_n \in M]) > 0.$$

The following is a generalisation of [**Key-New**]. The proof is an adaptation of Lemańczyk's proof of [**Key-New**] in [**Lem**]. See also [**Fu**].

8.3.2 THEOREM.

Suppose that T is ergodic and let $\varphi : X \to G$ be measurable, then $\mathfrak{P}(T_\varphi) = X \times G \Leftrightarrow \Pi(\varphi)$ is compact and $\exists \ \phi : X \to \Pi(\varphi)$ measurable and cohomologous with φ.

PROOF.

We first show $\Leftarrow$. Under the assumptions T_ϕ is a probability preserving transformation which is isomorphic to T_φ, whence $\mathfrak{P}(T_\varphi) = X \times G$.

We now prove $\Rightarrow$. If $\mathfrak{P}(T_\varphi) = X \times G$, then $\exists$ an ergodic T_φ-invariant probability P such that $P \circ \pi^{-1} = m$ where $\pi(x, y) = x$.

For each $a \in G$, $P \circ Q_a$ is also an ergodic T_φ-invariant probability, and therefore either $P \circ Q_a = P$ or $P \circ Q_a \perp P$. Define $H := \{a \in G : P \circ Q_a = P\}$. This is a closed subgroup of G.

Consider the Banach space $\mathfrak{b}$ of bounded measurable functions $X \times G \to \mathbb{R}$ equipped with the supremum norm. We need a separable subspace $\mathcal{A} \subset \mathfrak{b}$ with the properties that $f \in \mathcal{A} \implies f \circ Q_a \in \mathcal{A} \ \forall \ a \in G$, and

$$q, q' \in \mathcal{P}(X \times G), \ \int_{X \times G} f dq = \int_{X \times G} f dq' \ \ \forall \ f \in \mathcal{A} \implies q = q'.$$

To obtain such a subspace, fix a compact metric topology on X generating $\mathcal{B}$, then $\mathcal{A} = C(X) \otimes C_0(G)$ (where $C_0(G) := \{f \in C(G) : g(y) \underset{y \to \infty}{\to} 0\}$) is as needed.

Set

$$Y := \left\{ (x, y) \in X \times G : \ \frac{1}{n} \sum_{k=0}^{n-1} f \circ T_\varphi^k(x, y) \to \int_{X \times G} f dP \ \ \forall \ \ f \in \mathcal{A} \right\}.$$

Since $\mathcal{A}$ is a separable subspace of $\mathfrak{b}$, the set Y is determined by a countable subcollection of $\mathcal{A}$ whence $Y \in \mathcal{B}(X \times G)$, and by Birkhoff's ergodic theorem $P(Y) = 1$.

For $x \in X$, set $Y_x = \{y \in G : (x, y) \in Y\}$. We claim that Y_x is a coset of H whenever it is nonempty. Indeed, if $(x, y) \in Y$ and $a \in G$ then for $f \in \mathcal{A}$,

$$\frac{1}{n} \sum_{k=0}^{n-1} f \circ T_\varphi^k(x, y + a) = \frac{1}{n} \sum_{k=0}^{n-1} f \circ Q_a \circ T_\varphi^k(x, y) \to \int_{X \times G} f \circ Q_a dP$$

since $f \circ Q_a \in \mathcal{A}$. Thus, $(x, y + a) \in Y$ if and only if

$$\int_{X \times G} f \circ Q_a dP = \int_{X \times G} f dP \ \forall \ f \in \mathcal{A},$$

equivalently $a \in H$.

By the disintegration theorem, $\exists \ X_0 \in \mathcal{B}$, $m(X_0) = 1$ such that Y_x is a coset of $H \ \forall \ x \in X_0$, and $\exists$ a measurable map $x \mapsto P_x \quad (X_0 \to \mathcal{P}(G, \mathcal{B}(G)))$ such that $P_x(Y_x) = 1 \ \forall \ x \in X_0$ and

$$\int_{X \times G} u \otimes g dm dP = \int_X u(x) \int_G g dP_x dm(x) \ \ \forall \ u \otimes g \in \mathcal{A}.$$

It follows that for a.e. $x \in X_0$, $P_x(A + a) = P_x(A) \quad (a \in H, \ A \in \mathcal{B}(G))$.

By the analytic section theorem, $\exists \ h : X \to G$ measurable such that $h(x) \in Y_x$ for a.e. $x \in X$, whence $Y_x = h(x) + H$.

Now let $P'_x \in \mathcal{P}(G)$ be defined by $P'_x(A) := P_x(A - h(x))$. Clearly $P'_x(H) = 1$ and $P'_x(A + a) = P'_x(A) \quad (a \in H, \ A \in \mathcal{B}(G))$. Thus $P'_x = m_H$ and H is compact.

Defining $\Psi : X \times G \to X \times G$ by $\Psi(x, y) := (x, y + h(x))$, we have that $P \circ \Psi^{-1} = m \times m_H$. If $V := \Psi \circ T_\varphi \circ \Psi^{-1}$ then $m \times m_H \circ V = m \times m_H$ and $V = T_\phi$ where $\phi = \varphi + h \circ T - h$.

Since $(X \times G, \mathcal{B}(X \times G), m \times m_H, T_\phi)$ is an ergodic probability preserving transformation, we have that $\phi : X \to H = \Pi(\phi) = \Pi(\varphi)$. $\qquad\square$

8.3.3 COROLLARY.
Suppose that T is ergodic and let $\varphi : X \to G$ be measurable, then φ is a coboundary if and only if $\Pi(\varphi) = \{0\}$, and the distributions of $\{\varphi_n : n \geq 1\}$ are uniformly tight.

PROOF. This follows from theorems 8.3.1 and 8.3.2. $\qquad\square$

8.3.4 COROLLARY ([**Schm1**]).
Suppose that T is ergodic and let $\varphi : X \to \mathbb{R}^d$ be measurable, then φ is a coboundary if and only if the distributions of $\{\varphi_n : n \geq 1\}$ are uniformly tight.

PROOF. By theorem 8.3.1, the distributions of $\{\varphi_n : n \geq 1\}$ are uniformly tight iff $\mathfrak{P}(T_\varphi) = X \times \mathbb{R}^d$; which in turn is the case (by theorem 8.3.2) iff $\Pi(\varphi)$ is compact, and φ is cohomologous to a cocycle taking values in $\Pi(\varphi)$. The corollary follows since the only compact subgroup of $\mathbb{R}^d$ is $\{0\}$. $\qquad\square$

8.3.5 COROLLARY ([**Schm1**]).
Suppose that T is ergodic and let $\varphi : X \to G$ be measurable.

If $\Pi(\varphi) \neq \emptyset$ and the distributions of $\{\varphi_n + \Pi(\varphi) : n \geq 1\}$ are uniformly tight on $G/\Pi(\varphi)$, then $\exists \ \psi : X \to \Pi(\varphi)$ measurable, and T-cohomologous with φ.

REMARK. The hypothesis is satisfied whenever $G/\Pi(\varphi)$ is compact.

PROOF. Consider the measurable function $\tilde{\varphi} : X \to G/\Pi(\varphi)$ defined by $\tilde{\varphi} := \varphi + \Pi(\varphi)$.

We claim first that $\Pi(\tilde{\varphi}) = \{0\} \subset G/\Pi(\varphi)$.

To see this suppose that $a \in G$ and $a + \Pi(\varphi) \in \Pi(\tilde{\varphi})$. We must show that $a \in \Pi(\varphi)$. Let $\epsilon > 0$ and $A \in \mathcal{B}_+$, then $\exists \ \gamma \in \Pi(\varphi)$, $n \geq 1$ such that $m(B) > 0$ where $B := A \cap T^{-n}A \cap [\|\varphi_n - (a - \gamma)\| < \epsilon/2]$. Since $\gamma \in \Pi(\varphi)$, $\exists \ N \geq 1$ such that $m(B \cap T^{-N}B \cap [\|\varphi_n - \gamma\| < \epsilon/2]) > 0$. Since $B \cap T^{-N}B \cap [\|\varphi_n - \gamma\| < \epsilon/2] \subset A \cap T^{-(N+n)}A \cap [\|\varphi_{N+n} - a\| < \epsilon]$, we have that $a \in \Pi(\varphi)$.

Next, by corollary 8.3.3, $\exists \ h : X \to G/\Pi(\varphi)$ so that $\tilde{\varphi} = h - h \circ T$.

By the analytic section theorem, $\exists \ H : X \to G$ measurable such that $H + \Pi(\varphi) = h$ a.e., whence if $\psi := \varphi - H + H \circ T$, then

$$\psi + \Pi(\varphi) = \tilde{\varphi} - H + H \circ T = \Pi(\varphi)$$

and $\psi : X \to \Pi(\varphi)$. $\qquad\square$

§8.4 Skew products over Kronecker transformations

In this section, we present some ergodic skew products over Kronecker transformations. Together with ergodicity, we'll also be interested in squashability.

DEFINITION: SQUASHABLE.

A conservative, ergodic measure preserving transformation $(Y, \mathcal{A}, \mu, S)$ is said to be *squashable* if $\exists$ a non-singular, non measure preserving transformation $Q : Y \to Y$ commuting with S (i.e. $S \circ Q = Q \circ S$). In this case, it follows from theorem 1.5.6 that $m \circ Q^{-1} = cm$ for some $0 < c \leq \infty$.

Let $(X, \mathcal{B}, m, T)$ be a Kronecker transformation, let G be a locally compact, second countable, Abelian topological group, and let $\varphi : X \to G$ be measurable.

8.4.1 PROPOSITION.

If T_φ is ergodic, $Q : X \times G \to X \times G$ is non-singular and $Q \circ T_\varphi = T_\varphi \circ Q$, then there exist a group translation $S : X \to X$, a continuous, surjective group endomorphism $w : G \to G$ and a measurable map $f : X \to G$ such that

$$Q(x,y) = (Sx, f(x) + w(y)) \quad (x \in X, \ y \in G)$$

and

$$w \circ \varphi = \varphi \circ S + f - f \circ T.$$

PROOF. Write $Q = (S, F)$, where $S : X \times G \to X$ and $F : X \times G \to G$. We have $S \circ T_\varphi = T \circ S$ and $F \circ T_\varphi = \varphi \circ S + F$.

Defining $U : X \times G \to X$ by $U(x,h) = S(x,h) - x$ we see that $U \circ T_\varphi = U$, whence by ergodicity of T_φ $\exists x_1 \in X$ such that $U(x,h) = x_1$, and $S(x,g) = Sx = x + x_1$.

Consequently $F \circ T_\varphi(x,h) = \varphi(Sx) + F(x,h)$ and hence $\forall \ g \in G$

$$\left(F \circ Q_g - F \right) \circ T_\varphi(x,h) = F(T_\varphi(x, h+g)) - F(T_\varphi(x,h))$$

$$= F(x, h+g) - F(x,h)$$

$$= \left(F \circ Q_g - F \right)(x,h).$$

Again by ergodicity of T_φ $\exists \ w : G \to G$ such that $F \circ Q_g - F = w(g)$ $(g \in G)$. It follows that w is a measurable homomorphism and hence continuous.

Set $\phi(x,h) = F(x,h) - w(h)$. By the above, $\phi \circ Q_g = \phi \ \forall \ g \in G$ whence there exists a measurable $f : X \to G$ such that $\phi(x,h) = f(x)$ a.e., and

$$Q(x,g) = (Sx, f(x) + w(g)).$$

To see that $w : G \to G$ is surjective, we show that it is non-singular. This is because $m \times m_G \circ S_f^{-1} = m \times m_G$ and $\exists \ 0 < c \le \infty$ such that $m \times m_G \circ Q^{-1} = cm \times m_G$, whence $m_G \circ w^{-1} = cm_G$ since $\mathrm{Id.} \otimes w = S_f^{-1} \circ Q$.

The equation $w \circ \varphi = \varphi \circ S + f - f \circ T$ follows from $Q \circ T_\varphi = T_\varphi \circ Q$. $\square$

Irrational rotations of $\mathbb{T}$.

We consider skew products $T_{\alpha,\varphi} : \mathbb{T} \times \mathbb{R} \to \mathbb{T} \times \mathbb{R}$ of form $T_{\alpha,\varphi}(x,y) = (x + \alpha, y + \varphi(x))$.

We'll need some information about the continued fraction expansion

$$\alpha = \cfrac{1}{a_1 + \cfrac{1}{a_2 + \cfrac{1}{a_3 + \cfrac{1}{\ddots}}}}$$

of $\alpha \in \mathbb{T} \setminus \mathbb{Q}$. This can be found in [**Kh**].

The positive integers a_n are called the *partial quotients* of α. Set

$$q_0 = 1, \ q_1 = a_1, \ q_{n+1} = a_{n+1}q_n + q_{n-1} \quad p_0 = 0, \ p_1 = 1, \ p_{n+1} = a_{n+1}p_n + p_{n-1}.$$

The rationals p_n/q_n are called the *convergents* of α, the numbers q_n are called (principal) *denominators* and we have the inequalities

$$|\alpha - \frac{p_n}{q_n}| < \frac{1}{q_n q_{n+1}}, \quad \|q_n \alpha\| < \frac{1}{q_{n+1}}$$

where $\|x\| := \min_{n \in \mathbb{Z}} |x - n|$.

A function $f : \mathbb{T} \to \mathbb{R}$ is called *piecewise linear* if $\exists\, 0 \leq x_1 < x_2 < \cdots < x_N < 1$ such that $f|_{[0,x_1)}$, $f|_{[x_j,x_{j+1})}$ $(1 \leq j \leq N - 1)$ and $f|_{[x_N,1)}$ are linear.

The *jump* of f at x is given by

$$d_x := \left\{ \begin{array}{ll} f(x+) - f(x-) & x \neq 0, \\ f(0+) - f(1-) & x = 0. \end{array} \right.$$

Evidently, $d_x = 0$ if $x \neq x_j$ $(1 \leq j \leq N - 1)$. Also we have

$$0 = \int_0^1 df(t) + d_0 = \int_0^1 f'(t)dt + \sum_{j=1}^N d_{x_j}.$$

Any piecewise linear function $f : \mathbb{T} \to \mathbb{R}$ has bounded variation on $\mathbb{T}$, and so the Denjoy-Koksma inequality holds (see [**Kui-Ni**]):

$$|f_N(x) - N \int_0^1 f(t)dt| \leq D_N(\alpha) \bigvee_{\mathbb{T}} f$$

where

$$D_N(\alpha) := \sup_{x<y} \{|\#\{1 \leq j \leq N : j\alpha \in [x,y)\} - (y - x)|\}$$

and $\bigvee_{\mathbb{T}} f$ denotes the variation of f on $\mathbb{T}$. As is well known, $D_{q_n(\alpha)}(\alpha) = O(1)$ as $n \to \infty$.

8.4.2 PROPOSITION [**Pas**]. *Suppose that* $f : \mathbb{T} \to \mathbb{R}$, $\int_0^1 f(t)\, dt = 0$ *is piecewise linear, and* $\sum_{j=1}^N d_{x_j} \neq 0$, *then* $\forall\, \alpha \in \mathbb{T} \setminus \mathbb{Q}$, *the skew product* $T_{\alpha,f}$ *is ergodic, and non-squashable.*

The proof of proposition 8.4.2 given here is from [**A-Lem-Mau-Nak**] (see §6 there for generalisations).

The related case where $f : \mathbb{T} \to \mathbb{R}$ is a step function is treated in [**Co**] and [**Ore**] (see also [**Schm1**]).

We begin the proof with a lemma.

8.4.3 LEMMA [**Schm1**].

Suppose that T *is an invertible, ergodic probability preserving transformation of the standard probability space* $(X, \mathcal{B}, m)$, $\phi : X \to \mathbb{R}$ *is measurable and* $K \subset E(\phi)^c$ *is compact, then*

$$\forall\, A \in \mathcal{B}(X)_+, \,\exists\, B \in \mathcal{B}(A)_+ \text{ such that } m(B \cap T^{-n}B \cap [\phi_n \in K]) = 0\ \forall\, n \in \mathbb{Z}.$$

PROOF.

Suppose that $a \notin E(\phi)$. Call $\epsilon > 0$ *bad* for a if

$$\exists\, B \in \mathcal{B}(X)_+ \text{ such that } m(B \cap T^{-n}B \cap [\|\phi_n - a\| < \epsilon]) = 0\ \forall\, n \in \mathbb{Z}.$$

We first show that "badness is not localised". Suppose that $a \notin E(\phi)$, and $\epsilon > 0$ is bad for a. We claim that $\forall \ A \in \mathcal{B}(X)_+$, $0 < \epsilon' < \epsilon$, $\exists \ B \in \mathcal{B}(A)_+$ such that

$$m(B \cap T^{-n} B \cap [\|\phi_n - a\| < \epsilon']) = 0 \ \forall \ n \in \mathbb{Z}.$$

To see this, fix $A \in \mathcal{B}(X)_+$ and $\epsilon' = \epsilon - \delta < \epsilon$. Suppose $C \in \mathcal{B}(X)_+$ satisfies

$$m(C \cap T^{-n} C \cap [\|\phi_n - a\| < \epsilon]) = 0 \ \forall \ n \in \mathbb{Z}.$$

By ergodicity, $\exists \ q$ such that $m(A \cap T^{-q} C) > 0$, or, in other words, $\exists \ D \in \mathcal{B}(A)_+$ such that $T^q D \subset C$. By Lusin's theorem $\exists \ B \in \mathcal{B}(D)_+$ such that

$$\|\phi_q(x) - \phi_q(y)\| < \delta \ \forall \ x, y \in B.$$

Now suppose that $x \in B$, $n \in \mathbb{Z}$, are such that $T^n x \in B$ and $\|\phi_n(x) - a\| < \epsilon'$, then $y := T^q x \in C$, $T^n y = T^q T^n x \in C$, and

$$\begin{aligned}
\phi_n(y) &= \phi_{-q}(y) + \phi_n(x) + \phi_q(T^n x) \\
&= \phi_{-q}(T^q x) + \phi_n(x) + \phi_q(T^n x) = \phi_n(x) \pm \delta
\end{aligned}$$

whence $\|\phi_n(y) - a\| < \epsilon' + \delta = \epsilon$.

$$\therefore \ T^q \left(B \cap T^{-n} B \cap [\|\phi_n - a\| < \epsilon'] \right) \subset C \cap T^{-n} C \cap [\|\phi_n - a\| < \epsilon]$$

and

$$m(B \cap T^{-n} B \cap [\|\phi_n - a\| < \epsilon']) = 0 \ \forall \ n \in \mathbb{Z}.$$

This proves our claim.

To continue, since $K \cap E(\phi) = \emptyset$, $\forall \ b \in K \ \exists \ \epsilon_b > 0$ bad for b. By compactness, $\exists \ b_1, \ldots, b_k \in K$ such that

$$K \subset \bigcup_{j=1}^{k} B(b_j, \frac{\epsilon_{b_j}}{2}).$$

Repeated application of our claim gives sets $A_1 \supset A_2 \cdots \supset A_k \in \mathcal{B}(X)_+$ satisfying

$$m(A_j \cap T^{-n} A_j \cap [\phi_n \in B(bj, \frac{\epsilon_{bj}}{2})]) = 0 \ \forall \ n \in \mathbb{Z}, \ 1 \le j \le k,$$

whence

$$m(A_k \cap T^{-n} A_k \cap [\phi_n \in K]) = 0 \ \forall \ n \in \mathbb{Z}.$$

$\square$

PROOF OF PROPOSITION 8.4.2.

There is no loss of generality in assuming that $\sum_{j=1}^{N} d_{x_j} < 0$. Since f' is Riemann integrable, the ergodic theorem holds uniformly, so

$$\frac{1}{q} \sum_{j=0}^{q-1} f'(x + j\alpha) \to \int_0^1 f'(t) \, dt > 0$$

uniformly in x. Hence, we can find two constants $0 < C_1 < C_2$ such that for all q sufficiently large,

$$(*) \qquad C_1 q \le f_q'(x) \le C_2 q \ \forall \ x \in \mathbb{T}.$$

On the other hand, f_q is also piecewise linear with the discontinuity points of the form $x_i - j\alpha$, with jumps $d_{x_i - j\alpha} = d_{x_i}$, where $i = 1, \ldots, N$, $j = 0, \ldots, q-1$.

Now consider $q = q_n$ a denominator of α. Let ξ_n be the partition of $\mathbb{T}$ into intervals generated by $\{x_i + j\alpha : 1 \le i \le N, \ 0 \le j \le q_n - 1\}$.

Evidently $|\xi_n| \le \overline{N} q_n$ where $\overline{N} := N + 2$. Also each $J \in \xi_n$ has length $|J| \le \frac{1}{q_n}$ whence the sequence $\xi_1, \xi_2, \ldots$ *approximately generates* $\mathcal{B}$ in the sense that $\forall \ \eta > 0$, $A \in \mathcal{B}_+ \ \exists \ k_0 \ge 1$ such that $\forall \ k \ge k_0 \ \exists \ A' \in \sigma(\xi_k)$ with $m(A \triangle A') < \eta$.

Let $0 < c < 1$. Call $J \in \xi_n$ *c-long* if $|J| \ge \frac{c}{\overline{N} q_n}$ and *c-short* otherwise. Evidently

$$\sum_{J \in \xi_n \ c-\text{short}} |J| \le c$$

whence

$$\sum_{J \in \xi_n \ c-\text{long}} |J| \ge 1 - c.$$

Suppose now that $E(f) \ne \mathbb{R}$, then $E(f) = \lambda \mathbb{Z}$ for some $\lambda \ge 0$. For $\epsilon > 0$, consider the compact

$$K = K_\epsilon := \{r \in [-2 \bigvee_{\mathbb{T}} f, 2 \bigvee_{\mathbb{T}} f] : \ \text{dist}(r, \mathbb{Z}\lambda) \ge \epsilon\}.$$

Let $J \in \xi_n$ be c-long, then

$$|f_{q_n}(J)| \ge C_1 q_n |J| \ge \frac{C_1 c}{\overline{N}}$$

by (∗) and $f_{q_n}(J) \subset [-\bigvee_{\mathbb{T}} f, \bigvee_{\mathbb{T}} f]$ by the Denjoy-Koksma inequality, whence for

$$0 < \epsilon < \frac{C_1 c \lambda}{3 \overline{N} \bigvee_{\mathbb{R}} f},$$

$$m(f_{q_n}(J) \cap K_\epsilon) > \frac{2}{3} |f_{q_n}(J)|$$

and hence ($f_{q_n}|_J$ being linear)

$$m(J \cap [f_{q_n} \in K_\epsilon]) > \frac{2}{3} |J|.$$

We claim that $\forall \ A \in \mathcal{B}_+$, $\exists \ \epsilon > 0$ such that

$$\liminf_{n \to \infty} m(A \cap [f_{q_n} \in K_\epsilon]) \ge \frac{403}{960} m(A).$$

To see this fix $A \in \mathcal{B}$, $m(A) > 0$. Since the sequence $\xi_1, \xi_2, \ldots$ approximately generates $\mathcal{B}$, $\exists \ n_0 \ge 1$ such that $\forall \ n \ge n_0 \ \exists \ \mathfrak{c}_n \subset \xi_n$ such that $\sum_{J \in \mathfrak{c}_n} |J| > \frac{7 m(A)}{8}$ and such that $m(J \setminus A) < \frac{|J|}{8} \ \forall \ J \in \mathfrak{c}_n$.

Let $c = \frac{m(A)}{10}$ and let $\mathfrak{l}_n$ be the collection of c-long members of $\mathfrak{c}_n$. It follows that for $n \geq n_0$ and $0 < \epsilon < \frac{C_1 c \lambda}{3\overline{N} \, \mathsf{V}_{\mathbb{R}} f}$,

$$m(A \cap [f_{q_n} \in K_\epsilon]) \geq \sum_{J \in \mathfrak{l}_n} m(J \cap A \cap [f_{q_n} \in K_\epsilon])$$

$$\geq \sum_{J \in \mathfrak{l}_n} \left(m(J \cap [f_{q_n} \in K_\epsilon]) - \frac{|J|}{8} \right)$$

$$\geq \sum_{J \in \mathfrak{l}_n} \left(\frac{2}{3}|J| - \frac{|J|}{8} \right)$$

$$\geq \frac{13}{24} \left(\frac{7m(A)}{8} - c \right)$$

$$= \frac{403}{960} m(A)$$

and the claim is established.

To finish the proof of ergodicity, we'll obtain a contradiction from lemma 8.4.3 thereby showing that the assumption $E(f) \neq \mathbb{R}$ is untenable and that $T_{\alpha,f}$ is ergodic.

Fix $\epsilon > 0$ and set $K = K_\epsilon$. It follows from lemma 8.4.3 that
$\forall \, A \in \mathcal{B}(X)_+, \, \exists \, B \in \mathcal{B}(A)_+$ such that $m(B \cap T^{-q_n}B \cap [f_{q_n} \in K]) = 0 \ \forall \, n \geq 1$.
Since $m(B \Delta T^{-q_n} B) \to 0 \ \forall \, B \in \mathcal{B}$ it follows that

$$\forall \, A \in \mathcal{B}(X)_+, \, \exists \, B \in \mathcal{B}(A)_+ \text{ such that } m(B \cap [f_{q_n} \in K]) \to 0 \text{ as } n \to \infty;$$

whence (using the exhaustion lemma) $X = \bigcup_{k \in \mathbb{N}} B_k$ where $m(B_k \cap [f_{q_n} \in K]) \to 0 \ \forall \, k \geq 1$ with the conclusion that

$$m([f_{q_n} \in K]) \to 0$$

as $n \to \infty$ contradicting the claim.

If $T_{\alpha,f}$ is squashable, then by proposition 8.4.1 $\exists \, c > 0$, $c \neq 1$ and $\psi : \mathbb{T} \to \mathbb{R}$ measurable such that $g(x) := cf(x) - f(x + \beta) = \psi(x) - \psi(x + \alpha)$. However, this is impossible as the function $g : \mathbb{T} \to \mathbb{R}$ also satisfies the assumptions of the proposition, whence $T_{\alpha,g}$ is ergodic. $\square$

Odometers.

Odometers are generalisations of the adding machine of §1.2.

For $a_n \in \mathbb{N}$, $(n \in \mathbb{N})$, set

$$X := \prod_{n=1}^{\infty} \{0, \ldots, a_n - 1\}$$

equipped with the addition

$$(x + x')_n = x_n + x'_n + \epsilon_n \qquad \text{mod } a_n$$

where

$$\epsilon_1 = 0, \ \& \ \epsilon_{n+1} = \begin{cases} 0 & x_n + x'_n + \epsilon_n < a_n \\ 1 & x_n + x'_n + \epsilon_n \geq a_n. \end{cases}$$

Clearly (see [**Hew-Ros**]), X equipped with the product discrete topology is a compact Abelian topological group, with Haar measure

$$m = \prod_{n=1}^{\infty} (\frac{1}{a_n}, \ldots, \frac{1}{a_n}).$$

Also if $\tau = (1, 0, \ldots)$ then $X = \overline{\{n\tau\}}_{n\in\mathbb{Z}}$ whence $x \mapsto Tx := x + \tau$ is ergodic.

Set $q_1 = 1$, $q_{n+1} = \prod_{k=1}^{n} a_k$, then

$$(q_n\tau)_k = \begin{cases} 1 & k = n \\ 0 & k \neq n, \end{cases}$$

whence

$$T^{q_n}x = (x_1, \ldots, x_{n-1}, \tilde{T}_n(x_n, \ldots))$$

where $\tilde{T}_n : \prod_{k=n}^{\infty}\{0, \ldots, a_k-1\} \to \prod_{k=n}^{\infty}\{0, \ldots, a_k-1\}$ is defined by $\tilde{T}_n(x) = x + \tilde{\tau}_n$ where $\tilde{\tau}_n = (1, 0, \ldots)$. It follows that $T^{q_n} \overset{\mathcal{U}(L^2(m))}{\longrightarrow} \mathrm{Id}$.

The transformation $T : X \to X$ is called the *odometer with digits* $\{a_n : n \in \mathbb{N}\}$. Note that digits can be "grouped". If $1 = n_1 < n_2 < \cdots < n_k \uparrow \infty$ and $a_1' = a_1$, $a_k' := \frac{q_{n_k}}{q_{n_{k-1}}}$ $(k \geq 2)$, then

$$\prod_{n=1}^{\infty}\{0, \ldots, a_n - 1\} \cong \prod_{k=1}^{\infty}\{0, \ldots, a_k' - 1\}$$

as topological groups and so the digits of an odometer may always be assumed to be "as large as we like".

Let G be a second countable LCA group, and let (X, T) be an odometer. We consider cocycles $\varphi : X \to G$ of form

$$\varphi(x) := \sum_{n=1}^{\infty}[\beta_n((Tx)_n) - \beta_n(x_n)]$$

where $\beta_n : \{0, 1, \ldots, a_n - 1\} \to G$, this sum being a finite sum. Cocycles of this form are called *of product type*. We'll call the functions $\{\beta_k : k \in \mathbb{N}\}$ the *partial transfer functions* of φ.

8.4.4 THEOREM   [**A-Lem-Mau-Nak**].
$\exists$ *a cocycle* $\varphi : X \to G$ *of product type so that* T_φ *is ergodic and non-squashable.*

For $(X, \mathcal{B}, m, T)$ a Kronecker transformation and $\varphi : X \to G$ measurable, set

$$D(\varphi) := \{a \in G : \exists\ q_n \text{ such that } T^{q_n} \overset{\mathcal{U}(L^2(m))}{\longrightarrow} \mathrm{Id}, \ \& \ \varphi_{q_n} \overset{m}{\to} a\}.$$

LEMMA 8.4.5.
1) $D(\varphi)$ *is a subgroup of* $E(\varphi)$.
2) If $\overline{D(\varphi)} = G$, *then* T_φ *is ergodic and non-squashable.*

PROOF.

1) Suppose that $a, a' \in D(\varphi)$, and T^{q_n}, $T^{q'_n} \overset{\mathcal{U}(L^2(m))}{\longrightarrow}$ Id and that $\varphi_{q_n} \to a$, $\varphi_{q'_n} \to a'$ a.e., then

$$\varphi_{q_n - q'_n} = \varphi_{q_n} - \varphi_{q'_n} \circ T^{q_n} \overset{m}{\to} a - a'$$

and $a - a' \in D(\varphi)$. To see that $D(\varphi) \subset E(\varphi)$ let $a \in D(\varphi)$, $T^{q_n} \overset{\mathcal{U}(L^2(m))}{\longrightarrow}$ Id and $\varphi_{q_n} \overset{m}{\to} a$, then

$$m(A \cap T^{-q_n} A \cap [|\varphi_{q_n} - a| < \epsilon]) \to m(A)$$

as $n \to \infty$ $\forall A \in \mathcal{B}$, $\epsilon > 0$.

2) It follows from corollary 8.2.5 that T_φ is ergodic. Let $Q : X \times G \to X \times G$ be non-singular such that $Q \circ T_\varphi = T_\varphi \circ Q$. We show $m \times m_G \circ Q^{-1} = m \times m_G$.

Indeed, by proposition 8.4.1, $\exists$ a group translation $S : X \to X$, a continuous, surjective group endomorphism $w : G \to G$ and a measurable map $f : X \to G$ such that

$$Q(x, y) = (Sx, f(x) + w(y)) \quad (x \in X, \ y \in G)$$

and

$$w \circ \varphi = \varphi \circ S + f - f \circ T.$$

We show that $w \equiv \mathrm{Id}$.

It follows from $w \circ \varphi = \varphi \circ S + f - f \circ T$ that $D(w(\varphi) - \varphi \circ S) = \{0\}$. However, if $a \in D(\varphi)$, and

$$q_n \to \infty, \ T^{q_n} \overset{\mathcal{U}(L^2(m))}{\longrightarrow} \mathrm{Id}, \ \& \ \varphi_{q_n} \to a \text{ a.e. },$$

then

$$w(\varphi_{q_n}) - \varphi_{q_n} \circ S \to w(a) - a \text{ a.e.}$$

whence $w(a) - a \in D(w(\varphi) - \varphi \circ S) = \{0\}$ and $w(a) = a$ $\forall a \in D(\varphi)$ and hence $\forall a \in G$. $\qquad \square$

PROOF OF THEOREM 8.4.4.

We construct a cocycle $\varphi : X \to G$ of product type with $\overline{D}(\varphi) = G$. To this end, let $\Gamma \subset G$ be a countable set which generates a dense subgroup of G, and let the sequence

$$(\gamma_1, \gamma_2, \dots) \in \Gamma^{\mathbb{N}}$$

have the property that

$$|\{n \geq 1 : \gamma_n = g\}| = \infty \ \forall g \in \Gamma.$$

Let (X, T) be the odometer with digits $\{a_n : n \in \mathbb{N}\}$. Assume that $\sum_{n=1}^{\infty} \frac{1}{a_n} < \infty$ and define partial transfer functions $\beta_n : \{0, \dots, a_n - 1\} \to G$ $(n \geq 1)$ by

$$\beta_n(k) = \begin{cases} k\gamma_n & 0 \leq k \leq a_n - 2, \\ 0 & k = a_n - 1. \end{cases}$$

It follows that

$$
\varphi_{q_k}(x) = \sum_{n=1}^{\infty} [\beta_n((T^{q_k}x)_n) - \beta_n(x_n)]
$$

$$
= \sum_{n=1}^{\ell_k(x)-1} [\beta_{k+n}(0) - \beta_{k+n}(a_{k+n} - 1)]
$$

$$
+ \beta_{k+\ell_k(x)}(x_{k+\ell_k(x)} + 1) - \beta_{k+\ell_k(x)}(x_{k+\ell_k(x)}),
$$

where

$$
\ell_k(x) = \min\{n \geq 1 : x_{k+n} < a_{k+n} - 1\}.
$$

By our assumption on the digits, for a.e. $x \in X \; \exists \; N_x \geq 1$ such that $x_n < a_n - 5 \; \forall \; n \geq N_x$. It follows that for $k \geq N_x$, $\ell_k(x) = 1$ and

$$
\varphi_{q_k}(x) = \beta_{k+1}(x_{k+1} + 1) - \beta_{k+1}(x_{k+1}) = \gamma_{k+1}.
$$

Thus, if $y \in \Gamma$ and $\gamma_{n_k} = y \; \forall \; k \geq 1$ then $\varphi_{q_{n_k}-1} \to y$ a.e. and $y \in D(\varphi)$. $\square$

REMARK.

An ergodic Maharam $\mathbb{R}$-extension can be constructed in this way. Let

$$
G = \mathbb{R}, \; (\gamma_1, \gamma_2, \ldots) = (1, \sqrt{2}, 1, \sqrt{2}, \ldots), \; a_n = 2^n,
$$

$$
\beta_n(k) := \left\{ \begin{array}{ll} k\gamma_n & 0 \leq k \leq a_n - 2, \\ 0 & k = a_n - 1. \end{array} \right.
$$

Instead of considering the odometer (X, T) with Haar measure, we define a different product measure

$$
\mu := \prod_{n=1}^{\infty} p_n \text{ where } p_n \in \mathcal{P}(\{0, 1, \ldots, a_n - 1\}), \; p_n(k) := \frac{e^{\beta_n(k)}}{\sum_{j=0}^{a_n-1} e^{\beta_n(j)}}.
$$

It follows that $\mu \circ T \sim \mu$ and that $\log \frac{d\mu \circ T}{d\mu} = \varphi$. The Maharam $\mathbb{R}$-extension is shown to be ergodic as in the proof of theorem 8.4.4.

This is an example of a rather widespread technical analogue between the cohomologies of cocycles under probability preserving transformations, and logarithms of Radon-Nikodym derivatives of non-singular transformations under the same non-singular transformations (see [**A-Ham-Schm**]).

§8.5 Joinings of skew products

In this section, we show (*inter alia*) that if T is a Kronecker transformation, and $\varphi : X \to \mathbb{Z}^d$ is measurable with T_φ ergodic, then T_φ has a law of large numbers. This is done in the context of a study of the joinings of skew products.

DEFINITION: JOININGS, DILATIONS.

Let T_i $(i = 0, 1)$ be measure preserving transformations and let $0 < c_i \leq \infty$ $(i = 0, 1)$ not both infinite. A (c_0, c_1)-*joining* of T_0 and T_1 is a $T_0 \times T_1$-invariant measure $\mu : \mathcal{B}_{T_0} \otimes \mathcal{B}_{T_1} \to [0, \infty]$ such that $\mu(\pi_i^{-1}A) = c_i m_{T_i}(A)$ $(A \in \mathcal{B}_{T_i}, i = 0, 1)$. Note that the demand that $(c_0, c_1) \neq (\infty, \infty)$ ensures that the measure μ is σ-finite on $\mathcal{B}_{T_0} \otimes \mathcal{B}_{T_1}$.

The joining μ of T_0 and T_1 is called *ergodic* if the measure preserving transformation $(X_{T_0} \times X_{T_1}, \mathcal{B}_{T_0} \otimes \mathcal{B}_{T_1}, \mu, T_0 \times T_1)$ is ergodic.

A *self joining* of the measure preserving transformation T is a (c_0, c_1)-joining of T with itself where $c_0 < \infty$.

The *dilation* of a self joining of T is the constant $c = c_\mu := \frac{c_1}{c_0} \in (0, \infty]$.

The following proposition follows easily from the definitions.

8.5.1 PROPOSITION.

$\Delta_\infty(T) = \{c_\mu : \ \mu \text{ an ergodic self joining of } T\}$.

DEFINITION: GROUP JOININGS.

A *group joining* of the countable groups G_0 and G_1 is a subgroup $E \subset G_0 \times G_1$ such that $\forall \ x \in G_0 \ \exists y \in G_1$ such that $(x, y) \in E$ and $\forall \ y \in G_1 \ \exists x \in G_0$ such that $(x, y) \in E$.

Given a group joining $E \subset G_0 \times G_1$, define $p_i : E \to G_i \quad (i = 0, 1)$ by $p_i(x_0, x_1) = x_i \quad (i = 0, 1)$, and

$$c_i(E) := m_E(p_i^{-1}\{0\}) = |\operatorname{Ker} p_i| \quad (i = 0, 1).$$

A *group self joining* of the countable group G is a group joining of G with itself.

A group joining $E \subset G_0 \times G_1$ is called *admissible* if $0 < c_0(E) < \infty$, and in this case its *dilation* is $c_E := \frac{c_1(E)}{c_0(E)}$.

8.5.2 PROPOSITION (C.F. [A11]).

$E \subset G_0 \times G_1$ *is a group joining of G_0 and G_1 iff $\exists$ normal subgroups $K_i \lhd G_i \quad (i = 0, 1)$ and a group isomorphism $\pi : G_0/K_0 \to G_1/K_1$ such that $E = \bigcup_{g \in G_0} gK_0 \times \pi(gK_0)$.*

Here, $c_0(E) = |K_0|$ and $c_1(E) = |K_1|$.

PROOF.

Let $K_0 := \{x \in G_0 : \ (x, e) \in E\}$ and let $K_1 := \{y \in G_1 : \ (e, y) \in E\}$. It follows that $K_i \cong \operatorname{Ker} p_{1-i} \lhd E \quad (i = 0, 1)$, whence $K_i \lhd G_i \quad (i = 0, 1)$.

If (x, y), $(x, y') \in E$, then $(e, yy'^{-1}) \in E$ whence $yy'^{-1} \in K_1$ and $\exists$ a group homomorphism $\phi : G \to G/K_1$ such that $E = \bigcup_{g \in G}\{g\} \times \phi(g)$. Evidently $\phi(x) = \phi(x')$ if and only if $xx'^{-1} \in K_0$ whence $\operatorname{Ker} \phi = K_0$ and $\pi : G/K_0 \to G/K_1$ defined by $\pi(gK_0) := \phi(g)$ is the required group isomorphism. $\qquad \square$

EXAMPLES.

1) There are no admissible group self joinings of $\mathbb{Z}^d$ with non-unit dilation because

a) there are no finite subgroups other than $\{0\}$,
    and

b) there are no subgroups $K \subset \mathbb{Z}^d$ with $\mathbb{Z}^d \cong \mathbb{Z}^d/K$ other than $K = \{0\}$.

2) Let G be a countable group, set $G^\infty := \{x = (x_1, x_2, \dots) \in G^\mathbb{N} : \ x_n \underset{n \to \infty}{\to} e\}$ equipped with the multiplication $(x_1, x_2, \dots) \cdot (y_1, y_2, \dots) = (x_1 y_1, x_2 y_2, \dots)$ and define $A : G^\infty \to G^\infty$ by $A(x_1, x_2, \dots) := (x_2, x_3, \dots)$, then

$$E := \{(x, A(x)) : \ x \in G\}$$

is a group self joining of G^∞ with $c_0(E) = 1$ and $c_1(E) = |G|$.

Let T be an ergodic, probability preserving transformation of the standard probability space $(X, \mathcal{B}, m)$ and let G, H be countable, Abelian groups with counting measures m_G and m_H.

Suppose that $\varphi : X \to G$, $\phi : X \to H$ are measurable and that T_φ, T_ϕ are ergodic.

DEFINITION: ALGEBRAIC JOININGS.

An ergodic joining μ of T_φ and T_ϕ is called *algebraic* if $\exists$ a group joining E of G and H, measurable functions $\psi = (\alpha^{(0)}, \alpha^{(1)}) : X \to E$, $h = (h_0, h_1) : X \to G \times H$, a translation $S : X \to X$ such that

$$\pi \circ T_\psi = (T_\varphi \times T_\phi) \circ \pi \text{ and } m \times m_E \circ \pi^{-1} = \mu$$

where $\pi = (\pi_0, \pi_1) : X \times E \to (X \times G) \times (X \times H)$ is defined by $\pi_i(x, y_0, y_1) := (S^i x, y_i + h_i(x))$ $(i = 0, 1)$.

The dilation of an algebraic self joining is that of the associated group self joining E which is necessarily admissible; indeed

$$\pi_i : T_\psi \overset{c_i(E)}{\to} T_\varphi \quad (i = 0, 1).$$

8.5.3 PROPOSITION (C.F. [**A11**]).
Suppose that T is a Kronecker transformation, then any ergodic joining of T_φ and T_ϕ is algebraic.

PROOF. Let μ be an ergodic joining of T_φ and T_ϕ and consider the conservative, ergodic measure preserving transformation U given by

$$X_U = X_{T_\varphi} \times X_{T_\phi}, \ U = T_\varphi \times T_\phi, \ \& \ m_U = \mu.$$

Define factor maps $p_0 : X_U \to X_{T_\varphi}$ and $\phi_1 : X_U \to X_{T_\phi}$ by $p_i(y_0, y_1) = y_i$ $(i = 0, 1)$.

It is sufficient to show that U is isomorphic to an ergodic, admissible algebraic joining T_ψ of T_φ and T_ϕ.

Writing

$$X_U = X \times X \times G \times H = \{z = (x_0, x_1, y_0, y_1) : x_i \in X \ (i = 0, 1), \ y_0 \in G, \ y_1 \in H\},$$

we define $p : X_U \to X$ by $p(z) = x_1 - x_0$. Clearly $p \circ U = p$ whence $\exists$ a translation $S : X \to X$ such that for m_U-a.e. $z = (x_0, x_1, y_0, y_1) \in X_U$, $x_1 = S x_0$. Thus, up to isomorphism,

$$X_U = X \times G \times H, \ U(x, y_0, y_1) = (Tx, y_0 + \varphi(x), y_1 + \phi(Sx)) := T_\psi(x, y_0, y_1)$$

and $p_i(x, y_0, y_1) = (S^i, y_i)$ $(i = 0, 1)$.

For each $g \in G \times H$, $T_\psi \circ Q_g = Q_g \circ T_\psi$, whence $m_U \circ Q_g \circ T_\psi^{-1} = m_U \circ Q_g$. By ergodicity, either $m_U \circ Q_g \perp m_U$ or $m_U \circ Q_g \sim m_U$ and indeed $m_U \circ Q_g = a(g) m_U$ for some $a(g) \in \mathbb{R}_+$. Set $E := \{g \in G \times H : \ m_U \circ Q_g \sim m_U\}$, then E is a subgroup of $G \times H$ and $a : E \to \mathbb{R}_+$ is a multiplicative homomorphism.

$\exists \ c : X \times G \times H \to [0, \infty)$ measurable such that $dm_U(x, g) = c(x, g) dm(x)$. Since the measure class of m_U projects on to that of m under $(x, y_0, y_1) \mapsto x$, we have

$$\sum_{g \in G \times H} c(x, g) > 0 \text{ for a.e. } x \in X,$$

and consequently $\exists \ h = (h_0, h_1) : X \to G \times H$ measurable such that $c(x, h(x)) > 0$.

Now define $q : X \times G \times H \to X \times G \times H$ by $q(x, y_0, y_1) = (x, y_0 + h_0(x), y_1 + h_1(x))$, then $W := q^{-1} \circ U \circ q = q^{-1} \circ T_\psi \circ q = T_\Psi$ where $\Psi = \psi + h - h \circ T$. Note that $dm_W = dm_U \circ q^{-1}(x, g) = c'(x, g) dm(x)$ where $c'(x, g) = c(x, g + h(x))$.

We claim that $m_W = m \times m_E$. Let $m_1 \sim m$ be defined by $dm_1(x) := c(x, h(x))dm(x)$, then $m_W(A \times \{g\}) = dm_1(A)a(g) \ \forall \ g \in E$. For $g \notin E$, $m_W(\cdot \times \{g\}) \perp m_1$, but also $m_W(\cdot \times \{g\}) \ll m \sim m_1$ whence $m_W(\cdot \times \{g\}) \equiv 0$. It follows that $m_W \sim m \times m_E$, $dm_W(x, g) = a(g)dm_1(x)$ and that $\Psi : X \to E$. Since $m_W \circ T_\Psi = m_W$ we have $a \circ \Psi = 1$ a.e., whence by ergodicity of T_Ψ, $a \equiv 1$. The invariance of m_1 now shows that $m = m_1$ whence $m_W = m \times m_E$.

Evidently E is a group joining and so T_Ψ is an ergodic algebraic joining of T_φ via E, and μ is algebraic.

$\square$

8.5.4 THEOREM.
1) If $d \in \mathbb{N}$ and T_ϕ is an ergodic $\mathbb{Z}^d$-extension of a Kronecker transformation, then $\Delta_\infty(T_\phi) = \{1\}$.

2) If T_ϕ is an ergodic $\mathbb{Z}^\infty$-extension of a Kronecker transformation, then $\Delta(T_\phi) = \{1\}$.

PROOF.
By proposition 8.5.1, $\Delta_\infty(T_\phi)$ is the collection of dilations of ergodic self joinings of T_ϕ, which by proposition 8.5.2 is contained in the collection of dilations of group self joinings of Z^d $d \in \mathbb{N} \cup \{\infty\}$.

The theorem follows because there are no group self joinings of Z^d $d \in \mathbb{N}$ with non-unit dilation, and there are no group self joinings of Z^∞ with finite, non-unit dilation.

$\square$

8.5.5 COROLLARY. *Any ergodic $\mathbb{Z}^d$-extension of a Kronecker transformation has a law of large numbers.*

PROOF. Follows from theorem 3.2.5.

$\square$

REMARKS.
1) It follows from theorem 8.5.4 2) that if T_ϕ is an ergodic $\mathbb{Z}^\infty$-extension of a Kronecker transformation, then $\Delta_\infty(T_\phi) = \{1\}$ or $\{1, \infty\}$. Since all admissible group self joinings of $\mathbb{Z}^\infty$ are (up to isomorphism) of form $E = \{(g, A^j(g)) : \ g \in \mathbb{Z}^\infty\}$ where $A(n_1, n_2, \dots) = (n_2, n_3, \dots)$ we have by proposition 8.5.3 that if T_ϕ is non-squashable, then $\Delta_\infty(T_\phi) = \{1\}$. This can be achieved as in theorem 8.4.4.

We'll see in §8.6 that $\Delta_\infty(T_\phi) = \{1, \infty\}$ is also possible.

2) For rationally ergodic $\mathbb{Z}$-extensions of irrational rotations of $\mathbb{T}$, see [**A-Kea1**].

§8.6 Squashable skew products over odometers

In §8.4, we constructed ergodic, non-squashable skew products over odometers. In this section, we give (the more difficult) constructions of ergodic, squashable and completely squashable skew products.

DEFINITION: COMPLETELY SQUASHABLE.
A conservative, ergodic measure preserving transformation $(Y, \mathcal{C}, \mu, S)$ is said to be *completely squashable* if $\Delta_0(S) = \mathbb{R}_+$.

Note that any ergodic Maharam $\mathbb{R}$-extension is completely squashable.

In this section we construct a conservative, ergodic measure preserving transformation for which $\Delta_\infty = \{1, \infty\}$, and a completely squashable transformation which is not isomorphic to any Maharam $\mathbb{R}$-extension. The examples are group extensions of Kronecker transformations.

Squashability is achieved as follows. Let $L : G \to G$ be a surjective endomorphism such that $m_G \circ L^{-1} = d(L)m_G$ for some $0 < d(L) \leq \infty$, $d(L) \neq 1$. We construct $\varphi : X \to G$ such that T_φ is ergodic and $\exists$ a group translation $S : X \to X$ such that $\varphi \circ S$ is T-cohomologous to $L \circ \varphi$ (where $(X, \mathcal{B}, m, T)$ is a Kronecker transformation).

If $\varphi \circ S = L \circ \varphi + h - h \circ T$ where $h : X \to G$ is measurable, we define $Q : X \times G \to X \times G$ by $Q(x, y) = (Sx, L(y) + h(x))$ obtaining that $m \times m_G \circ Q^{-1} = d(L)m \times m_G$ and $T_\varphi \circ Q(x, y) = Q \circ T_\varphi(x, y)$.

We prove ergodicity of φ using the following lemma. Recall that a sequence of countable partial partitions $\alpha_1, \alpha_2, \ldots$ is said to *approximately generate* $\mathcal{B}$ if $\forall \eta > 0$, $A \in \mathcal{B}_+$ $\exists k_0 \geq 1$ such that $\forall k \geq k_0$ $\exists A' \in \sigma(\alpha_k)$ (the collection of unions of elements of α_k) with $m(A \triangle A') < \eta$.

It follows that if $\alpha_1, \alpha_2, \ldots$ approximately generate $\mathcal{B}$, then $\forall \eta > 0$, $A \in \mathcal{B}_+$ $\exists k_0 \geq 1$ such that $\forall k \geq k_0$ $\exists a \in \alpha_k$ with $m(a \setminus A) < \eta m(a)$.

8.6.1 LEMMA [**A-Lem-V**]. *Let $\varphi : X \to G$ be a cocycle, and let $\gamma \in G$.*

If $\forall$ $\epsilon > 0$ $\exists$ $\delta_k \to 0$ and a sequence of partitions α_k which approximately generate $\mathcal{B}$ with the property that
for every $k \geq 1$, for δ_k-a.e. $a \in \alpha_k$, $\exists$ $n = n(a) \geq 1$ such that

$$m(a \triangle T^{-n}a) < \delta_k m(a) \text{ and } m(a \cap [\varphi_n \in B(\gamma, \epsilon)]) > \frac{m(a)}{25},$$

then

$$\gamma \in \Pi(\varphi).$$

PROOF.
Let $0 < \epsilon < 1/4$, $A \in \mathcal{B}_+$. If α_k is as above, choose $k \geq 1$ such that $\delta_k < \frac{\epsilon}{25}$ and such that $\exists a \in \alpha_k$ with $m(a \setminus A) < \frac{\epsilon}{25}m(a)$. We have

$$m(A \cap T^{-n(a)}A \cap [\varphi_n \in B(\gamma, \epsilon)]) \geq m(a \cap T^{-n(a)}a \cap [\varphi_n \in B(\gamma, \epsilon)]) - \frac{2\epsilon}{25}m(a)$$

$$\geq m(a \cap [\varphi_n \in B(\gamma, \epsilon)]) - (\delta_k + \frac{2\epsilon}{25})m(a)$$

$$\geq \left(\frac{1}{25} - \delta_k - \frac{2\epsilon}{25}\right)m(a)$$

$$> 0.$$

$\square$

8.6.2 THEOREM [**A-Lem-V**]. *Suppose that $L : G \to G$ is a surjective, continuous endomorphism, then there is an odometer (X, T), an ergodic cocycle $\varphi : X \to G$ of product type, a measurable function $\psi : X \to G$, and a translation $S : X \to X$ satisfying*

(8.6.1) $$\varphi \circ S = L \circ \varphi + \psi \circ T - \psi.$$

PROOF. The partial transfer functions β_k of φ are defined by means of blocks. To $\underline{\gamma} = (\gamma_1, \gamma_2, \ldots, \gamma_m) \in G^m$, associate a *canonical difference block* $B_{\underline{\gamma}} = (b(0), b(1), \ldots, b(2^m - 1)) \in G^{2^m}$ defined by

$$b\left(\sum_{k=1}^{m} \epsilon_k 2^{k-1}\right) = \sum_{k=1}^{m} \epsilon_k \gamma_k \quad ((\epsilon_1, \ldots \epsilon_m) \in \{0, 1\}^m).$$

It is evident that for $0 \leq \nu \leq 2^m - 1$, $1 \leq j \leq m$ with $\epsilon_j(\nu) = 0$, we have $\nu + 2^{j-1} \leq 2^m - 1$, and $b(\nu + 2^{j-1}) - b(\nu) = \gamma_j$. It follows that $\forall\, 1 \leq j \leq m$, $\exists\, 1 \leq n = n(j) \leq 2^m$ such that

$$(8.6.2) \qquad \#\{1 \leq \nu \leq 2^m - 1 : \nu + n \leq 2^m - 1,\ b(\nu + n) - b(\nu) = \gamma_j\} \geq \frac{2^m}{2}.$$

Fix a countable set $\Gamma \subset G$ such that

$$\overline{\mathrm{Gp}(\Gamma)} = G,$$

and let $(\gamma_1, \gamma_2, \ldots) \in \Gamma^{\mathbb{N}}$ be such that

$$|\{n \geq 1 : \gamma_n = \gamma\}| = \infty \ \forall\, \gamma \in \Gamma.$$

For $k \geq 1$ let $\gamma_k(j)$ $(0 \leq j \leq 2^k)$ be such that $L^j(\gamma_k(j)) = \gamma_k$, set $\underline{\gamma}_k := (\gamma_k(0), \gamma_k(1), \ldots, \gamma_k(2^k - 1))$ and let $B_{\underline{\gamma}_k} \in G^{2^{2^k}}$ be the canonical difference block defined above. Set $N_k := 2^{k+2^k}$ and let

$$B_k = (b_k(0), b_k(1), \ldots, b_k(N_k - 1)) := \underbrace{B_{\underline{\gamma}_k} \ldots B_{\underline{\gamma}_k}}_{2^k\text{-times}}.$$

We have that

$$\forall\, 0 \leq j \leq 2^k - 1\ \exists\, n = n(j, k) \leq \frac{N_k}{2^k} \ni$$

$$|\{0 \leq \nu < N_k : b_k(\nu + n) - b_k(\nu) = \gamma_k(j)\}| > \frac{N_k}{2}.$$

Now let

$$a_k = 2^k N_k$$

and let (X, T) be the odometer with digits $\{a_n : n \in \mathbb{N}\}$. We specify a cocycle $\varphi : X \to G$ of product type, defining its partial transfer functions $\beta_k : \{0, \ldots, a_k - 1\} \to G$ by

$$\beta_k(j N_k + \nu) = L^j(b_k(\nu)) \quad (0 \leq j \leq 2^k - 1,\ 0 \leq \nu \leq N_k - 1).$$

Note that for $0 \leq j \leq 2^k - 1$,

$$|\{j N_k \leq \nu < (j+1)N_k : \beta_k(\nu + n(j, k)) - \beta_k(\nu) = \gamma_k\}|$$
$$= |\{0 \leq \nu < N_k : L^j\big(b_k(\nu + n(j, k)) - \beta_k(\nu)\big) = L^j(\gamma_k(j))\}|$$
$$\geq |\{0 \leq \nu < N_k : b_k(\nu + n(j, k)) - \beta_k(\nu) = \gamma_k(j)\}|$$
$$> \frac{N_k}{2}.$$

Let $S : X \to X$ be defined by

$$Sx = x + (N_1, N_2, \ldots).$$

To see that (8.6.1) is satisfied, note that by the Borel-Cantelli lemma, for a.e. $x \in X$, $\exists\, n_x \geq 1$ such that for $k \geq n_x$, $x_k < a_k - N_k$, whence $(Sx)_k = x_k + N_k$, and $\beta((Sx)_k) = L(\beta_k(x_k))$.

It follows that $\varphi \circ S - L(\varphi)$ is a product type cocycle, the sum of whose partial transfer functions converges (being a finite sum) and is hence a coboundary. Thus (8.6.1) is satisfied.

To check ergodicity of φ, we show, using lemma 8.6.1 that $\Gamma \subset E(\varphi)$. For $k \geq 1$, we let

$$\alpha_k = \{A((u_1, \ldots, u_{k-1}), j) : 0 \leq u_\nu < a_\nu,\ 1 \leq \nu \leq k-1,\ 0 \leq j \leq 2^k - 2\}$$

where

$$A((u_1, \ldots, u_{k-1}), j) = \{x \in X : x_\nu = u_\nu,\ 1 \leq \nu \leq k-1,\ \&\ jN_k \leq x_k < (j+1)N_k\}.$$

It follows that

$$m\left(\bigcup_{a \in \alpha_k} a\right) = 1 - \frac{1}{2^k}.$$

Also, if $n = n(j, k)q_k$, then writing $\underline{u} = (u_1, \ldots, u_{k-1})$,

$$m(A(\underline{u}, j) \Delta T^{-n} A(\underline{u}, j)) < \frac{1}{2^k} m(A(\underline{u}, j)),$$

and, by (8.6.2)

$$m(A(\underline{u}, j) \cap [\varphi_n = \gamma_k]) > (1 - \frac{1}{2^{k-1}}) m(A(\underline{u}, j)),$$

whence

$$m(A(\underline{u}, j) \cap T^{-n} A(\underline{u}, j) \cap [\varphi_n = \gamma_k]) > (1 - \frac{3}{2^k}) m(A(\underline{u}, j)).$$

Lemma 8.6.1 now shows that

$$\Gamma \subset E(\varphi).$$

$\square$

8.6.3 COROLLARY.

$\exists$ a conservative, ergodic, measure preserving transformation T with $\Delta_\infty(T) = \{1, \infty\}$.

PROOF.

Let $G = \mathbb{Z}^\infty$ and $L : G \to G$ be the shift $L(n_1, n_2, \ldots)_k := n_{k+1}$ $(k \geq 1)$, and let (X, T) be an odometer, $\varphi : X \to G$ measurable and $S : X \to X$ a group translation be such that T_φ is ergodic and $\varphi \circ S = L \circ \varphi - h + h \circ T$ where $h : X \to G$ is measurable.

Evidently $QT_\varphi = T_\varphi Q$ where $Q(x, y) = (Sx, L(y) + h(x))$ whence since $\infty \in \Delta_\infty(T)$.

On the other hand, by theorem 8.5.3, $\Delta(T) = \{1\}$. $\square$

8.6.4 THEOREM [A-Lem-V]. There is an odometer (X, T), and an ergodic cocycle $\varphi : X \to \mathbb{R}$ of product type such that T_φ is completely squashable, indeed $\forall\, c > 0$, $\exists$ a measurable function $\psi_c : X \to G$, and a translation $S_c : X \to X$ satisfying

$$\varphi \circ S_c = c\varphi + \psi_c \circ T - \psi_c.$$

PROOF.

The functions β_k are again defined by means of blocks.

As before, to $\underline{\gamma} = (\gamma_1, \gamma_2, \ldots, \gamma_m) \in \mathbb{R}^m$, associate a canonical difference block $B_{\underline{\gamma}} = (b(0), b(1), \ldots, b(2^m - 1)) \in \mathbb{R}^{2^m}$ defined by

$$b\left(\sum_{k=1}^m \epsilon_k 2^{k-1}\right) = \sum_{k=1}^m \epsilon_k \gamma_k \quad ((\epsilon_1, \ldots \epsilon_m) \in \{0,1\}^m)$$

and such that $\forall\, 1 \leq j \leq m,\ \exists\, 1 \leq n = n(j) \leq 2^m$ such that $\#\{1 \leq \nu \leq 2^m - 1 : \nu + n \leq 2^m - 1,\ b(\nu + n) - b(\nu) = \gamma_j\} \geq \frac{2^m}{2}$.

We'll need some control over the size of $|b(j)|,\ (0 \leq j \leq 2^m - 1)$ and to obtain this, we need the

balanced canonical difference block associated to $\underline{\gamma} = (\gamma_1, \gamma_2, \ldots, \gamma_m) \in \mathbb{R}^m$, defined by

$$B = (b(0), b(1), , \ldots, b(4^m - 1)) \in \mathbb{R}^{4^m}$$

where

$$b\left(\sum_{k=1}^m \epsilon_k 2^{k-1} + \sum_{\ell=1}^m \delta_\ell 2^{m+\ell-1}\right) = \sum_{k=1}^m (\epsilon_k - \delta_k)\gamma_k, \quad (\underline{\epsilon}, \underline{\delta} \in \{0,1\}^m).$$

Let B be the balanced canonical difference block associated to $(\gamma_1, \ldots, \gamma_m) \in \mathbb{R}^m$. Since B is also the canonical difference block associated to $(\gamma_1, \ldots, \gamma_m, -\gamma_1, \ldots, -\gamma_m) \in \mathbb{R}^{2m}$, we have by (8.6.2) that

$$(8.6.3) \qquad \#\{1 \leq \nu \leq 4^m - 1 : \nu + n \leq 4^m - 1,\ b(\nu + n) - b(\nu) = \gamma_j\} \geq \frac{4^m}{2}.$$

Also, we claim that

$$(8.6.4) \qquad \#\{0 \leq \nu \leq 4^m - 1 : |b(\nu)| \geq m^{\frac{3}{4}}\} \leq \max_{1 \leq j \leq m} |\gamma_j|^2 \frac{4^m}{\sqrt{m}}.$$

To see this

$$|\{0 \leq \nu \leq 4^m - 1 : |b(\nu)| \geq m^{\frac{3}{4}}\}| \leq \frac{1}{m^{\frac{3}{2}}} \sum_{\underline{\epsilon}, \underline{\delta} \in \{0,1\}^m} \left(\sum_{k=1}^m (\epsilon_k - \delta_k)\gamma_k\right)^2$$

$$= \frac{4^m}{m^{\frac{3}{2}}} \sum_{k=1}^m \frac{\gamma_k^2}{2}$$

$$\leq \frac{4^m}{\sqrt{m}} \max_{1 \leq j \leq m} |\gamma_j|^2.$$

We now construct the odometer and cocycle. We construct our cocycle φ to have $1, \frac{1}{\sqrt{2}} \in E(\varphi)$ thus ensuring ergodicity. Let $g_{2n} = 1$, and $g_{2n+1} = \frac{1}{\sqrt{2}}$.

For $k \geq 1$, choose natural numbers ν_k and μ_k satisfying

$$(8.6.5) \qquad\qquad \sum_{k=1}^\infty \frac{1}{\mu_k} < \infty,$$

$$(8.6.6) \qquad\qquad \sum_{k=1}^\infty \frac{e^{2\mu_k}}{\sqrt{\mu_k \nu_k}} < \infty,$$

and

(8.6.7)
$$\sum_{k=1}^{\infty} \frac{\mu_k}{\nu_k^{\frac{1}{4}}} < \infty.$$

For example:

$$\mu_k = k^2, \quad \text{and} \quad \nu_k = k^2 3^{4k^2}.$$

Set $m_k = \mu_k \nu_k$, and let

$$B_k = (b_k(0), b_k(1), \dots, b_k(4^{m_k} - 1))$$

be the balanced canonical difference block associated to $(\gamma_k(1), \dots, \gamma_k(m_k))$ where

$$\gamma_k(j) = g_k e^{-\frac{j-1}{\nu_k}}.$$

Now let

$$a_k = m_k 4^{m_k}$$

and let (X, T) be the odometer with digits $\{a_n : n \in \mathbb{N}\}$.

We specify a cocycle $\varphi : X \to \mathbb{R}$ of product type, defining its partial transfer functions $\beta_k : \{0, \dots, a_k - 1\} \to \mathbb{R}$ by

$$\beta_k(j4^{m_k} + \nu) = e^{\frac{j}{\nu_k}} b_k(\nu) \quad (0 \le j \le m_k - 1, \ 0 \le \nu \lambda e 4^{m_k} - 1).$$

By (8.6.4), for $0 \le j \le m_k - 1$

$$|\{j4^{m_k} \le \nu < (j+1)4^{m_k} : \beta_k(\nu + n(j,k)) - \beta_k(\nu) = g_k\}|$$
$$\ge |\{0 \le \nu < 4^{m_k} : b_k(\nu + n(j,k)) - \beta_k(\nu) = \gamma_k(j)\}|$$

(8.6.8)
$$\ge \frac{4^{m_k}}{2}.$$

To check ergodicity of φ, we show, using lemma 8.6.1 1, $\frac{1}{\sqrt{2}} \in E(\varphi)$. For $k \ge 1$, we let

$$\alpha_k = \{A((u_1, \dots, u_{k-1}), j) : 0 \le u_\nu < a_\nu, \ 1 \le \nu \le k-1, \ 0 \le j \le m_k - 2\}$$

where

$$A((u_1, \dots, u_{k-1}), j) = \{x \in X : x_\nu = u_\nu, \ 1 \le \nu \le k-1, \ \& \ jN_k \le u_k < (j+1)N_k\}.$$

It follows that

$$m\left(\bigcup_{a \in \alpha_k} a\right) = 1 - \frac{1}{m_k}.$$

Also, if $n = n(j,k)q_k$ where $n(j,k)$ is as in (8.6.8), then

$$m(A(\underline{u}, j) \Delta T^{-n} A(\underline{u}, j)) < \frac{1}{m_k} m(A(\underline{u}, j)),$$

and, by (8.6.8)

$$m(A(\underline{u}, j) \cap [\varphi_{n(j,k)} = \gamma_k]) \ge \frac{m(A(\underline{u}, j))}{2}.$$

Lemma 8.6.1 now shows that

$$1, \frac{1}{\sqrt{2}} \in E(\varphi).$$

To conclude, we show that $\forall\, c \in (1, e)$, $\exists S : X \to X$ such that $\varphi \circ S - c\varphi$ is a coboundary. Fix $c \in (1, e)$ and let

$$r_k = [\nu_k \log c] \leq \nu_k.$$

Let

$$S = (r_1 4^{m_1}, r_2 4^{m_2}, \dots).$$

We claim that $\forall\, k \geq 1$,

$$m(\{x \in X : |\beta_k(x_k)| \geq m_k^{\frac{3}{4}}\}) \leq \frac{e^{2\mu_k}}{\sqrt{m_k}}.$$

To see this

$$m(\{x \in X : |\beta_k(x_k)| \geq m_k^{\frac{3}{4}}\})$$
$$= \frac{1}{a_k} \#\{0 \leq \nu \leq a_k - 1 : |\beta_k(\nu)| \geq m_k^{\frac{3}{4}}\}$$
$$= \frac{1}{a_k} \sum_{j=0}^{m_k-1} \#\{0 \leq \nu \leq 4^{m_k} - 1 : |\beta_k(j4^{m_k} + \nu)| \geq m_k^{\frac{3}{4}}\}$$
$$= \frac{1}{a_k} \sum_{j=0}^{m_k-1} \#\{0 \leq \nu \leq 4^{m_k} - 1 : e^{\frac{j}{\nu_k}} |b_k(\nu)| \geq m_k^{\frac{3}{4}}\}$$
$$\leq \frac{1}{4^{m_k}} \#\{0 \leq \nu \leq 4^{m_k} - 1 : e^{\mu_k} |b_k(\nu)| \geq m_k^{\frac{3}{4}}\}$$
$$\leq \frac{e^{2\mu_k}}{\sqrt{m_k}}.$$

It now follows from (8.6.6), and the Borel-Cantelli lemma that for a.e. $x \in X, \exists\, k_x \geq 1$ such that $\beta_k(x_k) \leq m_k^{\frac{3}{4}}\ \forall\, k \geq k_x$.

It follows from (8.6.5) that

$$\sum_{k=1}^{\infty} m(\{x \in X : x_k \geq a_k - 2r_k 4^{m_k}\}) < \infty,$$

whence by the Borel-Cantelli lemma, for a.e. $x \in X, \exists\, K_x \geq 1$ such that $(Sx)_k = x_k + r_k N_k\ \forall\, k \geq K_x$.

It follows that a.e. $x \in X$, for $k \geq k_x, K_x$,

$$|\beta_k((Sx)_k) - c\beta_k(x_k)| = (c - e^{\frac{r_k}{\nu_k}})|\beta_k(x_k)|$$
$$\leq \frac{cm_k^{\frac{3}{4}}}{\nu_k} = \frac{c\mu_k^{\frac{3}{4}}}{\nu_k^{\frac{1}{4}}}$$

whence by (8.6.7),

$$\sum_{k=1}^{\infty} |\beta_k((Sx)_k) - c\beta_k(x_k)| < \infty \text{ for a.e. } x \in X$$

and $\varphi \circ S - c\varphi$ is a coboundary, being a product type cocycle, the sum of whose partial transfer functions converges.   □

8.6.5 COROLLARY [A-Lem-V].

∃ a completely squashable, conservative, ergodic measure preserving transformation which is not isomorphic to a Maharam transformation.

PROOF. Let T be an odometer, and let T_φ be an ergodic, completely squashable $\mathbb{R}$-extension of T as in theorem 8.6.4. We claim that T_φ is not isomorphic to a Maharam transformation.

If T_φ is isomorphic to some Maharam transformation, then there is a flow $\{Q_t : t \in \mathbb{R}\}$ on $X \times \mathbb{R}$ such that $Q_t T_\varphi = T_\varphi Q_t$ $\forall t$ such that $D(Q_t) = e^t$ $\forall\, t \in \mathbb{R}$. By proposition 8.4.1,

$$Q_t(x,y) = (S_t x, e^t y + g_t(x)).$$

where $S_t : X \to X$ is a group translation $S_t x = x + s(t)$. Clearly, the map $t \mapsto s(t)$ is a measurable homomorphism from $\mathbb{R} \to X$, whence by Banach's theorem, continuous. To see that $t \mapsto s(t)$ is in fact injective, suppose otherwise, that $S_a =$Id for some $a \neq 0$. Then $Q_a(x,y) = (x, e^a y + g_a(x))$, whence

$$\varphi = G - G \circ T \text{ where } G = \frac{g_a}{e^a - 1}$$

contradicting ergodicity of T_φ.

The fact that X is completely disconnected prevents the existence of such $t \mapsto s(t)$.   □

Bibliography

[A1]   J. Aaronson, *Rational ergodicity and a metric invariant for Markov shifts.*, Israel Journal of Math. **27** (1977), 93-123.

[A2]   ————, *On the ergodic theory of non-integrable functions and infinite measure spaces*, Israel Journal of Math. **27** (1977), 163-173.

[A3]   ————, *On the pointwise ergodic theory of transformations preserving infinite measures*, Israel Journal of Math. **32** (1978), 67-82.

[A4]   ————, *On the categories of ergodicity when the measure is infinite*, Ergodic Theory Proceedings, Oberwolfach 1978, Ed. M. Denker, SLN, vol. 729, 1978, pp. 1-9.

[A5]   ————, *Ergodic theory for inner functions of the upper half plane*, Ann. Inst. H. Poincaré (B) **XIV** (1978), 233-253.

[A6]   ————, *A remark on the exactness of inner functions*, Jour. L.M.S. **23** (1981), 469-474.

[A7]   ————, *The asymptotic distributional behavior of transformations preser- ving infinite measures*, J. D'Analyse Math. **39** (1981), 203-234.

[A8]   ————, *An ergodic theorem with large normalising constants*, Israel Journal of Math. **38** (1981), 182-188.

[A9]   ————, *The eigenvalues of non-singular transformations*, Israel Journal of Math. **45** (1983), 297-312.

[A10]   ————, *Random f-expansions*, Ann. Probab. **14** (1986), 1037-1057.

[A11]   ————, *The intrinsic normalising constants of transformations preserving infinite measures*, J. D'Analyse Math. **49** (1987), 239-270.

[A12]   ————, *Category theorems for some ergodic multiplier properties*, Israel Journal of Math. **51** (1985), 151-162.

[A-De1]   J. Aaronson, M. Denker, *Upper bounds for ergodic sums of infinite measure preserving transformations*, Trans. Amer. Math. Soc. **319** (1990), 101-138.

[A-De2]   ————, *Local limit theorems for Gibbs-Markov maps*, internet: http://www.math.tau.ac.il/∼aaro.

[A-De3]   ————, *The Poincaré series of* $\mathbb{C}\backslash\mathbb{Z}$, internet: http://www.math.tau.ac.il/∼aaro.

[A-De4]   ————, *Distributional limits for hyperbolic, infinite volume geodesic flows*, internet: http://www.math.tau.ac.il/∼aaro.

[A-De-Fi]   J. Aaronson, M. Denker, A. Fisher, *Second order ergodic theorems for ergodic transformations of infinite measure spaces*, Proc. Amer. Math. Soc. **114** (1992), 115-127.

[A-De-Ur] J. Aaronson, M. Denker, M. Urbański, *Ergodic theory for Markov fibred systems and parabolic rational maps*, Trans. Amer. Math. Soc. **337** (1993), 495-548.

[A-Ham-Schm] J. Aaronson, T. Hamachi, K. Schmidt, *Associated actions and uniqueness of cocycles*, Algorithms fractals and dynamics, Proceedings of the Hayashibara Forum '92, Okayama, Japan; and the Kyoto symposium. Ed.: Y. Takahashi, Plenum Publishing Company, New York, 1995, pp. 1-25.

[A-Kea1]   J. Aaronson, M.Keane, *The visits to zero of some deterministic random walks*, Proc. London Math. Soc., Ser III **44** (1982), 535-553.

[A-Kea2]   ————, *Isomorphism of random walks*, Israel J. of Maths. **87** (1994), 37-63.

[A-Le-Ma-Nak] J.Aaronson, M.Lemańczyk, C.Mauduit, H.Nakada, *Koksma'a inequality and group extensions of Kronecker transformations*, Algorithms fractals and dynamics, Proceedings of the Hayashibara Forum '92, Okayama, Japan; and the Kyoto symposium. Ed.: Y. Takahashi, Plenum Publishing Company, New York, 1995, pp. 27-50.

[A-Le-V]   J.Aaronson, M.Lemańczyk, D. Volný, *A salad of cocycles*, preprint, internet: http://www.math.tau.ac.il/~aaro.

[A-Lig-P]   J. Aaronson, T. Liggett, P. Picco, *Equivalence of renewal sequences and isomorphism of random walks*, Israel J. of Maths. **87** (1994), 65-76.

[A-Lin-W]   J. Aaronson, M. Lin and B. Weiss, *Mixing properties of Markov operators and ergodic transformations*, Israel Journal of Math. **33** (1979), 198-224.

[A-Nad]   J.Aaronson, M. Nadkarni, L^∞ *eigenvalues and* L^2 *spectra of non-singular transformations*, Proc. London Math. Soc. **55** (1987), 538-570.

[A-Su]   J. Aaronson, D. Sullivan, *Rational ergodicity of geodesic flows*, Ergod. Theory & Dynam. Syst. **4** (1984), 165-178.

[A-W1]   J. Aaronson and B. Weiss, *Generic distributional limits for measure preserving transform- ations*, Israel Journal of Math. **47** (1984), 251-259.

[A-W2]   _____, *A* $\mathbb{Z}^d$ *ergodic theorem with large normalising constants*, Convergence in ergodic theory and probability, Eds. V. Bergelson, P. March, J. Rosenblatt, O.S.U. Math. Res. Inst. Publ., vol. 5, de Gruyter, Berlin, 1996.

[A-W3]   _____, *On the asymptotics of a 1-parameter family of infinite measure preserving transformations*, internet: http://www.math.tau.ac.il/~aaro.

[Ad]   R. Adler, *F-expansions revisited*, Recent advances in topological dynamics, L.N.Math. 318, Springer, Berlin, Heidelberg, New York, 1973, pp. 1-5.

[Ad-W]   R. Adler, B. Weiss, *The ergodic, infinite measure preserving transformation of Boole*, Israel Journal of Math. **16** (1973), 263-278.

[Ahl]   L.V. Ahlfors, *Conformal invariants topics in geometric function theory*, McGraw-Hill, New York, 1973.

[Arn]   L. K. Arnold, *On σ-finite invariant measures*, Z. Wahrsch. u.v. Geb. **9** (1968), 85-97.

[At]   G. Atkinson, *Recurrence of co-cycles and random walks*, J. London math. Soc., II. Ser. **13** (1976), 486-488.

[Ban]   S. Banach, *Theorie des operations lineaires*, Chelsea, New York, 1932.

[Bar-Nin]   M. N. Barber, B. W. Ninham, *Random and restricted walks*, Gordon and Breach, New York, 1970.

[Bir]   G. D. Birkhoff, *Proof of the ergodic theorem*, Proc. Nat. Acad. Sci. USA **17** (1931), 656-660.

[Boo]   G. Boole, *On the comparison of transcendents with certain applications to the theory of definite integrals*, Philos. Transac. R. Soc. London **147** (1857), 745-803.

[Bow]   R. Bowen, *Invariant measures for Markov maps of the interval*, Comm. Math. Phys. **69** (1979), 1-17.

[Bra]   R. Bradley, *On the ψ-mixing condition for stationary random sequences*, Transac. Amer. Math. Soc. **276** (1983), 55-66.

[Bre]   L.Breiman, *Probability*, Addison-Wesley, Reading, Mass., U.S., 1968.

[Bru]   A. Brunel, *New conditions for the existence of invariant measures in ergodic theory*, Contributions to ergodic theory and probability, S.L.N. Math., vol. 160, 1970, pp. 7-17.

[Cha-Orn]   R. Chacon, D. Ornstein, *A general ergodic theorem*, Illinois J. Math. **4** (1960), 153-160.

[Cho-Ro]   Y.S. Chow, H.Robbins, *On sums of independent random variables with* ∞ *moments*, Proc. Nat. Acad. Sci. U.S.A. **47** (1961), 330-335.

[Chu]   K.L Chung, *Markov chains with stationary transition probabilities*, Springer, Heidelberg, 1960.

[Coh]   D. Cohn, *Measure theory*, Birkhauser, Boston, 1980.

[Con]   J.P.Conze,, *Ergodicite d'un flot cylindrique*, Bull. Soc. Mat. de France **108** (1980), 441-456.

[Cor-Sin-Fom] I. Cornfeld, S.V. Fomin, Ya. G. Sinai, *Ergodic theory*, Springer, New York, 1982.

[Cra]   M. Craizer, *The Bernoulli property of inner functions*, Ergodic Theory Dyn. Syst. **12** (1992), 209-215.

[Da-Kac]   D.A. Darling, M. Kac, *On occupation times for Markov processes*, Trans. Amer. Math. Soc. **84** (1957), 444-458.

[De]   A. Denjoy, *Fonctions contractent le cercle* $|Z| < 1$, C.R.Acad. Sci. Paris **182** (1926), 255-257.

[De-Gr-Sig] M. Denker, C. Grillenberger, K. Sigmund, *Ergodic theory on compact spaces*, Lecture notes in Mathematics vol. 527, Springer, Berlin, 1976.

[Do-Mañ]   C. Doering, R. Mañé, *The dynamics of inner functions*, Ensaios Matemáticos **3** (1991), Soc. Brasileira de Mat., 1-79.

[Doo]   J. Doob, *Stochastic processes*, Wiley, New York, 1953.

[E-Fr]   M. Ellis, N. Friedman, *On eventually weakly wandering sequences*, Studies in probability and ergodic theory, Ed. G. Rota, Adv. Math., Suppl. Stud., vol. 2, 1978, pp. 185-194.

[Fe1]   W. Feller, *A limit theorem for random variables with infinite moments*, Amer. J. Math. **68** (1946), 257-262.

[Fe2]   _____, *An introduction to probability theory and its applications, volume I*, John Wiley, New York, 1968.

[Fe3]   _____, *An introduction to probability theory and its applications, volume II*, Wiley, New York, 1966.

[Fo]   S. Foguel, *The ergodic theory of Markov processes*, van Nostrand, New York, 1969.

[Fo-Lin]   S. Foguel, M. Lin, *Some ratio limit theorems for Markov operators*, Z. Wahrsch. u.v. gebiete **23**, 55-66.

[Fr]   N. Friedman, *Introduction to ergodic theory*, van Nostrand, New York, 1970.

[Fu-W]   H. Furstenberg, B.Weiss, *The finite ergodic multipliers of infinite ergodic transformations*, Lecture notes in math. **668** (1978), Springer, Berlin, 127-132.

[Fu]   H. Furstenberg, *Recurrence in ergodic theory and combinatorial number theory*, Princeton university press, Princeton, N.J., 1981.

[Ga]   A. Garsia, *Topics in almost everywhere convergence*, Lectures in Advanced Mathematics : NO. 4, Markham, Chicago, 1970.

[Go-Sin]   V.I.Golodets, S.D.Sinel'shchikov, *Locally compact groups appearing as ranges of cocycles of ergodic* $\mathbb{Z}$-*actions*, Ergod. Th. and Dynam. Sys. **5** (1985), 45-57.

[G]   Y. Guivarc'h, *Propriétés ergodiques, en mesure infinie, de certains systèmes dynamiques fibrés*, Ergod. Th. and Dynam. Sys. **9** (1989), 433-453.

[Haj-It-Kak] A. Hajian, Y. Ito, S. Kakutani, *Invariant measures and orbits of dissipative transformations*, Adv. Math **9** (1972), 52-65.

[Haj-Kak1] A. Hajian, S. Kakutani, *Weakly wandering sets and invariant measures*, Trans. Amer. Math. Soc. **110** (1964), 136-151.

[Haj-Kak2] _____, *An example of an ergodic measure preserving transformation defined on an infinite measure space*, Contributions to ergodic theory and probability, S.L.N. Math., vol. 160, 1970, pp. 45-52.

[Half]   M. Halfant, *Analytic properties of Renyi's invariant density*, Israel Journal of Math. **27** (1977), 1-20.

[Halm1]   P. Halmos, *Measure theory*, van Nostrand, New York, 1950.

[Halm2]   _____, *Lectures on ergodic theory*, Chelsea, New York, 1956.

[Ham]   T. Hamachi, *The normaliser group of an ergodic automorphism of type III and the commutant of an ergodic flow*, Jour. Funct. Anal. **40** (1981), 387-403.

[Hed]   G. Hedlund, *Fuchsian groups and mixtures*, Ann. Math. **40** (1939), 370-383.

[Hei]   M. Heins, *On the finite angular derivatives of an analytic function mapping the disk onto itself*, J. London Math. Soc. **15** (1977), 239-254.

[Hew-Ros]   E. Hewitt, K. Ross, *Abstract harmonic analysis*, Springer Verlag, Berlin, 1979.

[Hof-Kel]   F. Hofbauer; G. Keller, *Ergodic properties of invariant measures for piecewise monotonic transformations.*, Math. Zeitschrift **180** (1982), 119-140.

[Hop1]   E. Hopf, *Ergodentheorie*, Ergeb. Mat., vol. 5, Springer, Berlin, 1937.

[Hop2]   _____, *The general temporally discrete Markov process*, J. Rat. Mech. Anal. **3** (1954), 13-45.

[Hop3]   _____, *Ergodic theory and the geodesic flow on surfaces of constant negative curvature*, Bull. Am. Math. Soc. **77** (1971), 863-877.

[Hos-Me-Par] B. Host, J-F. Melá, F. Parreau, *Nonsingular transformations and spectral analysis of measures*, Bull. Soc. Math. de France **119** (1991), 33-90.

[Hur]   W. Hurewicz, *Ergodic theorem without invariant measure*, Ann. Math. **45** (1944), 192-206.

[Io-Mar]   Ionescu-Tulcea, G. Marinescu, *Théorie ergodique pour des classes d'opérations non complètement continues*, Ann. Math. **47** (1946), 140-147.

[Kac]   M. Kac, *On the notion of recurrence in discrete stochastic processes*, Bull. Amer. Math. Soc. **53** (1947), 1002-1010.

[Kah-Sal]   J-P. Kahane, R. Salem, *Ensembles parfaits et series trigonometriques*, Hermann, Paris, 1963.

[Kak]   S. Kakutani, *Induced measure preserving transformations*, Proc. Imp. Acad. Sci. Tokyo **19** (1943), 635-641.

[Kar]   J. Karamata, *Sur un mode de croissance reguliere. Theoreme s fondamentaux*, Bull. Soc. Math. France **61** (1933), 55-62.

[Kal]   T.Kaluza, *Über die Koeffizienten reziproker Potenzreihen*, Math. Z. **28** (1928), 161-170.

[Kat]   Y. Katznelson, *An Introduction to Harmonic Analysis*, Dover Publ. inc., New York, 1967.

[Kec]   A. Kechris, *Classical descriptive set theory*, Graduate texts in mathematics : 156, Springer-Verlag, Berlin, 1995.

[Kel]   G. Keller, *Exponents, attractors and Hopf decompositions for interval maps*, Ergodic Theory Dyn. Syst. **10** (1990), 717-744.

[Ken]   D. G. Kendall, *Delphic semigroups*, Springer Lecture Notes in Math. **31** (1967), 147-175.

[Key-New]   H. Keynes, D. Newton, *The structure of ergodic measures for compact group extensions*, Israel J. Math. **18** (1974), 363-389.

[Kh]   A.Ya. Khinchin, *Continued Fractions*, University of Chicago Press, Chicago and London, 1964.

[Ki]   J.F.C.Kingman, *Regenerative Phenomena*, John Wiley, New York, 1972.

[Kol]   A.N. Kolmogorov, *Foundations of the theory of probability*, Chelsea, New York, 1956.

[Kom]   J. Komlos, *A generalization of a problem of Steinhaus*, Acta Math. Acad. Sci. Hung. **18** (1967), 217-229.

[Kre]   U. Krengel, *Ergodic theorems*, de Gruyter, Berlin, 1985.

[Kre-Suc]   U. Krengel, L. Sucheston, *On mixing in infinite measure spaces*, Z. Wahrsch. u.v. Geb. **13** (1969), 150-164.

[Kri]   W. Krieger, *On ergodic flows and isomorphism of factors*, Math. Annalen **223** (1976), 19-70.

[Kui-Ni]   L.Kuipers, H.Niederreiter, *Uniform Distribution of Sequences*, Wiley, N.Y., 1974.

[Kur]   K. Kuratowski, *Topology vols. 1 and 2*, Academic press, New York, 1966.

[La]   S. Lang, $SL_2(R)$, Addison-Wesley, Reading, Mass. USA., 1975.

[Lem]   M.Lemańczyk, *Ergodic compact Abelian group extensions*, Habilitation thesis, Nicholas Copernicus University, Toruń, 1990.

[Let]   G. Letac, *Which functions preserve Cauchy laws?*, Proc. Amer. Math. Soc. **67** (1977), 277-286.

[Lév]   P. Lévy, *Théorie de l'addition des variables aléatoires*, 2nd. edition, Gauthier-Villars, Paris, 1954.

[Li-Schw]   T-Y. Li, F. Schweiger, *The generalized Boole's transformation is ergodic*, Manuscr. Math. **25** (1978), 161-167.

[Lin]   M. Lin, *Mixing for Markov operators*, Z. Wahrsch. u.v. Geb. **19** (1971), 231-243.

[Mah]   D.Maharam, *Incompressible transformations*, Fund. Math. **56** (1964), 35-50.

[Mañ]   R. Mañé, *Ergodic theory and differential dynamics*, Springer-Verlag, Berlin, 1987.

[Me]   J. F. Melá, *Groupes de valeurs propres des systemes dynamiques et sous groupes satures du cercle*, C.R. Acad. Sci. Paris Ser. I Math **296** (1983), 419-422.

[Mo-Schm]   C.Moore, K.Schmidt, *Coboundaries and homomorphisms for nonsingular actions and a problem of H.Helson*, Proc. L.M.S. **40** (1980), 443-475.

[Myr]   P.J. Myrberg, *"Uber die Existenz der Greenschen Funktionen auf einer gegebenen Riemannschen Fläche*, Acta Math. **61** (1933), 39-79.

[Neu-v1]   von Neumann, J., *Allgemeine Eigenwerttheorie Hermitescher Funktionaloperatoren*, Math. Annalen **102** (1929), 49-131.

[Neu-v2]   _____, *Proof of the quasi ergodic hypothesis*, Proc. Nat. Acad. Sci. USA **18** (1932), 70-82.

[Neu-v3]   _____, *Zur Operatorenmethode in der Klassischen Mechanik*, Ann. Math **33** (1932), 587–642.

[Neu]   J. H. Neuwirth, *Ergodicity of some mappings of the circle and the line*, Israel Journal of Math. **31** (1978), 359-367.

[Nor]   E. A. Nordgren, *Composition operators*, Canad. J. Math. **20** (1968), 442-449.

[Nic]   P. Nicholls, *The ergodic theory of discrete groups*, London Math. Soc. Lecture Notes, vol. 143, Cambridge University Press, Cambridge, 1989.

[Ore]   I. Oren, *Ergodicity of cylinder flows arising from irregularities of distribution*, Israel J. Math. **44** (1983), 127-138.

[Os]   M. Osikawa, *Point spectra of non-singular flows*, Publ. R.I.M.S. Kyoto U. **13** (1977), 167-172.

[Parr1]   W. Parry, *Ergodic and spectral analysis of certain infinite measure preserving transformations*, Proc. Am. Math. Soc. **16** (1965), 960-966.

[Parr2]   _____, *Topics in ergodic theory*, Cambridge Tracts in Mathematics, Cambridge University Press, Cambridge, 1981.

[Part]   K. R. Parthasarathy, *Probability measures on metric spaces*, Academic press, New York, 1967.

[Pas]   D.A. Pask, *Skew products over the irrational rotation*, Israel J. Math. **69** (1990), 65-74..

[Pet]   K. Petersen, *Ergodic theory*, Cambridge University Press, Cambridge, 1983.

[Poi]   H. Poincaré, *Les méthodes nouvelles de la mécanique céleste*, vol. 3, Gauthier-Villars, Paris, 1899.

[Pol-Sh]   M. Pollicott, R, Sharp, *Orbit counting for some discrete groups acting on simply connected manifolds with negative curvature*, Invent. Math. **117** (1994), 275-302.

[Pom]   C. Pommerenke, *On the iteration of analytic functions in a half plane*, J. London Math. Soc. **19** (1979), 439-447.

[Ren1]   A. Rényi, *Representations for real numbers and their ergodic properties*, Acta. Math. Acad. Sci. Hung. **8** (1957), 477-493.

[Ren2]   _____, *Probability theory*, North-Holland, Amsterdam, 1970.

[Ree]   M. Rees, *Checking ergodicity of some geodesic flows with infinite Gibbs measure*, Ergodic Theory Dyn. Syst. **1** (1981), 107-133.

[Rie-Sz.N]   F. Riesz, B. Szokefalvi-Nagy, *Lecons d'analyse fonctionnelle*, Akademiai Kiado, Budapest, 1953.

[Ro1]   V.A. Rokhlin, *On the fundamental ideas of measure theory*, Mat. Sb. **25** (1949), 107-150; A.M.S.Transl. **71** (1952).

[Ro2]   _____, *Selected topics from the metric theory of dynamical systems*, Uspehi Mat. Nauk. **4** (1949), 57-125; A.M.S.Transl. Ser. 2 **49** (1966).

[Rudi]   W. Rudin, *Real and complex analysis*, Tata McGraw-Hill, New Delhi, India, 1974.

[Rudo]   D.J. Rudolph, *Fundamentals of measurable dynamics (ergodic theory on lebesgue spaces)*, Clarendon Press, Oxford, 1990.

[Rudo-Sil]   D. J. Rudolph, C. Silva, *Minimal self-joinings for nonsingular transformations*, Ergodic Theory Dyn. Syst. **4** (1989), 759-800.

[Rue]   D. Ruelle, *Thermodynamic formalism (the mathematical structures of classical equilibrium statistical mechanics)*, Encyclopedia of Mathematics and its applications, vol. 5, Addison-Wesley, Reading, Mass., 1978.

[Sach]   U. Sachdeva, *On category of mixing in infinite measure spaces*, Math. Syst. Theoty **5** (1971), 319-330.

[Schm1]   K. Schmidt, *Cocycles of Ergodic Transformation Groups*, Lect. Notes in Math. Vol. 1, Mac Millan Co. of India, 1977.

[Schm2]   _____, *Spectra of ergodic group actions*, Israel Journal of Math. **41** (1982), 151-153.

[Schm3]   _____ On recurrence, Wahrscheinlichkeitstheor. Verw. Geb. **68** (1984), 75-95.

[Schw1]   F. Schweiger, *Number theoretical endomorphisms with σ-finite invariant measures*, Israel Journal of Math. **21** (1975), 308-318.

[Schw2]   _____, *tan x is ergodic*, Proc. Am. Math. Soc. 71 (1978), 54-56.

[Schw3]   _____, *Ergodic theory of fibred systems and metric number theory*, Clarendon Press, Oxford, 1995.

[Sie]   C.L Siegel, *Some remarks on discontinuous groups*, Ann. Math. **46** (1945), 708-718.

[Sil]   C. Silva, *On μ-recurrent nonsingular endomorphisms*, Israel Journal of Math. **61** (1988), 1-13.

[So]   M. Souslin, *Sur une définition des ensembles measurable B sans nombres transfinis*, C. R. Acad. Sci. Paris **164** (1917), 88-91.

[St]   V. V. Stepanov, *Sur une extension du theoreme ergodique*, Compositio Math. **3** (1936), 239-253.

[Su]   D. Sullivan, *On the ergodic theory at infinity of an arbitrary discrete group of hyperbolic motions*, Riemann surfaces and related topics: Proc. 1978 Stony Brook Conf., Ann. Math. Stud., vol. 97, 1981, pp. 465-496.

[Ta]   D. Tanny, *A $0-1$ law for stationary sequences*, Z. Wahrsch. u.v. Geb. **30** (1974), 139-148.

[Te]   A. Tempelman, *Ergodic theorems for group actions (informational and thermodynamical aspects)*, Ser: Mathematics and its applications, vol. 78, Kluwer Academic, Dordrecht, The Netherlands, 1992.

[Tha1]   M. Thaler, *Estimates of the invariant densities of endomorphisms with indifferent fixed points*, Israel Journal of Math. **37** (1980), 303-314.

[Tha2]   _____, *Transformations on $[0,1]$ with infinite invariant measures*, Israel Journal of Math. **46** (1983), 67-96.

[Tha3]   _____, *A limit theorem for the Perron-Frobenius operator of transformations on $[0,1]$ with indifferent fixed points*, Israel Journal of Math. **91** (1995), 111-127.

[Tho]   J-P. Thouvenot, *Some properties and applications of joinings in ergodic theory*, Ergodic theory and its connections with harmonic analysis (Alexandria 1993 conference proceedings, Eds: K. Petersen, I. Salama), London Math. Soc. Lecture Notes, vol. 205, Cambridge University Press, Cambridge, 1995, pp. 207-235.

[Ts]   M. Tsuji, *Potential theory in modern function theory*, Maruzen Co. Ltd, Tokyo, 1959.

[vN1]   von Neumann, J., *Proof of the quasi ergodic hypothesis*, Proc. Nat. Acad. Sci. USA **18** (1932), 70-82.

[Var]   V. S. Varadarajan, *Groups of automorphisms of Borel spaces*, Trans. Amer. Math. Soc. **109** (1963), 191-220.

[vN2]   von Neumann, J., *Zur Operatorenmethode in der Klassischen Mechanik*, Ann. Math **33** (1932), 587–642.

[Wa]   P. Walters, *An introduction to ergodic theory*, Springer, New York, 1982.

[Wi]   N. Wiener, *The ergodic theorem*, Duke Math. J. **5** (1939), 1-18.

[Wo]   J. Wolff, *Sur l'iteration des fonctions dans une region*, C.R.Acad. Sci. Paris **182** (1926), 42-43.

[Yo-Kak]   K. Yosida, S. Kakutani, *Birkhoff's ergodic theorem and the maximal ergodic theorem*, Proc. Imp. Acad. Sci. Tokyo **15** (1939), 165-168.

[Yu]   M. Yuri, *Multi-dimensional maps with infinite invariant measures and countable state sofic shifts*, Indag. Math., New Ser. **6** (1995), 355-383.

[Zi]   R. J. Zimmer, *Ergodic theory and semisimple groups*, Birkhauser, Boston, 1984.

Index

numbers refer to sections   (i.e. a.b=§a.b)